Non-Perturbative Methods and Lattice QCD

Proceedings of the International Workshop

Non-Perturbative Methods and Lattice QCD

Editors

Xiang-Qian Luo
Eric B. Gregory
Zhongshan University, China

World Scientific
Singapore • New Jersey • London • Hong Kong

Published by

World Scientific Publishing Co. Pte. Ltd.
5 Toh Tuck Link, Singapore 596224
USA office: 27 Warren Street, Suite 401-402, Hackensack, NJ 07601
UK office: 57 Shelton Street, Covent Garden, London WC2H 9HE

British Library Cataloguing-in-Publication Data
A catalogue record for this book is available from the British Library.

NON-PERTURBATIVE METHODS AND LATTICE QCD

ISBN-13 978-981-02-4595-5
ISBN-10 981-02-4595-5

PREFACE

Lattice field theory is the most reliable tool for investigating non-perturbative phenomena in particle physics. It has also become a cross-discipline overlapping with other physics and computer sciences. The International Workshop on Non-Perturbative Methods and Lattice QCD was held at Zhongshan University in Guangzhou, China, from May 15 through 20, 2000, with approximately 50 attendees from more 10 different countries.

Guangzhou, also known as Canton, is a city with a history that stretches back over two millennia. The city has long been a center of international trade and it continues in this role today in this period of rapid development in China. Perhaps though, Guangzhou, is most famous for its cuisine, as Cantonese food has become popular throughout the world. Dr. Sun Yat-Sen, a great figure in China's modern history, founded Zhongshan University in 1924. To commemorate its founder, the university was renamed Zhongshan, Dr. Sun's courtesy name, in 1926. It developed into the most famous university in south China. The lattice QCD group is an active one, equipped with recently constructed 20 CPU Beowulf cluster for parallel computing.

During those six days in May, physicists were able to meet colleagues — old and new acquaintances — from around the world, exchange ideas, plan future collaborations, and share cultural differences and similarities all in a relaxed intimate setting. Together we also explored the sights, history, and cuisine of south China's largest and oldest city and its environs. A day trip took us to see the unique landscape of the Seven Star Crags of Zhaoqing. An evening dinner on the Pearl River displayed the colorful lights of modern Guangzhou, and a tour of the Chen ancestral temple and the 2000 year old tomb of the southern Yue king gave a glimpse of the city's history.

This book contains the written version of 30 presentations at this conference by active physicists. It covers new developments in algorithms, statistical physics, parallel computers and quantum computation, recent advances in standard model and beyond, QCD vacuum, glueball, hadron and quark masses, finite temperature and density, chiral fermions, SUSY, and heavy quark effective theory. We hope it serves as both a useful reference of the current work in these fields and as a reminder of successful workshop and a pleasant stay in Guangzhou.

We thank the National Science Fund for Distinguished Young Scientists, National Natural Science Foundation, Guangdong Provincial Natural Science Fund, Guangdong National Communication Network Ltd., Ministry of Education and Zhongshan University for partial financial support. We also thank

Prof. Jia-Rui Xu, the vice president of Zhongshan University for his opening address. Chun-Hua Luo and Jin-Xiao Luo, parents of Xiang-Qian Luo, contributed encouragement and many hours of tireless help with organization and logistics prior to and throughout the workshop. They have our heartfelt gratitude. Our thanks also go to the many undergraduate and graduate ZSU physics students who provided invaluable assistance. We are very grateful to all the attendees whose high quality contributions made the workshop a very successful one.

Xiang-Qian Luo and Eric B. Gregory
Zhongshan University
May 2000

CONTENTS

*Denotes Speaker

1 Chiral Fermions and Perfect Action

Conference participants.

APPROXIMATE GINSPARG-WILSON FERMIONS FOR QCD

W. BIETENHOLZ
NORDITA, Blegdamsvej 17
DK-2100 Copenhagen, Denmark
E-mail: bietenho@nordita.dk

Lattice fermions obeying the Ginsparg-Wilson relation do correctly represent the physical properties related to chirality. This can be achieved by local fermions, which involve an infinite number of couplings, however. For practical purposes, it is useful to first construct approximate Ginsparg-Wilson fermions within a short range. We report on a successful construction in QCD at $\beta = 6$. The good quality of the approximation is observed from the spectrum, which is situated close to a Ginsparg-Wilson circle. These fermions also provide an excellent approximation to rotational symmetry and they are promising for a good scaling, since they arise from the perfect action framework. Their insertion into the overlap formula renders the Ginsparg-Wilson relation exact. It leads to an improved overlap fermion with a high level of locality. This insertion is statistically on safe grounds at $\beta \gtrsim 5.6$.

1 Ginsparg-Wilson fermions

In a slightly simplified form, the famous Nielsen-Ninomiya theorem [1] states that a local lattice fermion without species doublers cannot be chiral in the sense that the lattice Dirac operator D anti-commutes with γ_5. Locality means here that the couplings in D decay at least exponentially in the separation between ψ and $\bar{\psi}$. [a] Hence it is an obvious idea to break the chiral symmetry by an irrelevant term, so that it should be restored in the continuum limit. The simplest way to do so is to set $\frac{1}{2}\gamma_5\{D,\gamma_5\}$ equal to some local term of $O(a^2)$, such as the Wilson term $\frac{1}{2}\Delta$, where Δ is a discretized Laplacian. [b] However, this type of chiral symmetry breaking on the lattice is rather violent; it causes quite some trouble such as additive mass renormalization, $O(a)$ scaling artifacts, renormalization of currents, mixing of matrix elements, etc.

On the other hand, it turned out to be harmless to introduce a non-vanishing anti-commutator as

$$\frac{1}{2}\gamma_5\{D^{-1},\gamma_5\} = R \tag{1}$$

where R is a *local* term with $\{R,\gamma_5\} \neq 0$. The superiority of this relation can be understood intuitively from the fact that R doesn't shift the poles in D^{-1}.

[a] Actually the proof in Ref. [1] still holds for an even weaker form of locality.

[b] Here a is the lattice spacing, but in general we will refer to a hypercubic lattice of unit spacing in Euclidean space.

In the form

$$\{D_{x,y}, \gamma_5\} = 2(D\gamma_5 RD)_{x,y} \tag{2}$$

it is known as the *Ginsparg-Wilson relation* (GWR) [2].

1.1 Virtues

Amazingly, it seems that all physical properties related to chirality are represented correctly by a lattice fermion obeying the GWR (a GW fermion). The mass and the vector current (as well as the flavor non-singlet axial vector current) are not renormalized and weak matrix elements do not mix [3]. Moreover, the chiral anomalies [2,4,5] as well as global anomalies [6], and the soft pion theorems are reproduced correctly [7]. Even the construction of chiral gauge theories on the lattice is feasible based on GW fermions [8].

For the understanding of these properties, it is a key observation that GW fermions have an exact — though lattice modified — chiral symmetry at finite lattice spacing [4].

It is instructive to consider the spectrum of a GW fermion. For simplicity, we assume $D^\dagger = \gamma_5 D \gamma_5$ and $R_{x,y} = \frac{1}{2\mu}\delta_{x,y}$ ($\mu > 0$), hence the GWR reads $\mu(D + D^\dagger) = D^\dagger D$. If we introduce the operator $A = D - \mu$, the GWR simplifies further to

$$A^\dagger A = \mu^2. \tag{3}$$

Therefore we know that the spectrum of a GW Dirac operator is — with the above assumptions — always situated on a circle in the complex plane, with center and radius μ. This confirms the absence of additive mass renormalization, and it also rules out "exceptional configurations" [3]. Moreover, it provides a well-defined index (since the zero eigenvalues are exact), and together with the index theorem we obtain a sensible definition of the topological charge of a lattice configuration.

1.2 Limitations

After celebrating the impressive properties of GW fermions, we now have to address their limitations. One point is that the GWR guarantees a correct chiral behavior, but it does hardly imply anything about other properties, which are also essential for a formulation of lattice fermions, in particular the scaling behavior.

A second point concerns locality: the relaxation of the condition $\{D, \gamma_5\} = 0$ to the GWR allows the lattice fermion to be local in the sense that the couplings decay exponentially with the lattice distances. This is sufficient from a conceptual point of view, since there is a decay length of a finite number of lattice spacings, which ensures the right continuum limit. However, for applications one would like to have even "ultralocality", which means that the couplings drop to zero beyond a finite number of lattice spacings. Unfortunately, GW fermions cannot have this property, not even in the case of free fermions [10]. For example, if we want to insert the free Wilson Dirac operator and solve for R, then we obtain a pseudo-GW kernel R which decays as $R_{x,y} \propto |x-y|^{-4}$ in $d=2$, and like $R_{x,y} \propto |x-y|^{-6}$ in $d=4$. This is non-local, and therefore the Wilson fermion does not obey any GWR, not even in the free case (in the interacting case this is also clear from the mass renormalization).

Of course, in practice one cannot work with couplings over infinite distances. In a finite volume with certain boundary conditions, the GWR — with these boundary conditions implemented — can be solved, but this still requires the coupling of sites (and links) over all distances in this volume.

1.3 Exact and approximate solutions

Regarding the first limitation, there exists a class of lattice actions called "perfect actions" which deserves its name by yielding a scaling identical to the continuum at any lattice spacing. At the same time, perfect actions solve the GWR [2], but unfortunately their construction is about as difficult as solving directly the model under consideration, since it requires a functional integration extrapolated to the continuum.

The construction of "classically perfect actions" [11] is much easier, though still difficult. They also solve the GWR [3], and their scaling is still excellent. However, a successful construction and application for interacting fermions could only be achieved in $d=2$ so far [12]. From the second limitation we know that such actions need some truncation. Here we truncate to a "hypercube fermion", HF (with couplings not only to nearest neighbor sites, but to all sites inside a unit hypercube), which is actually applicable in QCD simulations [13]. We first consider the truncation for the free fermion. There the perfect lattice Dirac operator can be constructed, and the term R occurs in the block variable transformation. Locality is optimal — that is, the exponential decay of the couplings is optimally fast — for $R_{x,y} = \frac{1}{2}\delta_{x,y}$, [14] which we denote as the "standard GW kernel". Hence the spectrum of that perfect fermion is situated on a unit circle with center 1. We now truncate by evaluating the perfect couplings in a small volume of 3^4 sites, and then we use the same couplings in any volume [15]. We obtain a lattice Dirac operator of the form

$$D(x,y) = \rho_\mu(x-y)\gamma_\mu + \lambda(x-y) , \tag{4}$$

where the support of $\rho_\mu(x-y)$, $\lambda(x-y)$ is restricted to $|x_\nu - y_\nu| \leq 1$, ($\nu =$

1...4). (These couplings are given in Ref. [15], Table 1.) Of course, in a larger volume it is not exactly perfect any more, and the GWR is violated a little. To probe this truncation effect, we look at the spectrum of the hypercube fermion on a 20^4 lattice. Fig. 1 shows that it is indeed very close to a GW circle. Also

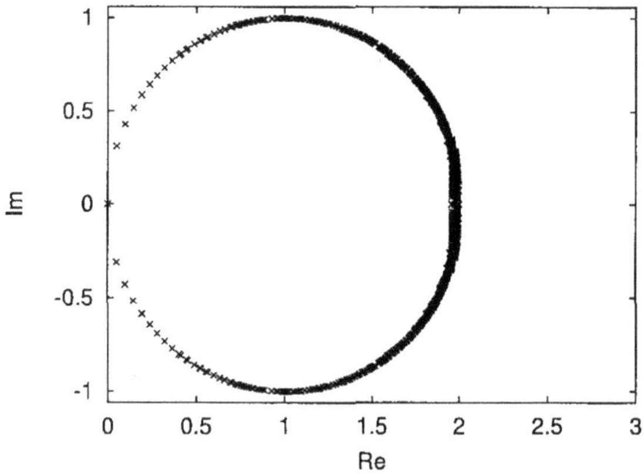

Figure 1: *The spectrum of a truncated perfect, free HF on a* 20^4 *lattice (plotted in* $\mathbb{C}$*).*

the scaling artifacts in this truncated perfect free fermion are very small, as we see from the fermion dispersion relation and thermodynamic scaling quantities [15,16].

The same construction can be done in $d = 2$, but there we prefer to use a "scaling-optimal hypercube fermion" (SO-HF) [17], which is similar to the truncated perfect one, but still a bit improved (with respect to scaling). Again the spectrum for the free fermion is close to a GW unit circle, and its scaling is excellent. To simulate this lattice fermion in the 2-flavor Schwinger model, we performed a "minimal gauging" by hand: we attached the free couplings to the shortest lattice paths only, in equal parts where several shortest paths exist. We also added a clover term with coefficient 1, and used the standard plaquette gauge action. Of course, this simple gauging brings in additional artifacts and a further deviation from the GWR. Hence the eigenvalues spread more and more around the unit circle as β decreases, but for instance at $\beta = 2$ the circle is still approximated more or less, see Fig. 2.

As a scaling test we measured the dispersion relations of the two meson-type states, a massless triplet and a massive singlet [18], which we denote as π and η. Indeed, the scaling is dramatically improved over the Wilson fermion (at critical hopping parameter), see Fig. 3. Amazingly, this SO-HF reaches a similar scaling quality as the classically perfect action, which was truncated only very mildly to 123 independent couplings per site [12]. In contrast, the

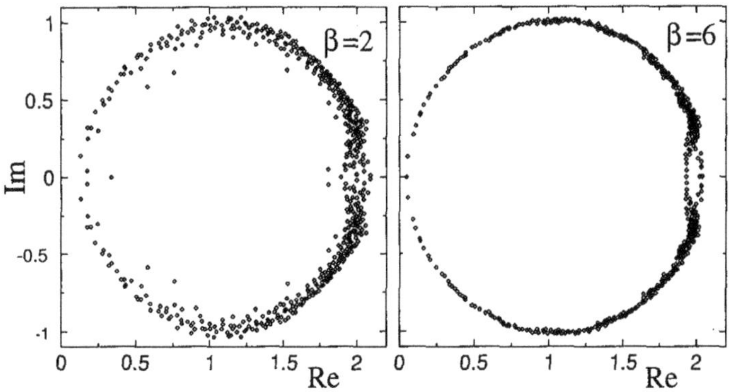

Figure 2: *The spectra of the 2d scaling optimal hypercube fermion (SO-HF) for typical configurations at strong resp. weak coupling, approximating a GW unit circle.*

SO-HF involves only 6 independent couplings per site, hence this approach has the potential to be extended and applied in $d = 4$.

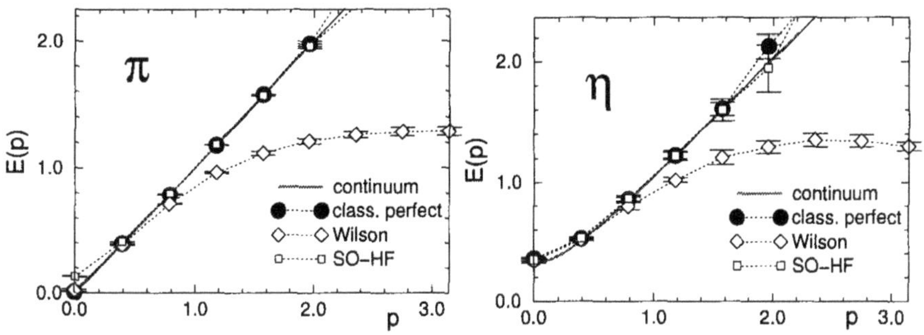

Figure 3: *The "meson" dispersion relations for three types of lattice fermions in the 2-flavor Schwinger model at $\beta = 6$.*

Also the rotational symmetry is approximated very well for such HFs. We tested this for free and for interacting fermions in $d = 2$ [17] and $d = 4$ [15]. As an example, we show the "speed of light" for the minimally gauged truncated perfect HF in QCD at $\beta = 5$, and compare it to the Wilson fermion in Fig. 4. What is not visible from that figure, however, is the dramatic mass renormalization: what is supposed to be the pion mass amounts to $M = 3.0$. Also in the Schwinger model this problem is serious, as Figs. 2 and 3 show: at $\beta = 6$ we obtain a π mass of 0.13. This is the one unpleasant feature of the otherwise successful HFs with minimal gauging. In the next section we discuss a possibility to eliminate this effect completely. In Sec. 4 we are going to consider further methods to approach the chiral limit.

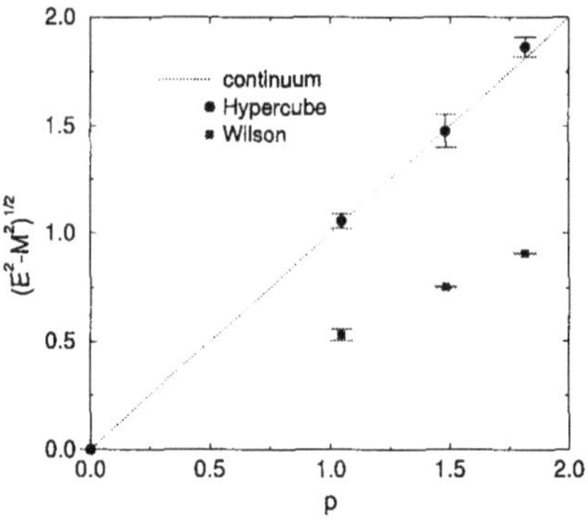

Figure 4: *The "speed of light" $c = p/\sqrt{E^2 - M^2}$ (M = "pion mass") in QCD at $\beta = 5$.* [15]

2 Improved overlap fermions

Let us start from some lattice Dirac operator D_0, which obeys the conditions of the Nielsen-Ninomiya theorem (local, no doublers etc.). We assume again $D_0^\dagger = \gamma_5 D_0 \gamma_5$, and we recall that the GWR for $R_{x,y} = \frac{1}{2\mu}\delta_{x,y}$ is equivalent to $A_0^\dagger A_0 = \mu^2$, for $A_0 = D_0 - \mu$, $(\mu > 0)$. In general this will not hold, of course, but we can simply enforce it by the overlap formula

$$A_{ov} = \frac{\mu A_0}{\sqrt{A_0^\dagger A_0}} \; ; \quad D_{ov} = A_{ov} + \mu \; . \tag{5}$$

D_{ov} represents a GW fermion. A prototype was proposed by H. Neuberger [19], starting from the Wilson fermion $D_0 = D_W$ and $\mu = 1$,

$$D_{Ne} = 1 + \frac{A_W}{\sqrt{A_W^\dagger A_W}} \; ; \quad A_W = D_W - 1 \; . \tag{6}$$

In the free case and in a smooth gauge background, D_{Ne} is free of doublers and local. The latter has been established analytically assuming a constraint on each plaquette variable [20,21], and numerically for QCD at $\beta = 6$ [20]. For first simulations in quenched QCD, see Refs. [22,23].

However, this is just one example in a large class of overlap fermions, which is obtained by varying D_0 and μ. [9] In particular, if it happens that D_0 represents a GW fermions already, then it is just reproduced under the overlap formula, $D_{ov} = D_0$ (for a fixed GW kernel R). Therefore, also perfect and classically perfect fermions are special types of overlap fermions. In practice we do not have an exact GW operator D_0 at hand, but we can construct an approximation, as discussed in Secs. 1.3 and 4. Then the square root will keep close to the constant μ, and $D_{ov} \approx D_0$. We may say that the overlap formula provides a "GW correction" of D_0; in particular it removes the additive mass renormalization.

Based on this property, we suggest the following concept [9]: *start from a truncated perfect fermion (or something similar), which scales well and approximates the GWR, and insert it as D_0 into the overlap formula.* The resulting D_{ov} has exact GW chirality (in particular we are in the chiral limit) *and* a good scaling behavior can also be expected, due to $D_{ov} \approx D_0$. The latter also suggests that an approximate rotational symmetry of D_0 is essentially inherited by D_{ov}, and that D_{ov} is very local in the sense of a fast exponential decay: remember that D_0 is ultralocal, and — due to the modest modification — the long-range couplings will be turned on just a little bit in D_{ov}. [c]

To summarize, this concept aims at combining all desirable properties of a lattice fermion formulation. It has been tested comprehensively in the two-flavor Schwinger model, and the above expectations are observed to hold in a very satisfactory way. In rest of this Section we summarize our 2d results from Ref. [17].

We first checked the scaling quality of the free overlap fermions (fermionic dispersion, thermodynamic scaling ratios) and it turns out that the improvement of the SO-HF over the Wilson fermion persists under the overlap formula, i.e. the overlap SO-HF is strongly improved over the Neuberger fermion. By comparing the spectrum of a fixed configuration for the SO-HF and for the overlap SO-HF, we could then literally see that the alteration due to the overlap formula is small; the eigenvalues are moved almost radially onto the unit circle. Next we tested the scaling in the interacting case, and the drastic improvement of the overlap SO-HF over the Neuberger fermion is confirmed again, see Fig. 5. We have used standard operators in both cases, hence the comparison is fair. Of course, both types of overlap fermions may improve if one consistently improves the operators, which can be tedious, however. Then one expects also the Neuberger fermion to scale better [24,23] (in particular $O(a)$ artifacts are excluded [25]), but our results imply that improved overlap fermions can be used very successfully even with the simple standard operators (as it was also observed for the classically perfect action [11,12]).

Finally we also tested our prediction regarding the degree of locality, and we found that indeed the exponential decay is much faster for the overlap

[c] Also D_W is ultralocal, but not an approximate GW fermion, so its change to D_{Ne} is rather drastic and the argument for a high level of locality does not apply to that case.

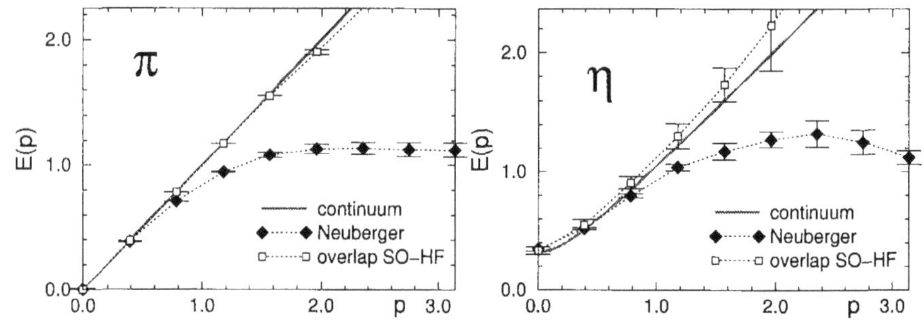

Figure 5: *The "meson" dispersion relations for different types of overlap fermions in the 2-flavor Schwinger model at $\beta = 6$.*

SO-HF than for the Neuberger fermion, see Fig. 6 and Table 1. [d]

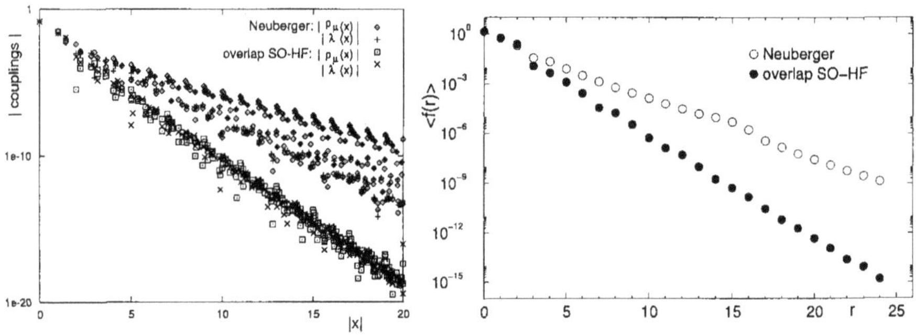

Figure 6: *The level of locality for the Neuberger fermion compared to the overlap SO-HF: the decay of the free couplings in all directions (left) and the decay of the "maximal correlation"* [20] *over a distance r at $\beta = 6$ (right). The width of the "cones" in the left figure also shows how well rotation invariance is approximated.*

3 Overlap fermions and the doubling problem

As we mentioned in Sec. 2, overlap fermions are free of doublers at least in a smooth gauge background. It should be clarified, however, that the overlap formula itself does not remove any doublers. They have to be removed before, in D_0, by something like a Wilson term; in our ansatz $D_0 = \rho_\mu \gamma_\mu + \lambda$, the scalar term λ is crucial for that purpose. Then the overlap formula cures the

[d] Ref. [24] suggests to use non-standard operators (with $\langle R \rangle$ being subtracted from $\langle D^{-1} \rangle$) even for the free fermion, which does improve e.g. the scaling of the free Neuberger fermion. However, one then deals with a non-local fermion, even though it is used only in an indirect way, which raises conceptual questions concerning the continuum limit. An exception is the perfect fermion at $R = 0$, where for instance the axial anomaly is reproduced correctly [26].

		perfect	overlap-HF	Neuberger
$d = 2$	r_ρ	1.268	1.255	1.930
$d = 2$	r_λ	0.871	0.888	1.248
$d = 4$	r_ρ	1.635	1.519	2.530
$d = 4$	r_λ	1.187	1.109	1.708

Table 1: *The characteristic radius* $r_\lambda = (\sum_x |\lambda(x)|x^2)/(\sum_x |\lambda(x)|)$ *and (analogously)* r_ρ *for various types of free GW fermions.*

chiral symmetry again in the sense of the GWR, while doublers are supposed not to be re-introduced.

We keep on referring to a point-like GW kernel $R_{x,y} = \frac{1}{2\mu}\delta_{x,y}$, where the mass parameter $\mu > 0$ can be chosen. It is the center of the circle through zero that the spectrum of the Dirac operator is mapped on by the overlap formula (which also involves the parameter μ). Let us focus on the eigenvalues of D_0 close to the real axis: the small (almost) real eigenvalues have to be mapped on (the vicinity of) zero, whereas the eigenvalues with large real parts have to be mapped on the opposite arc, i.e. on (the vicinity of) 2μ. For weak up to moderate coupling, there is a (statistically) safe interval where μ can be chosen such that this separation is achieved. Inside this interval, μ may be optimized with respect to various criteria; the optimal μ tends to rise with increasing coupling strength — since it has to adapt to the mass renormalization — as has been observed by optimizing locality [20] or minimizing the mapping effect [17]. For instance, in the Schwinger model such a safe interval still exists at $\beta = 2$, as Fig. 2 illustrates.

But how about really strong coupling ? At some point, the eigenvalues of typical configurations spread all over inside a certain area for simple (short-ranged) operators D_0. This is the case for the Wilson fermion and for the HF in the Schwinger model at $\beta = 1$, and for QCD at $\beta = 5$, [27] as Fig. 7 shows. Wherever we choose μ, mappings to the "wrong" arc will occur frequently; the possibility of "wrong" projections implies that both, the doubling problem (too many projections to the left arc) as well as mass renormalization (too many projections to the right arc) are back. Hence at really strong coupling, both of these problems return for the overlap fermions based on some simple D_0, as was also revealed by a strong coupling expansion [28] [e]. Probably the only way out of this problem would be an excellent approximation to a (classically)

[e] The question if also locality breaks down at some point of relatively strong coupling, where the overlap formula is still applicable, is currently under investigation [29]. For instance, one can check this for the Neuberger fermion in the Schwinger model at $\beta \lesssim 2$, or in QCD at $\beta \approx 5.6$.

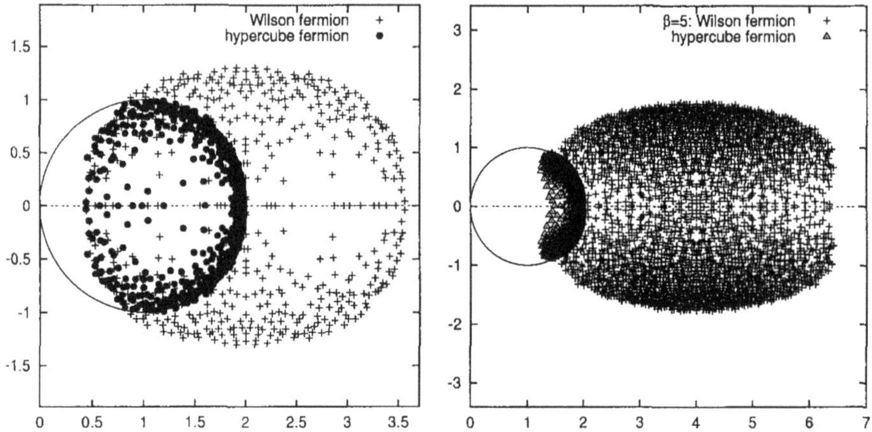

Figure 7: *Typical spectra of minimally gauged HFs and Wilson fermions (with the hopping parameter of the free fermion) in the Schwinger model at $\beta = 1$ (left) and in QCD at $\beta = 5$ (right). This illustrates situations where the overlap projection is not applicable any more.*

perfect fermion, which is, however, very difficult to construct and implement.

It is now of interest to get an idea where this transition occurs. Fig. 8 shows the Wilson spectra for typical (quenched) QCD configurations at $\beta = 5.4$ and $\beta = 5.6$. We see that the existence of a safe interval for μ seems to set in in between. [f] For the HF the situation is similar, just a little better. Hence $\beta = 6$ can be regarded a really safe regime with respect to this issue — Fig. 9 shows that an isolated left arc is present — and also with respect to locality [20], as we mentioned before.

We therefore concentrate on QCD at $\beta = 6$ in the next Section, and we show that we can construct HFs, which approximate the GWR well at that coupling strength. This is the crucial ingredient needed to carry on the program of improved overlap fermions — described in Sec. 3 — to QCD.

Note that in $d = 4$ the notorious square root has to be evaluated by some iterative method, and starting from a good GW fermion approximation instead of the Wilson fermion is also highly profitable with that respect: starting in the right vicinity speeds up the convergence a lot, as we see for instance from the perturbative expansion of the square root around μ. [17]

[f] From our experience, even lattices as small as 4^4 provide a reliable insight into such questions as the applicability of the overlap formula and the level of approximation to the GW circle. A disadvantage is that the left arc close to zero is missing — not due to the lattice action but just due to small volume. Therefore we also evaluated this subset of the spectra (the eigenvalues with smallest real parts) on 8^4 lattices, see below.

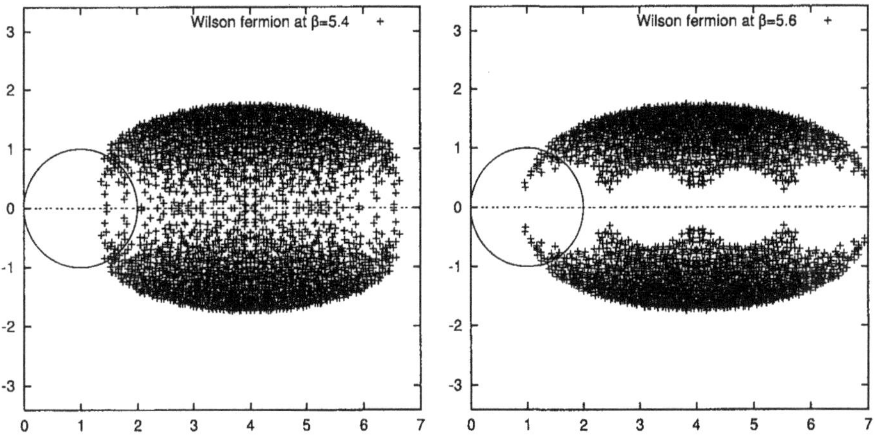

Figure 8: *Typical spectra of the Wilson fermion in QCD on a* 4^4 *lattice at* $\beta = 5.4$ *(left) and* $\beta = 5.6$ *(right). We recognize the transition to the regime where the overlap projection is statistically safe (for* $1.5 \lesssim \mu \lesssim 2$*).*

4 Approximate Ginsparg-Wilson Fermions for QCD

Our approach to construct a short-ranged approximate GW fermion for QCD is to stay with the couplings of the truncated perfect free fermion and gauge it by hand, using just very few new parameters to go beyond the "minimal gauging". This concept was successful in $d = 2$, and in $d = 4$ we already know that the free HF is doing well in scaling, approximating the GWR and approximating rotational invariance (the latter is also checked at strong coupling), see Sec. 1.3. Alternatively, one may try to minimize the GWR violation directly within a limited set of parameters [17,30], or undertake a new effort to parameterize an (approximate) classically perfect action [31].

We are confident that our free HF couplings already provide a good scaling, so the issue is to find a suitable gauging in the sense that the GWR violation is small. As our criterion, we compute the spectra on small lattices and try to arrange for them to be close to a GW unit circle (for typical quenched configurations at $\beta = 6$).

As we see from Fig. 9, the minimally gauged HF suffers from mass renormalization almost as much as the Wilson fermion. On the other hand, the right arc is excellent (see Fig. 7), but less important. Our first step beyond minimal gauging is the use of fat links: each link in a given configurations is substituted as

$$link \;\rightarrow\; (1-\alpha)\; link + \frac{\alpha}{6}\;[\; \sum \; staples \;] \qquad (\alpha \in \mathbb{R}). \qquad (7)$$

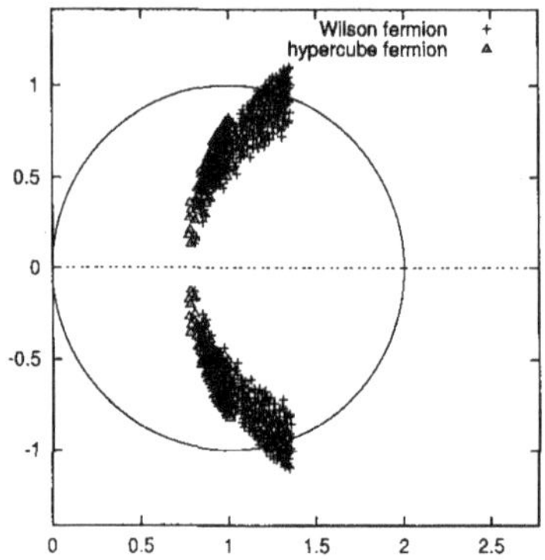

Figure 9: *The left arc of typical spectra in QCD at $\beta = 6$ on an 8^4 lattice for the Wilson fermion and the minimally gauged HF. We show the 600 resp. 300 eigenvalues with smallest real parts, and we see that here the left arc is manifestly isolated.*

This is computationally cheap, and since we perform just one such fattening step we do not need to project back onto the gauge group (in contrast to Ref. [32]). The results for different α are shown in Fig. 10 (left). As we observed already in $d = 2$, a strongly negative α is required if one wants to remove the mass renormalization in this way [17]. However, then the spectrum moves far away from the GW circle, so we do not recommend this way to approach the chiral limit. Positive α increases the pion mass further, but it makes the shape of the spectrum more circle-like. We are going to take advantage of that.

Next we attach an amplification factor $1/u$ ($u \lesssim 1$) to each link to compensate the mean suppression by the gauge field. This is related in spirit to tadpole improvement [33], but it can also be viewed as directly generalizing the tuning of the Wilson hopping parameter. For our value of $\beta = 6$, we reach criticality for the minimally gauged HF at $u \simeq 0.8$, see Fig. 10 (right). Once u is fixed, its inclusion is computationally for free, and it does lead already to a decent approximation of the GW circle.

In a next step, we include fat links with positive staples — $\alpha = 0.3$ is a good value — and use again the critical link amplification parameter, which now amounts to $u \simeq 0.76$. Indeed, this helps to move the eigenvalues closer to the GW circle, as Fig. 11 (left) shows.

Still one would like to further reduce the imaginary part of the eigenvalues; in particular the upper (and lower) arc still calls for improvement. Considering the structure of $D_{HF}(x, y, U) = \rho_\mu(x, y, U)\gamma_\mu + \lambda(x, y, U)$, we recognize that ρ_μ is responsible for the imaginary part, so we multiply a damping factor $v \lesssim 1$ on each link *only* in the vector term ρ_μ. The scalar term λ, which controls the left and right arc already successfully, remains untouched. The optimal

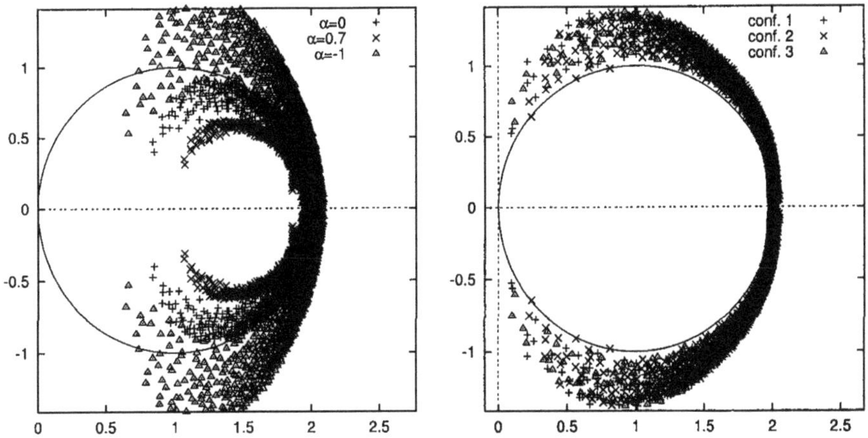

Figure 10: *The effect of fat links (left, $u = 1$) and link amplification (right, $\alpha = 0, u = 0.8$) for a HF spectrum in QCD at $\beta = 6$ on a 4^4 lattice.*

value for the new parameter is $v \simeq 0.9$ without fat links, and $v \simeq 0.92$ at $\alpha = 0.3$, always at critical u (which remains practically unchanged). Now also the upper arc follows the GW circle closely, and the fat link helps the eigenvalues to spread less around the circle, see Fig. 11 (right). This is the best approximation achieved so far, and we are confident that this is about the optimum that can be achieved with $O(10)$ independent parameters. In fact, even if we include fat links we are using just 10 independent parameters.

Of course one might still try further parameters, but they should only be included if they really lead to a significant progress. One could include terms with a new Dirac structure, and we are currently testing the clover term: its inclusion (with a positive coefficient) helps to improve the physically important arc near zero a little, but it distorts the opposite arc (in the sense that the eigenvalues fluctuate stronger around the right half-circle). Generally the clover term tends to attract the eigenvalues closer to the real axis, which is also known from the Wilson fermion [34]. Hence the optimal value of v increases a little: for instance at clover coefficient 0.15 (and $\alpha = 0.3$) it amounts to $v \simeq 0.94$, while $u \simeq 0.767$ is critical.

The consideration of the "magnetic mass" m_B suggests that also a term $\propto \gamma_\mu \gamma_\nu \gamma_\rho$ could be useful [15] (we refer to the Pauli term $\vec{\sigma}\vec{B}/2m_B$ in the low energy expansion).

What is computationally simple and perhaps promising is an extension of the fat link to include also (selected) paths of length 5. All this is currently under investigation and the results will be reported in Ref. [29].

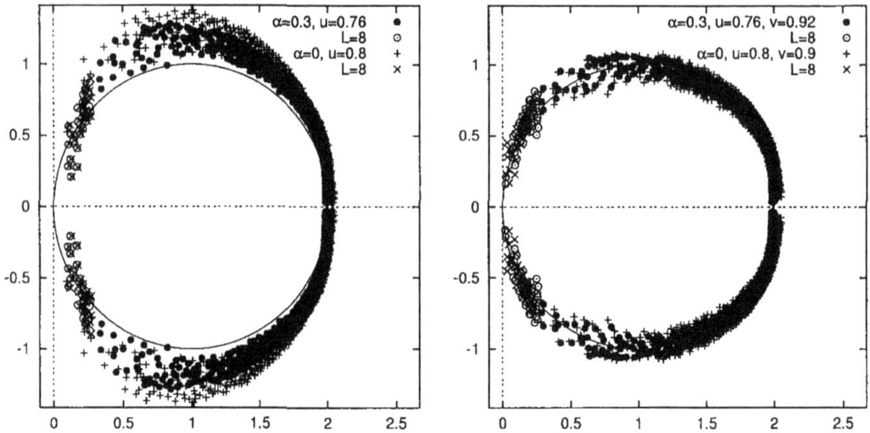

Figure 11: *QCD spectra on* 4^4 *lattices for the HF at critical link amplification parameter u: with and without fat links (left) and typical results if we further include the kinetic damping parameter v at its optimal value (right). We also show the "continuation" near 0 in each case on an* 8^4 *lattice with the same parameters, everything at* $\beta = 6$.

5 Summary and Outlook

Our concept is to first construct a short-ranged approximate GW fermion with a good scaling behavior and quasi rotational invariance. Inserting this fermion into the overlap formula makes the chirality exact. Since the alteration is small, the good scaling and approximate rotational symmetry is essentially preserved, so that we obtain an improved overlap fermion. For the same reason, such an overlap fermion has a high level of locality and its iterative evaluation is fast. This program was tested extensively in the Schwinger model [17]. Furthermore, for very good approximate GW operators the interval (for the mass parameter μ) where the overlap formula can be used safely is extended, and such an interval exists up to stronger coupling compared to the Neuberger fermion (which inserts the Wilson fermion).

We add that the same mechanism applies to domain wall fermions [35]: inserting an improved 4d fermion may induce the same advantages [9,36].

For QCD, we use a truncated perfect free HF as our building block, and construct a good gauging in the sense of a small GWR violation. At $\beta = 6$ this is achieved, as we observed from the spectrum: it is close to a GW circle, hence the fermion approximately obeys the GWR with a point-like kernel R.

In our construction, we start form the minimally gauged HF, remove the additive mass renormalization by a critical amplification factor for each link, and suppress the vector term only in order to damp the imaginary part of the eigenvalues, so that the spectrum follows the shape of a GW unit circle.

Fat links help to reduce the fluctuations of the eigenvalues around that circle, and hence provide an approximation on a satisfactory level. It seems quite optimal for a set of about 10 parameters. Perhaps extended staple terms lead to some further progress, and the clover term helps a little with respect to the fine resolution close to zero, which we now focus on. In any case, a good approximation to a GW fermion in QCD has been accomplished, and it is ready to be inserted into the overlap formula (5) [29].

Acknowledgments

I am very much indebted to Ivan Hip for our ongoing collaboration, which was the bases of this talk, and for reading the manuscript. I also thank him, Norbert Eicker and Thomas Lippert for their work on implementing hypercube fermions in QCD, and we were glad about the assistance by Christoph Best. Next I would like to thank the organizers of this workshop, in particular Xiang-Qian Luo and Eric Gregory. Finally I enjoyed stimulating discussions related this subject with Ruedi Burkhalter, Philippe de Forcrand, Yoshio Kikukawa, Martin Lüscher, Gerrit Schierholz and Urs Wenger.

References

1. H.B. Nielsen and M. Ninomiya, *Phys. Lett.* B **105**, 219 (1981); *Nucl. Phys.* B **185**, 20 (1981) [Erratum: B **195** 541 (1982)].
2. P.H. Ginsparg and K.G. Wilson, *Phys. Rev.* D **25**, 2649 (1982).
3. P. Hasenfratz, V. Laliena and F. Niedermayer, *Phys. Lett.* B **427**, 125 (1998). P. Hasenfratz, *Nucl. Phys.* B **525**, 401 (1998).
4. M. Lüscher, *Phys. Lett.* B **428**, 342 (1998).
5. K. Fujikawa, *Nucl. Phys.* B **546**, 480 (1999). Y. Kikukawa and A. Yamada, *Nucl. Phys.* B **547**, 413 (1999). H. Suzuki, *Prog. Theor. Phys.* **102**, 141 (1999). D. Adams, hep-lat/9812003. T.-W. Chiu and T.-H. Hsieh, hep-lat/9901011. T. Reisz and H.J. Rothe, *Phys. Lett.* B **455**, 246 (1999).
6. O. Bär and I. Campos, hep-lat/0001025.
7. S. Chandrasekharan, *Phys. Rev.* D **60**, 074503 (1999).
   Y. Kikukawa and A. Yamada, *Nucl. Phys.* **B547** (1999) 413.
8. M. Lüscher, *Nucl. Phys.* B **538**, 515 (1999); *Nucl. Phys.* B **549**, 295 (1999); *Nucl. Phys.* B **568**, 162 (2000); *JHEP* **0006**, 028 (2000). H. Suzuki, *Prog. Theor. Phys.* **101**, 1147 (1999); hep-lat/0002009.
9. W. Bietenholz, *Eur. Phys. J.* C **6**, 537 (1999).
10. I. Horváth, *Phys. Rev. Lett.* **81**, 4063 (1998). I. Horváth, C. Balwe and R. Mendris, hep-lat/0006027. W. Bietenholz, hep-lat/9901005.
11. P. Hasenfratz and F. Niedermayer, *Nucl. Phys.* B **414**, 785 (1994).

12. C.B. Lang and T.K. Pany, *Nucl. Phys.* B **513**, 645 (1998).
13. K. Orginos et al., *Nucl. Phys. [Proc. Suppl.]* **63**, 904 (1998). T. DeGrand, *Phys. Rev.* D **58**, 094503 (1998); *Phys. Rev.* D **60**, 094501 (1999). W. Bietenholz, N. Eicker, A. Frommer, Th. Lippert, B. Medeke, K. Schilling and G. Weuffen, *Comput. Phys. Commun.* **119**, 1 (1999).
14. W. Bietenholz and U.-J. Wiese, *Nucl. Phys.* B **464**, 319 (1996).
15. W. Bietenholz, R. Brower, S. Chandrasekharan and U.-J. Wiese, *Nucl. Phys. [Proc. Suppl.]* **53**, 921 (1997).
16. W. Bietenholz and U.-J. Wiese, *Phys. Lett.* B **426**, 114 (1998).
17. W. Bietenholz and I. Hip, *Nucl. Phys.* B **570**, 423 (2000); *Nucl. Phys. [Proc. Suppl.]* **83-84**, 600 (2000).
18. I. Sachs and A. Wipf, *Helv. Phys. Acta* **65**, 652 (1992).
19. H. Neuberger, *Phys. Lett.* B **417**, 141 (1998); *Phys. Lett.* B **427**, 353 (1998).
20. P. Hernández, K. Jansen and M. Lüscher, *Nucl. Phys.* B **552**, 363 (1999).
21. H. Neuberger, *Phys. Rev.* D **61**, 085015 (2000).
22. R.G. Edwards, U.M. Heller and R. Narayanan, *Phys. Rev.* D **59**, 094510 (1999). P. Hernández, K. Jansen and L. Lellouch, *Phys. Lett.* B **469**, 198 (1999).
23. S.J. Dong, F.X. Lee, K.F. Liu and J.B. Zhang, hep-lat/0006004.
24. S. Capitani et al., *Phys. Lett.* B **468**, 150 (1999); hep-lat/0007004.
25. F. Niedermayer, *Nucl. Phys. [Proc. Suppl.]* **73**, 105 (1999).
26. W. Bietenholz and U.-J. Wiese, *Phys. Lett.* B **378**, 222 (1996).
27. W. Bietenholz, *in* "Lattice Fermions and the Structure of the Vacuum" V. Mitrjushkin and G. Schierholz (eds.), Kluwer Academic C 553 (2000) p. 77 (hep-lat/0001001).
28. R. Brower and B. Svetitsky, *Phys. Rev.* D **61**, 114511 (2000).
29. W. Bietenholz, N. Eicker, I. Hip, Th. Lippert and K. Schilling, *in preparation.*
30. C. Gattringer and I. Hip, *Phys. Lett.* B **480**, 112 (2000).
31. P. Hasenfratz, S. Hauswirth, K. Holland, Th. Jörg, F. Niedermayer and U. Wenger, hep-lat/0003013.
32. T. DeGrand, A. Hasenfratz and T. Kovács, *Nucl. Phys.* B **520**, 301 (1998). M. Stephenson, C. DeTar, T. DeGrand and A. Hasenfratz, hep-lat/9910023.
33. G.P. Lepage and P.B. Mackenzie, *Phys. Rev.* D **48**, 2250 (1993).
34. C. Gattringer and I. Hip, *Nucl. Phys.* B **541**, 305 (1999).
35. D. Kaplan, *Phys. Lett.* B **288**, 342 (1992). Y. Shamir, *Nucl. Phys.* B **406**, 90 (1993).
36. Y. Shamir, hep-lat/0003024.

LOW ENERGY STRUCTURE OF DOMAIN WALL FERMIONS

Y. KIKUKAWA

Department of Physics, Nagoya University, 464-8602, Japan
E-mail: kikukawa@eken.phys.nagoya-u.ac.jp

Chiral symmetry and locality property of low energy effective action of domain-wall fermion are discussed.

1 Introduction

Locality properties of Neuberger's lattice Dirac operator [1], which is derived from the overlap formalism [2] and satisfies the Ginsparg-Wilson relation [3], has been examined by Hernándes, Jansen and Lüscher [4]. For a certain class of gauge fields with small lattice field strength, exponential bounds have been proved rigorously on the kernels of the Dirac operator and its differentiations with respect to the gauge field. These properties assure that the index theorem holds true on the lattice [5,2]. The index theorem implies the topological properties of chiral anomalies. It plays the crucial role in the recent construction of lattice chiral gauge theories [6].

The purpose of this paper is to argue the locality of the low energy effective action of the domain-wall fermion.[7,8,9 10] It has been known that the partition function of the domain-wall fermion, in the anti-periodic subtraction scheme, reduces to a single determinant of an effective four-dimensional Dirac operator [11],

$$D_{\text{eff}}^{(N)} = \frac{1}{2a}\left(1 + \gamma_5 \tanh \frac{N}{2} a_5 \widetilde{H}\right). \tag{1.1}$$

Here N and a_5 denote the lattice size and the lattice spacing of the fifth dimension, respectively. $\widetilde{H}$ is defined through the transfer matrix of the five-dimensional Wilson fermion

$$T = e^{-a_5 \widetilde{H}} = \begin{pmatrix} \frac{1}{B} & -\frac{1}{B}C \\ -C^\dagger \frac{1}{B} & B + C^\dagger \frac{1}{B} C \end{pmatrix}, \tag{1.2}$$

where

$$C = a_5\, \sigma_\mu \frac{1}{2}\left(\nabla_\mu + \nabla_\mu^*\right), \quad B = 1 + a_5\left(-\frac{a}{2}\nabla_\mu \nabla_\mu^* - \frac{m_0}{a}\right). \tag{1.3}$$

The limit $N \to \infty$ is defined well as long as $\widetilde{H}^2 > 0$. The effective Dirac operator Eq. (1.1) then reduces to Neuberger's lattice Dirac operator with $\widetilde{H}$,

$$D_{\text{eff}} = \frac{1}{2a}\left(1 + \gamma_5 \frac{\widetilde{H}}{\sqrt{\widetilde{H}^2}}\right), \tag{1.4}$$

and turns out to satisfy the Ginsparg-Wilson relation. Moreover, the propagator of the light fermion field introduced by Furman and Shamir [9],

$$q(x) = \psi_L(x,1) + \psi_R(x,N), \qquad \bar{q}(x) = \bar{\psi}_L(x,1) + \bar{\psi}_R(x,N), \tag{1.5}$$

can be expressed in terms of the effective Dirac operator [12]:

$$\langle q(x)\bar{q}(y)\rangle = \frac{a_5}{a^4}\left(\frac{1}{a} {D_{\text{eff}}^{(N)}}^{-1} - \delta(x,y)\right). \tag{1.6}$$

In view of this direct relation, it seems reasonable to argue locality in the domain-wall fermion approach through the locality properties of the effective Dirac operators Eq. (1.1) and Eq. (1.4). It is expected that a similar exponential bound could be established under certain conditions, like the result obtained by Hernández, Jansen and Lüscher [4]. In fact all the properties proved rigorously for Neuberger's Dirac operator can be established for the effective Dirac operator of domain wall fermion.[13]

This paper is organized as follows. In section 2, we briefly review the domain-wall fermion in order to fix our notation. In section 3, we describe how to derive the integral representation for the effective Dirac operator. We also discuss how the Ginsparg-Wilson relation for the effective Dirac operator follows in this integral representation. In section 4, we discuss the positivity of the five-dimensional Wilson-Dirac operator with the anti-periodic boundary condition. With this result, we consider exponential bounds for the effective Dirac operator in section 5.

2 Domain-wall fermion in the anti-periodic subtraction scheme

In this section, we review the domain-wall fermion [7,8,9] and fix our notation. The domain-wall fermion is defined by the five-dimensional Wilson-Dirac fermion with the Dirichlet boundary condition.

$$S_{\text{DW}} = \sum_{t=1}^{N} a^4 \sum_{x} \bar{\psi}(x,t) D_{5\text{w}} \psi(x,t), \tag{2.1}$$

$$D_{5\text{w}} = \gamma_\mu \frac{1}{2}\left(\nabla_\mu + \nabla_\mu^*\right)\delta_{st} + P_L M_{st} + P_R M_{st}^\dagger. \tag{2.2}$$

We assume the lattice size of the fifth dimension N is even. For $N = 6$, the mass matrix reads

$$M_{st} = \frac{1}{a_5}\begin{pmatrix} B & -1 & 0 & 0 & 0 & 0 \\ 0 & B & -1 & 0 & 0 & 0 \\ 0 & 0 & B & -1 & 0 & 0 \\ 0 & 0 & 0 & B & -1 & 0 \\ 0 & 0 & 0 & 0 & B & -1 \\ 0 & 0 & 0 & 0 & 0 & B \end{pmatrix}. \tag{2.3}$$

The chiral transformation is introduced as vector-like one so that the symmetry breaking is minimized [9]:

$$\delta\psi(x,t) = -\psi(x,t) \qquad t \leq \frac{N}{2}, \tag{2.4}$$

$$\delta\psi(x,t) = +\psi(x,t) \qquad t \geq \frac{N}{2} + 1. \tag{2.5}$$

The partition function of the domain-wall fermion may be defined with the subtraction of the Pauli-Villars fields, which is subject to the anti-periodic boundary condition in the fifth dimension.

$$Z_{\rm DW} = \frac{\det D_{\rm 5w}}{\det \overline{D}_{\rm 5w}}, \tag{2.6}$$

where

$$\overline{D}_{\rm 5w} = \gamma_\mu \frac{1}{2}\left(\nabla_\mu + \nabla_\mu^*\right)\delta_{st} + P_L \overline{M}_{st} + P_R \overline{M}^\dagger_{st}, \tag{2.7}$$

$$\overline{M}_{st} = \frac{1}{a_5}\begin{pmatrix} B & -1 & 0 & 0 & 0 & 0 \\ 0 & B & -1 & 0 & 0 & 0 \\ 0 & 0 & B & -1 & 0 & 0 \\ 0 & 0 & 0 & B & -1 & 0 \\ 0 & 0 & 0 & 0 & B & -1 \\ 1 & 0 & 0 & 0 & 0 & B \end{pmatrix}, \qquad (N=6) \tag{2.8}$$

For later convenience, we perform a chirally asymmetric parity transformation in the fifth dimension:

$$\psi(x,t) = (P_R + P_L P)_{ts}\, \psi'(x,s), \tag{2.9}$$

$$\bar{\psi}(x,t) = \bar{\psi}'(x,s)\,(P_R P + P_L)_{st}, \tag{2.10}$$

where

$$P_{st} = \begin{pmatrix} 0\,0\,0\,0\,0\,1 \\ 0\,0\,0\,0\,1\,0 \\ 0\,0\,0\,1\,0\,0 \\ 0\,0\,1\,0\,0\,0 \\ 0\,1\,0\,0\,0\,0 \\ 1\,0\,0\,0\,0\,0 \end{pmatrix} \qquad (N=6). \tag{2.11}$$

Accordingly, the five-dimensional Dirac operators are transformed as follows:

$$D'_{5\mathrm{w}} = \gamma_\mu \frac{1}{2}\left(\nabla_\mu + \nabla^*_\mu\right) + M^H_{st}, \tag{2.12}$$

$$\overline{D}'_{5\mathrm{w}} = \gamma_\mu \frac{1}{2}\left(\nabla_\mu + \nabla^*_\mu\right) + \overline{M}^H_{st}, \tag{2.13}$$

where, for $N = 6$,

$$M^{\mathrm{H}}_{st} = \frac{1}{a_5}\begin{pmatrix} 0 & 0 & 0 & 0 & -1 & B \\ 0 & 0 & 0 & -1 & B & 0 \\ 0 & 0 & -1 & B & 0 & 0 \\ 0 & -1 & B & 0 & 0 & 0 \\ -1 & B & 0 & 0 & 0 & 0 \\ B & 0 & 0 & 0 & 0 & 0 \end{pmatrix}, \quad \overline{M}_{st} = \frac{1}{a_5}\begin{pmatrix} 0 & 0 & 0 & 0 & -1 & B \\ 0 & 0 & 0 & -1 & B & 0 \\ 0 & 0 & -1 & B & 0 & 0 \\ 0 & -1 & B & 0 & 0 & 0 \\ -1 & B & 0 & 0 & 0 & 0 \\ B & 0 & 0 & 0 & 0 & 1 \end{pmatrix}. \tag{2.14}$$

In this basis, the chiral transformation adopted by Shamir and Furman [9] can be expressed as follows:

$$\delta\psi'_s(x) = \left(\Gamma_5\right)_{st}\psi'_t(x), \tag{2.15}$$

where Γ_5 is given (for $N = 6$) by

$$\left(\Gamma_5\right)_{st} = \begin{pmatrix} -\gamma_5 & 0 & 0 & 0 & 0 & 0 \\ 0 & -\gamma_5 & 0 & 0 & 0 & 0 \\ 0 & 0 & -\gamma_5 & 0 & 0 & 0 \\ 0 & 0 & 0 & \gamma_5 & 0 & 0 \\ 0 & 0 & 0 & 0 & \gamma_5 & 0 \\ 0 & 0 & 0 & 0 & 0 & \gamma_5 \end{pmatrix} \qquad (N=6). \tag{2.16}$$

With this definition of the chiral transformation, $D'_{5\mathrm{w}}$ and $\overline{D}'_{5\mathrm{w}}$ satisfy the following identities, respectively,

$$\left\{\Gamma_5 D'_{5\mathrm{w}} + D'_{5\mathrm{w}}\Gamma_5\right\}_{st} = \frac{2}{a_5}\gamma_5\delta_{s\frac{N}{2}}\delta_{t\frac{N}{2}}, \tag{2.17}$$

$$\left\{\Gamma_5 \overline{D}'_{5\mathrm{w}} + \overline{D}'_{5\mathrm{w}}\Gamma_5\right\}_{st} = \frac{2}{a_5}\gamma_5\delta_{s\frac{N}{2}}\delta_{t\frac{N}{2}} + \frac{2}{a_5}\gamma_5\delta_{sN}\delta_{tN}. \tag{2.18}$$

The chiral symmetry breaking occurs at $t = \frac{N}{2}$ in the five-dimensional Dirac operator for the domain-wall fermion. On the other hand, it occurs both at $t = \frac{N}{2}$ and at $t = N$ for the Pauli-Villars field, because of the anti-periodic boundary condition.

3 Effective four-dimensional Dirac operator

3.1 An integral representation of the effective four-dimensional Dirac operator

The functional determinant of the domain-wall fermion, in the anti-periodic subtraction scheme, reduces to a single determinant of a four-dimensional Dirac operator,

$$\frac{\det D_{5\mathrm{w}}}{\det \overline{D}_{5\mathrm{w}}} = \det aD_{\mathrm{eff}}^{(N)}. \tag{3.1}$$

In this section, we reproduce this result and derive an integral representation for the effective four-dimensional Dirac operator.

We may write the partition function as follows:

$$\frac{\det D_{5\mathrm{w}}}{\det \overline{D}_{5\mathrm{w}}} = \frac{\det D'_{5\mathrm{w}}}{\det \overline{D}'_{5\mathrm{w}}} = \det \left(D'_{5\mathrm{w}} \left\{ \overline{D}'_{5\mathrm{w}} \right\}^{-1} \right). \tag{3.2}$$

Then, we note a simple relation between two five-dimensional Wilson-Dirac operators:

$$D'_{5\mathrm{w}} = \overline{D}'_{5\mathrm{w}} - \frac{1}{a_5} \delta_{sN} \delta_{Nt}. \tag{3.3}$$

This relation implies that

$$D'_{5\mathrm{w}} \left\{ \overline{D}'_{5\mathrm{w}} \right\}^{-1} = \delta_{st} - \frac{1}{a_5} \delta_{sN} \left\{ \overline{D}'_{5\mathrm{w}} \right\}^{-1}_{1t}. \tag{3.4}$$

Since this matrix is lower triangle in the lattice indices of the fifth dimension, we can easily see that its determinant reduces to a single four-dimensional determinant:

$$\det \left(D'_{5\mathrm{w}} \left\{ \overline{D}'_{5\mathrm{w}} \right\}^{-1} \right) = \det \left(1 - \frac{1}{a_5} \left\{ \overline{D}'_{5\mathrm{w}} \right\}^{-1}_{NN} \right). \tag{3.5}$$

From this result, we may set

$$aD_{\mathrm{eff}}^{(N)} = 1 - \frac{1}{a_5} \left\{ \overline{D}'_{5\mathrm{w}} \right\}^{-1}_{NN} \tag{3.6}$$

$$= 1 - \frac{1}{a_5}\Big(P_R\left\{\overline{D}_{5\mathrm{w}}\right\}^{-1}_{NN} P_L + P_L\left\{\overline{D}_{5\mathrm{w}}\right\}^{-1}_{11} P_R$$
$$+P_R\left\{\overline{D}_{5\mathrm{w}}\right\}^{-1}_{N1} P_R + P_L\left\{\overline{D}_{5\mathrm{w}}\right\}^{-1}_{1N} P_L\Big). \quad (3.7)$$

Thus the effective four-dimensional Dirac operator can be expressed in terms of the inverse of the five-dimensional Wilson-Dirac operator *with the anti-periodic boundary condition.* Since the gauge field is four-dimensional, the inverse of this five-dimensional Wilson-Dirac operator may be expressed as follows:

$$\left\{a_5\overline{D}_{5\mathrm{w}}\right\}^{-1}_{st} = \frac{1}{N}\sum_p e^{ip(s-t)}\left\{i\gamma_5\sin p + 1 - \cos p + a_5\left(D_{\mathrm{w}} - \frac{m_0}{a}\right)\right\}^{-1}, \quad (3.8)$$

where the summation is taken over the discrete momenta $p = \frac{2\pi}{N}(k - \frac{1}{2})$ $(k = 1, 2, \cdots, N)$ and D_{w} is the four-dimensional Wilson-Dirac operator.

3.2 The Ginsparg-Wilson relation

Next we discuss the Ginsparg-Wilson relation for the effective Dirac operator in the integral representation. As we have seen in the previous subsection, the effective Dirac operator, D_{eff}, is defined by

$$aD_{\mathrm{eff}} = 1 - \frac{1}{a_5}\left\{\overline{D}'_{5\mathrm{w}}\right\}^{-1}_{NN} \qquad (N = \infty). \quad (3.9)$$

If it would satisfies the Ginsparg-Wilson relation

$$\gamma_5 D_{\mathrm{eff}} + D_{\mathrm{eff}}\gamma_5 = 2aD_{\mathrm{eff}}\gamma_5 D_{\mathrm{eff}}, \quad (3.10)$$

then the following identity must hold true in the limit of $N \to \infty$:

$$\gamma_5\left\{\bar{D}'_{5\mathrm{w}}\right\}^{-1}_{NN} + \left\{\bar{D}'_{5\mathrm{w}}\right\}^{-1}_{NN}\gamma_5 = \frac{2}{a_5}\left\{\bar{D}'_{5\mathrm{w}}\right\}^{-1}_{NN}\gamma_5\left\{\bar{D}'_{5\mathrm{w}}\right\}^{-1}_{NN} \qquad (N = \infty). \quad (3.11)$$

We may compare this identity with Eq. (2.18) which express the chiral property of $\bar{D}'_{5\mathrm{w}}$ under the chiral transformation introduced by Furman and Shamir Eq. (2.15). The latter we may write

$$\left\{\Gamma_5\left\{\overline{D}'_{5\mathrm{w}}\right\}^{-1} + \left\{\overline{D}'_{5\mathrm{w}}\right\}^{-1}\Gamma_5\right\}_{st} = \frac{2}{a_5}\left\{\overline{D}'_{5\mathrm{w}}\right\}^{-1}_{s\frac{N}{2}}\gamma_5\left\{\overline{D}'_{5\mathrm{w}}\right\}^{-1}_{\frac{N}{2}t}$$
$$+\frac{2}{a_5}\left\{\overline{D}'_{5\mathrm{w}}\right\}^{-1}_{sN}\gamma_5\left\{\overline{D}'_{5\mathrm{w}}\right\}^{-1}_{Nt}. \quad (3.12)$$

Setting $s = t = N$, we obtain

$$\gamma_5 \left\{\overline{D}'_{5\mathrm{w}}\right\}^{-1}_{NN} + \left\{\overline{D}'_{5\mathrm{w}}\right\}^{-1}_{NN} \gamma_5 = \frac{2}{a_5} \left\{\overline{D}'_{5\mathrm{w}}\right\}^{-1}_{N\frac{N}{2}} \gamma_5 \left\{\overline{D}'_{5\mathrm{w}}\right\}^{-1}_{\frac{N}{2}N} + \frac{2}{a_5} \left\{\overline{D}'_{5\mathrm{w}}\right\}^{-1}_{NN} \gamma_5 \left\{\overline{D}'_{5\mathrm{w}}\right\}^{-1}_{NN}. \tag{3.13}$$

Then we can see that Eq. (3.11) is equivalent to the following condition in the limit of $N \to \infty$:

$$\left\{\overline{D}'_{5\mathrm{w}}\right\}^{-1}_{N\frac{N}{2}} = 0 \qquad (N \to \infty). \tag{3.14}$$

4 Positivity of the square of the five-dimensional Wilson-Dirac operator with anti-periodic boundary condition

From the results in the previous section we see that for the effective four-dimensional Dirac operator to be defined consistently, it is required that the five-dimensional Wilson-Dirac operator with the anti-periodic boundary condition should be non-singular and invertible. To examine this requirement, we evaluate the square of the five-dimensional Wilson-Dirac operator. Setting $a_5 = a$ for simplicity, we have

$$\begin{aligned}&\{i\gamma_5 \sin p + 1 - \cos p + (aD_{\mathrm{w}} - m_0)\}^\dagger \{i\gamma_5 \sin p + 1 - \cos p + (aD_{\mathrm{w}} - m_0)\} \\ &= 4\sin^2(p/2)\left(1 - m_0 - \frac{a^2}{2}\nabla_\mu \nabla^*_\mu\right) + (aD_{\mathrm{w}} - m_0)^\dagger (aD_{\mathrm{w}} - m_0).\end{aligned} \tag{4.1}$$

Then we see that the positivity of $a^2\overline{D}^\dagger_{5\mathrm{w}}\overline{D}_{5\mathrm{w}}$ is closely related to the positivity of the four-dimensional Wilson-Dirac operator square, $(aD_{\mathrm{w}} - m_0)^\dagger(aD_{\mathrm{w}} - m_0)$. In fact, according to the result by Hernándes et al.[4], if the plaquette variables $U(p)$ are uniformly bounded as

$$\| 1 - U(p) \| < \epsilon, \tag{4.2}$$

we obtain

$$a^2\overline{D}^\dagger_{5\mathrm{w}}\overline{D}_{5\mathrm{w}} \geq \left\{(1-30\epsilon)^{\frac{1}{2}} - |1 - m_0|\right\}^2 \quad \text{if} \quad 1 - 30\epsilon > |1 - m_0|^2. \tag{4.3}$$

Then we may assume the positive lower and upper bounds of the square of the five-dimensional Wilson-Dirac operator with the anti-periodic boundary condition as

$$0 < \tilde{\alpha} \leq a^2\overline{D}^\dagger_{5\mathrm{w}}\overline{D}_{5\mathrm{w}} \leq \tilde{\beta}, \tag{4.4}$$

under the following condition,

$$\| 1 - U(p) \| < \epsilon, \qquad \epsilon < \frac{1}{30}\left(1 - |1 - m_0|^2\right). \tag{4.5}$$

5 Expansion with Chebycheff polynomials and an exponential bound

Given the bounds on the square of the five-dimensional Wilson- Dirac operator with the anti-periodic boundary condition, we can obtain an expansion of its inverse, using the generating function of the Chebycheff polynomials

$$\frac{1}{1 - 2tz + t^2} = \sum_{k=0}^{\infty} t^k U_k(z), \qquad \|U_k(z)\| \leq U_k(1) = k. \tag{5.1}$$

Following [4], we set

$$z = \frac{\tilde{\beta} + \tilde{\alpha} - 2a^2 \bar{D}_{5\mathrm{w}}^{\dagger} \bar{D}_{5\mathrm{w}}}{\tilde{\beta} - \tilde{\alpha}}, \qquad t = e^{-\tilde{\theta}}, \qquad \cosh\theta = \frac{\tilde{\beta} + \tilde{\alpha}}{\tilde{\beta} - \tilde{\alpha}}. \tag{5.2}$$

Then we obtain

$$\frac{1}{a^2 \bar{D}_{5\mathrm{w}}^{\dagger} \bar{D}_{5\mathrm{w}}} = \frac{4t}{\tilde{\beta} - \tilde{\alpha}} \sum_{k=0} t^k U_k(z). \tag{5.3}$$

This defines an expansion in terms of the square of the five-dimensional Wilson-Dirac operator with only nearest-neighbor and next-to-nearest-neighbor couplings. In order to contribute to the kernel of the operator Eq. (5.3) between two lattice sites (x, t) and (y, s) of the lattice distance $d_5(x, t; y, s) = |x - y| + \min(|s - t|, N - s - t)$, the order of the polynomials $U_k(x)$ in the expansion must be greater than $\frac{d_5(x,t;y,s)}{2a}$:

$$k \geq \frac{d_5(x, t; y, s)}{2a} \tag{5.4}$$

Then for the given distance $d_5(x, t; y, s)$, the series expansion Eq. (5.3) can be arranged as follows:

$$\frac{4t}{\tilde{\beta} - \tilde{\alpha}} \exp\left\{ -\frac{\tilde{\theta}}{2a} d_5(x, t; y, s) \right\} \cdot \sum_{k=0} t^k U_{k+d/2a}(z). \tag{5.5}$$

Noting the bound on the polynomials, $\|U_k(z)\| \leq k$, we obtain

$$\left\| \frac{1}{a^2 \bar{D}_{5\mathrm{w}}^{\dagger} \bar{D}_{5\mathrm{w}}} (x, t; y, s) \right\|$$

$$\leq \frac{4t}{\tilde{\beta}-\tilde{\alpha}}\left(\frac{1}{1-t}\frac{d_5(x,t;y,s)}{2a}+\frac{t}{(1-t)^2}\right)\exp\left\{-\frac{\tilde{\theta}}{2a}d_5(x,t;y,s)\right\}. \tag{5.6}$$

From this bound, it follows immediately that

$$\lim_{N\to\infty}\left\{\overline{D}'_{5\mathrm{w}}\right\}^{-1}_{N\frac{N}{2}}=0. \tag{5.7}$$

This completes the proof of the Ginsparg-Wilson relation under the condition on the plaquette variables Eq. (4.5). The bound then implies that the effective four-dimensional Dirac operator is exponentially bounded in the four-dimensional lattice and thus local. Note again that this proof does not refer to the transfer matrix and it applies for any value of $m_0 \in [0,2]$.

Acknowledgments

The author would like to thank X.-Q. Luo for the kind hospitality at the International Workshop on Nonperturbative Methods and Lattice QCD, Guangzhou, China.

References

1. H. Neuberger, Phys. Lett. **B417** (1998) 141; Phys. Lett. **B427** (1998) 353.
2. R. Narayanan and H. Neuberger, Nucl. Phys. **B412** (1994) 574; Phys. Rev. Lett. **71** (1993) 3251; Nucl. Phys. **B443** (1995) 305.
3. P. H. Ginsparg, K. G. Wilson, Phys. Rev. **D25** (1982) 2649.
4. P. Hernándes, K. Jansen, M. Lüscher, Nucl.Phys. B552 (1999) 363-378, `hep-lat/9808010`.
5. P. Hasenfratz, V. Laliena, F. Niedermayer, Phys. Lett. B427 (1998) 125.
6. M. Lüscher, plenary talk at the International Symposium on Lattice Field Theory, Pisa, Italy, June 29-July 3, 1999.
7. D. B. Kaplan, Phys. Lett. **B288** (1992) 342.
8. Y. Shamir, Nucl. Phys. **B406** (1993) 90.
9. V. Furman, Y. Shamir, Nucl. Phys. **B439** (1995) 54.
10. P. Vranas, plenary talk at the International Symposium on Lattice Field Theory, Bangalore, India, August 17-22, 2000, `hep-lat/0011066`
11. H. Neuberger, Phys. Rev. **D57** (1998) 5417.
12. Y. Kikukawa, T. Noguchi, `hep-lat/9902022`.
13. Y. Kikukawa, Nucl. Phys. B584 (2000) 511.

AN ALMOST PERFECT LATTICE ACTION FOR INFRARED QCD

T. SUZUKI
Institute for Theoretical Physics, Kanazawa University,
Kanazawa 920-1192, Japan
E-mail: suzuki@hep.s.kanazawa-u.ac.jp

A block-spin transformation on the dual lattice leads us to an almost perfect lattice action for monopoles and strings in QCD. The perfect operator for a static quark potential is fixed when we compare the above action with the perfect action obtained analytically after infinite-step block-spin transformations in a simple case. The continuum rotational invariance is restored and the physical value of the string tension is reproduced fairly well. Gauge independence of the abelian and the monopole scenario is discussed. This talk is based on our recent works.[1,2,3,4,5]

1 Introduction

Low-energy effective theory of QCD is important for analytical understanding of hadron physics. One of approaches is to search for relevant dynamical variables and to construct an effective theory for these variables.

From this point of view, the idea proposed by 't Hooft [6] is very promising. The confinement of quarks can be explained as the dual Meissner effect which is due to condensation of these monopoles after an abelian projection.

Many numerical results support the dual superconductor picture of confinement [7] in the Maximal abelian (MA) gauge [8] in the framework of lattice QCD. We expect hence, after integrating out all degrees of freedom other than the monopoles, an effective theory described by the monopoles works well in the IR region of QCD.

The effective monopole action on the MA projection of $SU(2)$ lattice QCD was first obtained by Shiba and Suzuki [9] using an inverse Monte-Carlo method [10]. See also Ref.[11].

The purpose of this talk is to review briefly the results in our recent publications [1,2,3,4,5].

2 An (almost) perfect monopole action

The method to derive the monopole action is the following.

1 We generate link fields $\{U(s,\mu)\}$ using the simple Wilson action for SU(2) and SU(3) QCD. We consider 24^4 and 48^4 hyper-cubic lattice for $\beta =$

2.0 ~ 2.8 (SU(2)) and for $\beta = 5.6 \sim 6.4$ (SU(3)).

2 Next we perform the abelian projection in the MA gauge to separate abelian link variables $\{u(s,\mu)\}$.

3 Monopole currents $k_\mu(s)$ can be defined from abelian link variables following DeGrand and Toussaint[12].

4 We determine the set of couplings G_i from the monopole current ensemble $\{k_\mu(s)\}$ with the aid of an inverse Monte-Carlo method first developed by Swendsen[10] and extended to closed monopole currents by Shiba and Suzuki [9]. Here the monopole action can be written as a linear combination of some operators $\mathcal{S}[k] = \sum_i G_i \mathcal{S}_i[k]$.

Practically, we have to restrict the number of interaction terms. The form of actions adopted here is 27 quadratic interactions and 4-point and 6-point interactions.

5 We perform a block-spin transformation in terms of the monopole currents on the dual lattice to investigate the renormalization flow in the IR region. We adopt $n = 1, 2, 3, 4, 6, 8$ extended monopole currents as an n blocked operator[13]:

$$K_\mu(s^{(n)}) = \sum_{i,j,l=0}^{n-1} k_\mu(ns^{(n)} + (n-1)\hat{\mu} + i\hat{\nu} + j\hat{\rho} + l\hat{\sigma}) \qquad (1)$$

The renormalized lattice spacing is $b = na(\beta)$ and the continuum limit is taken as the limit $n \to \infty$ for a fixed physical length b.

6 The physical length $b = na(\beta)$ is taken in unit of the physical string tension $\sqrt{\sigma_{phys}}$.

3 Numerical results

1. The couplings are fixed clearly. We see the scaling $\mathcal{S}[k_\mu, n, a(\beta)] \to \mathcal{S}[k_\mu, b = na(\beta)]$ for fixed $b = na(\beta)$ looks good. The continuum limit is taken as $a \to 0,\quad n \to \infty$ for $b = n \cdot a$ fixed.
2. The four- and six-point interactions become negligibly small for IR region. Two-point interactions are relatively dominant for large b region.
3. We see the direction dependence of the current action from the data. For example, two nearest-neighbor interactions G_2 and G_3 are quite different for small b region.

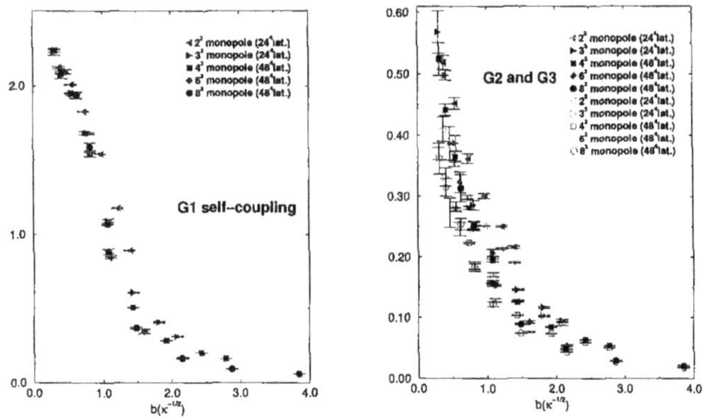

Figure 1: The dominant couplings of quadratic interactions versus physical length b in SU(2).

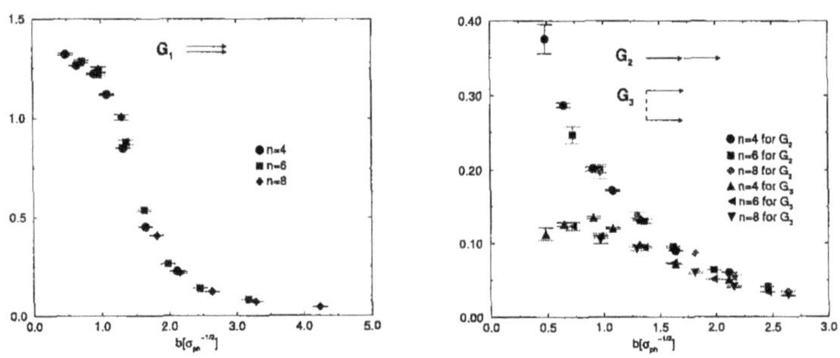

Figure 2: The dominant couplings of quadratic interactions versus physical length b in SU(3).

4. The $SU(3)$ case has three types of monopole currents $\{k_\mu^{(a)}(s), a = 1 \sim 3\}$ with one constraint $\sum_a k_\mu^{(a)}(s) = 0$. But the behaviors of the effective action are similar to those of the $SU(2)$ case especially for large b region.

4 A perfect operator for a physical quantities

In QCD, the string tension from the static potential is an important physical quantity. A naive abelian Wilson loop operator on the coarse lattice is not good [14], because the cut-off effect is of order of the lattice spacing of the coarse lattice. We should use an improved operator on the coarse lattice in order to get the correct values of the physical observables. An operator giving a cut-off independent value on RT is called **perfect operator**.

4.1 The method

The abelian monopole action $\mathcal{S}[k]$ which is obtained numerically is well approximated by quadratic interactions alone. We can perform the analytic block spin transformation along the flow projected on the quadratic coupling constant plane. When we define an operator on the fine a lattice, we can find a perfect operator along the projected flow in the $a \to 0$ limit for fixed b. We adopt the perfect operator on the projected space as an approximation of the correct operator for the action $\mathcal{S}[k]$ on the coarse b lattice.

4.2 The operator for the abelian static potential

First let us consider the following abelian gauge theory of the generalized Villain form on a fine lattice with a very small lattice distance:

$$\mathcal{S}[\theta,n] = \frac{1}{4\pi^2}\sum_{s,s';\mu>\nu}(\partial_{[\mu}\theta_{\nu]}(s) + 2\pi n_{\mu\nu}(s))(\Delta D_0)(s-s')(\partial_{[\mu}\theta_{\nu]}(s') + 2\pi n_{\mu\nu}(s')),$$

where $\theta_\mu(s)$ is a compact abelian gauge field and integer-valued tensor $n_{\mu\nu}(s)$ stands for Dirac string. Both of variables are defined on the original lattice. Since we are considering a fine lattice near to the continuum limit, we assume the direction symmetry of D_0. In this model, it is natural to use an abelian Wilson loop $W(C) = \exp i\sum_C(\theta_\mu(s), J_\mu(s))$ for particles with fundamental abelian charge, where $J_\mu(s)$ is abelian integer-charged electric current. The expectation value of $W(C)$ is written as

$$\langle W(C)\rangle = \left\langle \exp\left\{i\sum_{s,\mu} J_\mu(s)\theta_\mu(s)\right\}\right\rangle = Z[J]/Z[0], \tag{2}$$

$$Z[J] \equiv \int_{-\pi}^{\pi}\prod_{s;\mu} d\theta_\mu(s) \sum_{n_{\mu\nu}(s)=-\infty}^{+\infty} \exp\left\{-\mathcal{S}[\theta,n] + i\sum_{s,\mu} J_\mu(s)\theta_\mu(s)\right\}, \tag{3}$$

When use is made of BKT transformation[15], the area law term from the partition function (3) is equivalent to that from the following monopole expression

$$\langle W_m(C)\rangle = \frac{1}{Z}\sum_{\substack{k_\mu(s)=-\infty \\ \partial'_\mu k_\mu(s)=0}}^{\infty} \exp\left\{-\sum_{s,s',\mu} k_\mu(s)D_0(s-s')k_\mu(s') + 2\pi i\sum_{s,\mu} N_\mu(s)k_\mu(s)\right\},$$

$$N_\mu(s,S_J)=\sum_{s'}\Delta_L^{-1}(s-s')\frac{1}{2}\epsilon_{\mu\alpha\beta\gamma}\partial_\alpha S^J_{\beta\gamma}(s'+\hat\mu), \tag{4}$$

where $S^J_{\beta\gamma}(s'+\hat\mu)$ is a plaquette variable satisfying $\partial'_\beta S^J_{\beta\gamma}(s)=J_\gamma(s)$.

4.3 The perfect operator for the static potential on the coarse lattice

We perform a block spin transformation (1) of the monopole currents. Let us start from (4). The cutoff effect of the operator (4) is $O(a)$ by definition. The δ-function renormalization group transformation (1) can be done analytically. Taking the continuum limit $a\to 0$, $n\to\infty$ (with $b=na$ is fixed) finally, we obtain the expectation value of the operator on the coarse lattice with spacing $b=na$:

$$\langle W(C)\rangle=\frac{1}{Z}\langle W_m(C)\rangle_{cl}\exp\left\{\pi^2\sum_{\substack{s^{(n)},s^{(n)\prime}\\ \mu,\nu}}B_\mu(s^{(n)})D_{\mu\nu}(s^{(n)}-s^{(n)\prime})B_\nu(s^{(n)\prime})\right\}$$

$$\times\sum_{\partial'_\mu K_\mu=0}\exp\left\{-S[K_\mu(s^{(n)})]+2\pi i\sum B_\mu(s^{(n)})D_{\mu\nu}(s^{(n)}-s^{(n)\prime})K_\nu(s^{(n)\prime})\right\} \tag{5}$$

where $B_\mu(s^{(n)})$ is the renormalized electric source term. $S[K_\mu(s^{(n)})]$ denotes the effective action defined on the coarse lattice:

$$S[K_\mu(s^{(n)})]=\sum_{s^{(n)},s^{(n)\prime}}\sum_{\mu,\nu}K_\mu(s^{(n)})D_{\mu\nu}(s^{(n)}-s^{(n)\prime})K_\nu(s^{(n)\prime}). \tag{6}$$

Here $\langle W_m(C)\rangle_{cl}$ is defined by

$$\langle W_m(C)\rangle_{cl}=\exp\left\{-\pi^2\int_{-\infty}^{\infty}d^4xd^4y\sum_\mu N_\mu(x)D_0^{-1}(x-y)N_\mu(y)\right\}. \tag{7}$$

Since we take the continuum limit analytically, the operator (5) does not have no cutoff effect. For details, see Ref.[1].

Performing BKT transformation on the coarse lattice, we can get the loop operator for the static potential in the framework of the string model:

$$\langle W_m(C)\rangle=\sum_{\substack{\sigma_{\mu\nu}(s)=-\infty\\ \partial_{[\alpha}\sigma_{\mu\nu]}(s)=0}}^{\infty}\exp\left\{-\pi^2\sum_{\substack{s,s'\\ \mu\neq\alpha\\ \nu\neq\beta}}\sigma_{\mu\alpha}(s)\partial_\alpha\partial'_\beta D^{-1}_{\mu\nu}\Delta_L^{-2}(s-s')\sigma_{\nu\beta}(s')\right.$$

$$\left.-2\pi^2\sum_{\substack{s,s'\\ \mu,\nu}}\sigma_{\mu\nu}(s)\partial_\mu\Delta_L^{-1}(s-s')B_\nu(s')\right\}\times\langle W_m(C)\rangle_{cl}. \tag{8}$$

Table 1. The calculated atring tension

$SU(2)$				$SU(3)$			
b	2.1	2.9	3.8	b	2.64	3.27	4.23
$\sqrt{\frac{\sigma_{cl}}{\sigma_{phys}}}$	1.64	1.56	1.45	$\sqrt{\frac{\sigma_{cl}}{\sigma_{phys}}}$	1.15	1.22	1.21

4.4 Parameter fitting

In order to compare our analysis with the inverse Monte-Carlo method, we expand $D_0(s-s')$ in the monopole action (4) as $\alpha\delta_{s,s'} + \beta\Delta_L^{-1}(s-s') + \gamma\Delta_L(s-s')$, where α, β and γ are free parameters. Then by matching the set of numerically obtained coupling constants of the monopole action $\{G_i(b)\}$ with $D_{\mu\nu}(s^{(n)} - s^{(n)'})$ we fix the free parameters as shown in Ref.[2].

5 Analytic evaluation of physical quantities

5.1 The string tension

Let us evaluate the string tension using the perfect operator (8). The plaquette variable $S_{\alpha\beta}$ in Eq.(4) for the static potential $V(Ib,0,0)$ is expressed by

$$S_{\alpha\beta}(z) = \delta_{\alpha 1}\delta_{\beta 4}\delta(z_2)\delta(z_3)\theta(z_1)\theta(Ib-z_1)\theta(z_4)\theta(Tb-z_4). \tag{9}$$

The string model is in the strong coupling region for large b. Therefore, we evaluate Eq.(8) by the strong coupling expansion. As shown in Ref.[2], the (classical) string tension coming from Eq. (7) is dominant and it becomes in SU(2)

$$\sigma_{cl} = \frac{\pi\kappa}{2}\ln\frac{m_1}{m_2}. \tag{10}$$

Using the optimal values κ, m_1 and m_2, we get the string tension $\sqrt{\sigma_{cl}/\sigma_{phys}}$ as shown in Table 1.

5.2 On the continuum rotational invariance

We here comment on the continuum rotational invariance of the quark-antiquark static potential. We get the static potential $V(Ib,Ib,0)$ in SU(2) can be written as

$$V(Ib,Ib,0) = \frac{\sqrt{2}\pi\kappa Ib}{2}\ln\frac{m_1}{m_2}. \tag{11}$$

The potentials from the classical part take only the linear form and the rotational invariance is recovered completely even for the nearest $I = 1$ sites.

6 Gauge independence

If the color confinement mechanism is due to the condensation of the monopoles, it should be gauge independent, since the monopole condensation in some special gauge means only abelian charge confinement which is quite different from color confinement.

Recently Ogilvie has developped a character as well as a strong coupling expansions for Abelian projection and found that gauge fixing is unnecessary, i.e., Abelian projection yields string tensions of the underlying non-Abelian theory even *without* gauge fixing[17,7]. Hence at least gauge independence of abelian dominance seems to be realized. In the following, we show numerical analyses of gauge dependence for a general class of gauge between MA gauge and no-gauge fixing. Then we next prove if abelian dominance is gauge independent, gauge independence of monopole dominance is derived.

6.1 *Numerical analyses in a general gauge*

We consider the Langevin equation with stochastic gauge fixing term[18,19]:

$$U_\mu(x, \tau + \Delta\tau) = \omega^\dagger(x, \tau)\exp(if_\mu^a t^a)U_\mu(x, \tau)\omega(x + \hat{\mu}, \tau),$$

$$f_\mu^a = -\frac{\partial S}{\partial A_\mu^a}\Delta\tau + \eta_\mu^a(x, \tau)\sqrt{\Delta\tau},$$

$$\omega(x, \tau) = \exp(i\beta\Delta^a(x, \tau)t^a\Delta\tau/2N_c\alpha).$$

We set $\Delta(x, \tau) = i[\sigma_3, X(x, \tau)]$, where X is the operator to be diagonalized in MA gauge. $\alpha = 0$ ($\alpha = \infty$) corresponds to the MA gauge fixing (no gauge fixing).

We have performed numerical simulations on $8^3 \times 12$ and $16^3 \times 24$ lattices with improved Iwasaki action. See Ref [4] for details. We get the results shown in Fig.3. Abelian amd monopole dominances are seen for a wide range of α.

6.2 *Gauge independence of monopole dominance*

Assume gauge independence of abelian dominance is realized. Then we expect the existence of an abelian effective action. We express the abelian action in

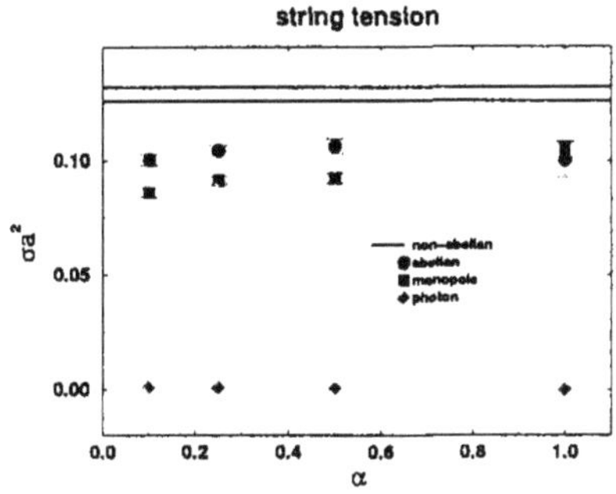

Figure 3. Gauge dependence of the string tension

terms of the Villain form $Z = \int_{-\pi}^{\pi} D\theta \sum_{n\in \mathbf{Z}} \mathrm{e}^{-F[d\theta+2\pi n]}$. The general Villain action can be expressed as follows:

$$\mathrm{e}^{-(d\theta+2\pi n, D(d\theta+2\pi n)) - F'[d\theta+2\pi n]}$$
$$= e^{-F'[-i\delta/\delta B]} \mathrm{e}^{-(d\theta+2\pi n, D(d\theta+2\pi n)) + i(B, d\theta+2\pi n)} \Big|_{B=0},$$

where $[D, d] = [D, \delta] = 0$ are satisfied in the large β scaling region.

The abelian Wilson loop $\mathrm{e}^{i(\theta, J)}$ is estimated with this action, where J is the electric current. When we use the BKT transformation[15], we get an action in terms of monopole currents:

$$Z(J) = e^{-F'[-i\delta/\delta B]} \sum_{k\in \mathbf{Z}, dk=0} \mathrm{e}^{-\frac{1}{4}(\delta B, (\Delta D)^{-1}\delta B)}$$
$$\times\, \mathrm{e}^{\frac{1}{2}(iB, 4\pi\delta\Delta^{-1}k + i(\Delta D)^{-1}dJ)} \mathrm{e}^{-4\pi^2(k, \Delta^{-1}Dk)} \mathrm{e}^{2\pi i(\delta\Delta^{-1}k, S)} \mathrm{e}^{-\frac{1}{4}(J, (\Delta D)^{-1}J)} \Big|_{B=0}$$

1)Electric-electric current $J - J$ interactions (with no monopole k) come from the exchange of regular photons and have no line singularity leading to a linear potential. Hence the linear potential of abelian Wilson loops is due to the monopole contribution alone. Monopole dominance is proved from abelian dominance. 2)The linear potential comes only from $\exp(2\pi i(\delta\Delta^{-1}k, S))$. The surface independence of the static potential is assured due to the 4-d linking number.

The author is grateful to S.Kato, S.Fujimoto, M. Chernodub and M.Polikarpov for useful discussions and collaboration. This work is supported by the Supercomputer Project (No.98-33, No.99-47) of High Energy Accelerator Research Organization (KEK) and the Supercomputer Project of the Institute of Physical and Chemical Research (RIKEN). The author acknowledges also the financial support from JSPS Grant-in Aid for Scientific Research (B) (No.10440073 and No.11695029).

References

1. S.Fujimoto, S.Kato and T.Suzuki, *Phys. Lett.* **B476** (2000) 437.
2. M.Chernodub *et al.*, to appear in Phys. Rev. D.
3. K. Yamagishi *et al.*, to appear in JHP.
4. F. Shoji *et al.*, *Phys. Lett.* **B476** (2000) 199.
5. T. Suzuki, to appear in *Nucl. Phys.*(Proc. Suppl.).
6. G.'t Hooft, *Nucl. Phys.* **B190** (1981) 455.
7. T. Suzuki and I. Yotsuyanagi, *Phys. Rev.* **D42** (1990) 4257; *Nucl. Phys.* B *(Proc. Suppl.)* **20** (1991) 236; S. Hioki *et al.*, *Phys. Lett.* **B272** (1991) 326 and references therein.
8. A.S. Kronfeld *et al.*, *Phys. Lett.* **198B** (1987) 516; A.S. Kronfeld, G. Schierholz and U.J. Wiese, *Nucl. Phys.* **B293** (1987) 461.
9. H. Shiba and T. Suzuki, *Phys. Lett.* **B343** (1995) 315, *Phys. Lett.* **B351** (1995) 519 and references therein.
10. R.H. Swendsen, *Phys. Rev. Lett.* **52** (1984) 1165; *Phys. Rev.* **B30** (1984) 3866,3875.
11. S. Kato, S. Kitahara, N. Nakamura and T. Suzuki, *Nucl. Phys.* **B520** (1998) 323.
12. T.A. DeGrand and D. Toussaint, Phys. Rev. **D22** (1980) 2478.
13. T.L. Ivanenko, A.V. Pochinskii and M.I. Polikarpov, Phys. Lett. **B 252** (1990) 631.
14. S. Fujimoto et al, Nucl.Phys.B(Proc.Suppl)73(1999) 533.
15. V.L. Beresinskii, *Sov. Phys. JETP* **32** (1970) 493. J.M. Kosterlitz and D.J. Thouless, *J. Phys.*,**C6** (1973) 1181.
16. M. Teper, hep-th/9812187
17. M.C. Ogilvie,Phys.Rev. D59 (1999) 074505.
18. D.Zwanziger, Nucl. Phys. **B192** (1981) 259.
19. M.Mizutani and A.Nakamura, Vistas in Astronomy, **37**, 305, 1993, Pergamon Press; Nucl.Phys. B(Proc.Suppl.)34, 253, 1994.

2 Hadron and Glueball Masses

On the deck of the Pearl River cruise boat.

FULL QCD LIGHT HADRON SPECTRUM AND QUARK MASSES

R. BURKHALTER
Center for Computational Physics and Institute of Physics, University of Tsukuba, Tsukuba, Ibaraki 305-8577, Japan
E-mail: burkhalt@rccp.tsukuba.ac.jp

We present recent results of the light hadron spectrum and quark masses in QCD with two flavors of dynamical quarks from the CP-PACS computer. To compensate the increase in computer time, we use a renormalization-group improved gauge action and a tadpole-improved clover quark action. A comparison with quenched results with the same improved action and with the standard action shows clearly the existence of sea quark effects in meson and quark masses.

1 Introduction

The reproduction of the experimental light hadron spectrum from ab initio non-perturbative calculations in QCD has always been one of the major goals of lattice QCD. If this approach is successful, it will not only be a confirmation of QCD as the correct theory for the strong interaction at low energies but also a validation of the lattice method itself.

In order to attain this goal a number of sources of systematic errors have to be brought under control. One of them, the quenched approximation of ignoring quark pair creation and annihilation, has until recently been widely used in light hadron spectrum calculations. Within the quenched approximation, a recent calculation by the CP-PACS Collaboration [1] attempted to control all remaining systematic errors and to obtain the quenched spectrum with an error of a few %, improving significantly upon the previous best effort in this direction by the GF11 Collaboration [2].

The results of this calculation, reproduced in Fig. 1, show that the quenched spectrum is consistent with experiment within 10%. However, they also clearly show the limitations of the quenched approximation: there is a statistically significant disagreement from experiment. One of the differences is that when using the K meson as input to fix the strange quark mass, the K^* and ϕ mesons turn out to be lighter than experiment. In fact, there is no choice of input to fix the strange quark mass so that all strange hadron masses are consistent with experiment.

The last remaining task towards a realistic determination of the light hadron spectrum is to perform calculations which include the effects of sea quarks. A first step in this direction, consisting of a systematic calculation of

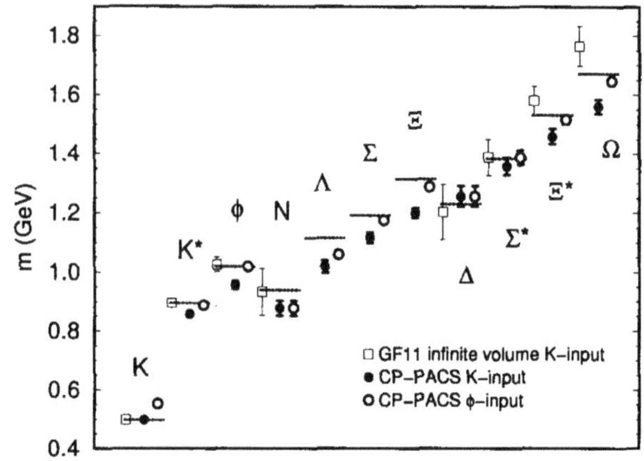

Figure 1: Quenched spectrum with the standard action. Experimental values of π and ρ meson masses have been used to fix the light quark mass and the lattice scale, the ones of the remaining particles are shown as dotted lines. The strange quark mass has been fixed using as input either the K meson (filled circles) or the ϕ meson (open circles). For comparison, previous results of the GF11-Collaboration [2] are also shown (open squares).

the spectrum with two flavors of light sea quarks, has been taken recently by the CP-PACS Collaboration [a]. In these proceedings I will report on the results on the light hadron spectrum and quark masses from this calculation. Earlier reports can be found in Reference [3].

2 Full QCD simulations

The quenched calculation mentioned in the introduction used a set of carefully chosen parameters. In order to keep finite size effects small a spatial volume with a physical lattice size of 3 fm has been used, systematic extrapolations to physical quark masses were made from a range of pseudoscalar to vector meson mass ratio $m_{PS}/m_V = 0.75$–0.4 and extrapolations to the continuum limit from the range of lattice spacings $a = 0.1$–0.05 fm. Since the simulation of full QCD is computationally much more demanding, it is at present not possible to attempt a completely parallel calculation with the same parameter ranges.

This obstacle can be partly overcome by the use of improved actions. They are designed to reduce cutoff errors and can therefore be used for simulations at coarser lattices. We use a combination of a renormalization-group improved

[a]CP-PACS Collaboration: A. Ali Khan, S. Aoki, G. Boyd, R. B., S. Ejiri, M. Fukugita, S. Hashimoto, N. Ishizuka, Y. Iwasaki, K. Kanaya, T. Kaneko, Y. Kuramashi, T. Manke, K. Nagai, M. Okawa, H. P. Shanahan, A. Ukawa, T. Yoshié

Table 1: Parameters of two-flavor QCD simulations. The scale a is set using M_ρ as input.

β	$L^3 \times T$	c_{SW}	a[fm]	La[fm]	m_{PS}/m_V for sea quarks # HMC trajectories			
1.80	$12^3 \times 24$	1.60	0.215(2)	2.58(3)	0.807(1)	0.753(1)	0.694(2)	0.547(4)
					6250	5000	7000	5250
1.95	$16^3 \times 32$	1.53	0.155(2)	2.48(3)	0.804(1)	0.752(1)	0.690(1)	0.582(3)
					7000	7000	7000	5000
2.10	$24^3 \times 48$	1.47	0.108(1)	2.58(3)	0.806(1)	0.755(2)	0.691(3)	0.576(3)
					4000	4000	4000	4000

gauge action [4] and a mean-field improved SW-clover action [5]. In an initial study[6] we found that this particular combination has a clearly improved scaling behavior already at $a \approx 0.2$ fm and therefore is suitable for our simulations in the range $a \approx 0.2$–0.1 fm. We also make an additional compromise and reduce the physical lattice size to 2.5 fm which allows us to make simulations at much smaller lattices. Finally, we reduce the range of quark masses to $m_{PS}/m_V \approx 0.8$–0.5 for valence quarks and $m_{PS}/m_V \approx 0.8$–0.6 for sea quarks.

In Table 1 we show an overview of the simulation parameters used for creating gauge configurations with two flavors of sea quarks. On these configurations we determine hadron masses using a standard correlated fit to hadron propagators. Errors are estimated using the jackknife procedure with a bin size of 50 trajectories, derived from an analysis of autocorrelations.

3 The quenched spectrum revisited with improved actions

Our main interest in this study is to look for sea quark effects and to examine whether they can reduce the observed discrepancy in the quenched hadron spectrum. To this end it will be useful to have additional quenched results which are obtained with the same improved action and with the same range of lattice spacings as the one used in the full QCD calculation. This will enable us to make a direct comparison between quenched and full QCD results.

We made such kind of additional quenched simulations choosing 10 values of the bare coupling β so that the string tension matches that of two-flavor full QCD for each value of the sea quark mass and for the chiral limit at $\beta = 1.95$ and 2.1. The simulation parameters are summarized in Table 2.

A natural question to ask is whether this new quenched simulation gives results consistent with the ones in Ref[1]. While doing this one has to keep in mind that the two calculations differ in some of the details. Due to our choice of parameters, some of the lattices in the improved case have a smaller physical

Table 2: Parameters of quenched QCD simulations with improved actions. The lattice spacing a is set using M_ρ as input.

$L^3 \times T = 16^3{\times}32$				$L^3 \times T = 24^3{\times}48$			
β	c_{SW}	a [fm]	La[fm]	β	c_{SW}	a [fm]	La[fm]
2.187	1.439	0.2004(20)	3.206(32)	2.416	1.379	0.1446(18)	3.470(43)
2.214	1.431	0.1903(19)	3.045(30)	2.456	1.370	0.1328(13)	3.187(31)
2.247	1.422	0.1807(18)	2.891(29)	2.487	1.363	0.1284(14)	3.082(34)
2.281	1.412	0.1765(20)	2.824(32)	2.528	1.355	0.1206(13)	2.894(31)
2.334	1.398	0.1632(16)	2.611(26)	2.575	1.345	0.1130(11)	2.712(26)
corresponds to $\beta_{\rm full} = 1.95$				corresponds to $\beta_{\rm full} = 2.1$			

size than in the standard case and therefore finite size effects might affect the results. Furthermore, the range of quark masses do not entirely match and the functional forms used for chiral extrapolations are different. For the work in Ref. [1] which employs the standard action, formulae from Quenched Chiral Perturbation Theory have been used whereas for the improved action polynomial functions, matching the ones employed for full QCD, have been used. We checked, however, that changing the functional form for the improved case to the one used in the standard case gives consistent results.

For all these differences we find a good agreement of the two calculations of the quenched spectrum in the continuum limit, as is shown in Figure 2, for both mesons and baryons. In particular, we find the K^* and ϕ meson masses to be lighter than experiment in both calculations. This is a nice confirmation of the universality of the continuum limit in quenched QCD and a corroboration of the pattern of disagreement of the quenched spectrum with experiment.

4 Sea quark effects in the light hadron spectrum

A new feature in two-flavor QCD, which is not present in the quenched case, is that hadron masses depend on both the sea and the valence quark mass. We take this into account by fitting vector meson and decuplet baryon masses using a combined quadratic ansatz

$$M_{V,D} = A + B_s\mu_{sea} + B_v\mu_{val} + C_s\mu_{sea}^2 + C_v\mu_{val}^2 + C_{sv}\mu_{sea}\mu_{val}, \qquad (1)$$

where $\mu_{val} = (\mu_1 + \mu_2)/2$ represents the average of the pseudo scalar meson mass squared $\mu_i = m_{PS}^2(\kappa_{val,i}, \kappa_{val,i}; \kappa_{sea})$ made of a degenerate quark-antiquark pair of the valence hopping parameter $\kappa_{val,i}$, and $\mu_{sea} = m_{PS}^2(\kappa_{sea}, \kappa_{sea}; \kappa_{sea})$. For octet baryons we have to use a more complicated formula with 12 parameters.

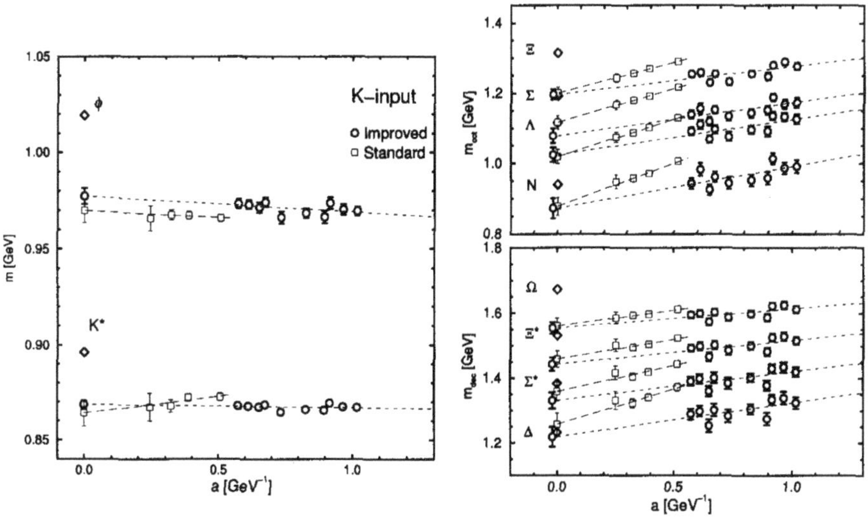

Figure 2: Continuum extrapolation of meson and baryon masses in quenched QCD with standard action (squares) and improved action (circles). Experimental values are displayed as diamonds. The K meson mass has been used for fixing the strange quark. In both cases linear extrapolations to the continuum limit have been employed.

Using Eq. 1 we can extrapolate to the physical points by keeping μ_{sea} at its value for the light (up and down) quark and putting μ_{val} to either the light or the strange point. Doing this we use as input the experimental values of the π and ρ meson masses and additionally either the K or ϕ meson mass. The meson and baryon masses obtained in this way are shown together with the quenched results in Fig. 3.

In comparing quenched and full QCD results the most notable difference is seen in meson masses. The K^* and ϕ meson masses extrapolate to a value in the continuum limit which is much closer, in fact almost consistent, with the experimental value. We would like to stress that this very encouraging conclusion would not have been possible without a systematic approach which includes a continuum extrapolation. Since full and quenched results appear to have a different slope even when using the same actions, one could, when performing the calculation only at one lattice spacing, easily be misled to conclude that there are no sea quark effects.

Conclusions for baryons are not so clear-cut. We find the nucleon and the Δ to have values higher than experiment, whereas some of the strange baryons are consistent with experiment. It might be, however, that our lattice size of

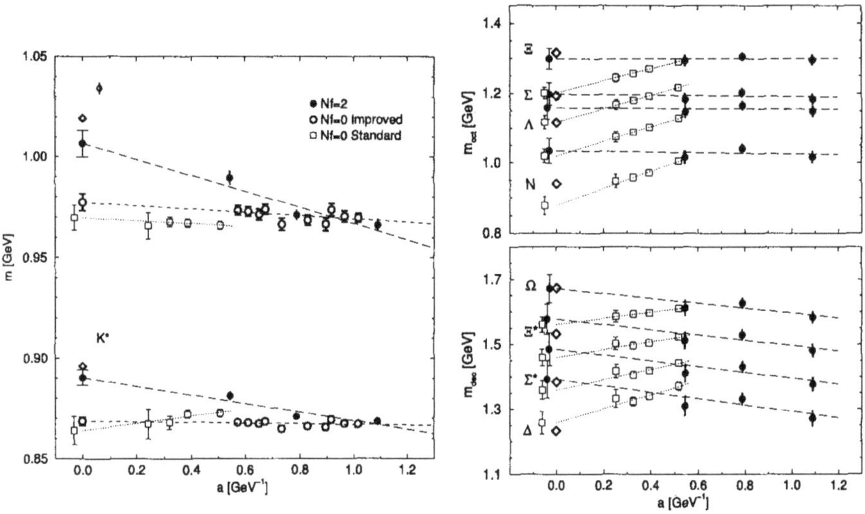

Figure 3: Comparison of continuum extrapolations in two-flavor (filled symbols) and quenched (open symbols) QCD. The K meson mass has been used for fixing the strange quark. In all cases fits linear in the lattice spacing have been employed.

2.5 fm is too small to completely avoid finite size effects for baryons.

5 Light quark masses

While for the light hadron spectrum lattice calculations can serve as a tool for the verification of QCD, they can make real, independent predictions for quark masses. The reason is that because of confinement, quark masses can not be determined directly by experiment. The lattice approach is at the moment the only known first-principle method in calculating quark masses.

Such calculations have seen considerable progress in the last few years [7]. However, except for a recent attempt in full QCD [8], they have been restricted to the quenched approximation. Our systematic calculations in two-flavor QCD therefore allow us to take a significant step forward in the determination of light quark masses [9].

Quark masses from lattice simulations can be defined in various ways. One uses the axial vector Ward identity $\nabla_\mu A_\mu(x) = 2am_q^{\rm AWI}P(x)$. Another can be derived from the Ward identity for vector currents and reads $m_q^{\rm VWI} = (1/\kappa - 1/\kappa_c)/2a$. For the latter definition there exists the additional ambiguity of choosing κ_c in full QCD. A natural choice is to use the definition $m_{PS}(\kappa_{val} =$

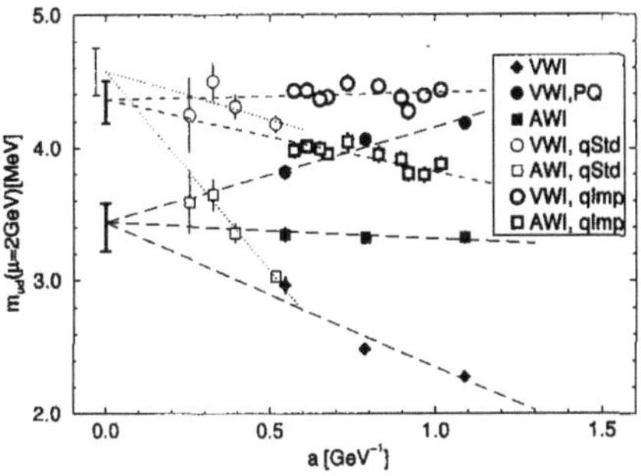

Figure 4: Continuum extrapolation of the average up and down quark mass m_{ud} for full QCD (filled symbols) and quenched QCD (thick open symbols) obtained with the improved action, and quenched results with the standard action (thin open symbols). The lines show combined fits linear in a.

$\kappa_{sea} = \kappa_c) = 0$, i.e. to choose the hopping parameter at which the PS meson vanishes when valence and sea hopping parameters are varied simultaneously. Another possibility is to keep κ_{sea} fixed at the physical point κ_{ud} for u and d quarks and to define the partially quenched critical hopping parameter $\kappa_c^{\rm PQ}$ to be the point where the PS meson vanishes as function of κ_{val} only, i.e. $m_{PS}(\kappa_{val} = \kappa_c^{\rm PQ}; \kappa_{sea} = \kappa_{ud}) = 0$. Since different definitions lead to different results at finite lattice spacings, we use all of them and check for consistency in the continuum limit.

In Fig. 4 we show results for the average up and down quark mass in the $\overline{MS}$ scheme renormalized at the scale $\mu = 2$ GeV for various definitions in quenched and full QCD with standard and improved actions. For our calculations with the improved action the leading scaling violation is expected to be $O(g^2 a)$ where g^2 is the renormalized coupling $g^2_{\overline{MS}}(\mu)$.

We find that the three definitions in full QCD tend to converge toward the continuum limit. A similar trend for two definitions can also be seen for both quenched simulations. Indeed, applying linear extrapolations in the lattice spacing to each definition separately, we find the values in the continuum limit to be consistent within two standard deviations. This finding leads us to perform a combined linear fit to all definitions with a common continuum limit to obtain our central value. The spread of results of the separate fits is used as an estimate of systematic error from the continuum extrapolation. We also make estimates of systematic errors from chiral fits and higher order contributions of renormalization factors.

Adding the statistical error and all systematic errors quadratically, we obtain the final result $m_{ud}^{\overline{MS}}(2\,\mathrm{GeV}) = 3.44^{+0.14}_{-0.22}$ MeV. This is about 25%

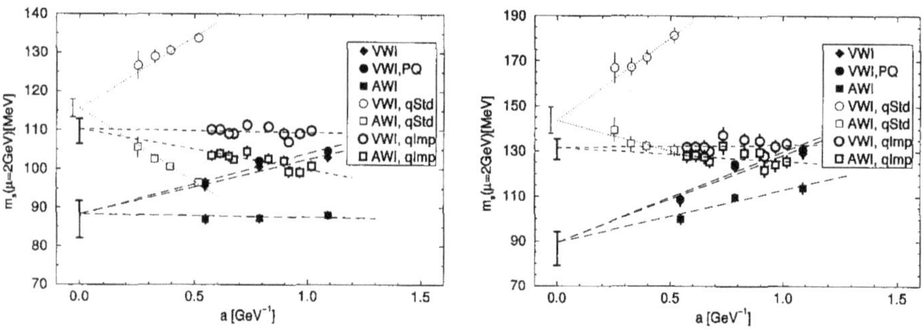

Figure 5: Continuum extrapolation of the strange quark mass m_s using M_K (left figure) or M_ϕ (right figure) as input. Symbols have the same meaning as in Fig. 4.

smaller than quenched results for which we obtain the mutually agreeing values 4.57(18) MeV (standard action [1]) and $4.36^{+0.14}_{-0.17}$ MeV (improved action).

In Fig. 5 we show results for the strange quark mass using either the K meson mass or the ϕ meson mass as input. Extrapolating in the same way as for m_{ud} we obtain the results $m_s^{\overline{MS}}(2\,\text{GeV}) = 88^{+4}_{-6}$ MeV (K input) and 90^{+5}_{-11} MeV (ϕ input), compared to 116(3) MeV (K input) and 144(6) MeV for the standard action [1] and 110^{+3}_{-4} MeV (K input) and 132^{+4}_{-6} MeV (ϕ input) for the improved action.

A very encouraging result is that the uncertainty in the strange quark mass in quenched QCD, reflected in the dependence on the experimental input, is reduced to a level well contained in the error bar of about 10% in full QCD with two flavors of sea quarks.

6 Conclusions

Our systematic investigations of quenched and full QCD have shown that the inclusion of two flavors of sea quarks leads to clearly detectable effects in meson masses and light quark masses in the continuum limit. We find that the discrepancies of mesons masses in quenched QCD and experiment are significantly reduced in full QCD. Light quark masses are smaller by about 25% in full QCD, resulting in a strange quark mass of about 90 MeV. This value is smaller than previous phenomenological estimates, which might have consequences for the analysis of CP violations.

Acknowledgments

I would like to thank the members of the CP-PACS collaboration for a very fruitful collaboration and constant support. In particular I thank Hugh Shanahan and Akira Ukawa for useful suggestions on the manuscript. This work was supported in part by the Grant-in-Aid No. 11640250 of the Ministry of Education of Japan.

References

1. S. Aoki *et al.* (CP-PACS Collaboration), Phys. Rev. Lett. **84**,238(2000); Nucl. Phys. B(Proc. Suppl.)**73**,189(1999).
2. F. Butler, *et al.*, Nucl. Phys. **B430**, 179 (1994).
3. For earlier preliminary reports, see, R. Burkhalter, Nucl. Phys. B(Proc. Suppl.)**73**,3(1999); S. Aoki *et al.* (CP-PACS Collaboration), Nucl. Phys. B(Proc. Suppl.)**73**,192(1999).
4. Y. Iwasaki, Univ. of Tsukuba Report UTHEP-118(1983).
5. B. Sheikholeslami and R. Wohlert, Nucl. Phys. **B259**, 572(1985).
6. S. Aoki *et al.* (CP-PACS Collaboration), Phys. Rev. D **60**, 114508 (1999).
7. Early studies were reviewed in, A. Ukawa, Nucl. Phys. B (Proc. Suppl.) **30**, 3 (1993). A more recent review can be found in, R. Gupta and T. Bhattacharya, Nucl. Phys. B (Proc. Suppl.) **63**, 95 (1998); Phys. Rev. D **55**, 7203 (1997). Recent progress has been reported in, B. J. Gough *et al.*, Phys. Rev. Lett. **79**, 1622 (1997); D. Becirevic *et al.*, Phys. Lett. B**444**, 401 (1998); S. Aoki *et al.* (JLQCD Collaboration), Phys. Rev. Lett. **82**, 4392 (1999); V. Giménez *et al.*, Nucl. Phys. **B540**, 472 (1999); S. Capitani *et al.*, Nucl. Phys. **B544**, 669 (1999); S. Aoki *et al.* (CP-PACS Collaboration), Phys. Rev. Lett. **84**,238(2000); J. Garden *et al.*, Nucl. Phys. **B571**, 237 (2000); M. Göckeler *et al.*, hep-lat/9908005.
8. N. Eicker *et al.* (SESAM Collaboration), Phys. Lett. B**407**,290(1997).
9. A. Ali Khan *et al.* (CP-PACS Collaboration), hep-lat/0004010.

EXCITED HEAVY BARYON MASSES FROM THE $1/N_c$ EXPANSION OF HQET

JONG-PHIL LEE

Department of Physics, Seoul National University, Seoul 151-742, Korea
E-mail: jplee@phya.snu.ac.kr

CHUN LIU[a]

Institute of Theoretical Physics, CAS, Beijing 100080, China
E-mail: liuc@itp.ac.cn

H. S. SONG

Department of Physics, Seoul National University, Seoul 151-742, Korea
E-mail: hssong@physs.snu.ac.kr

The mass spectra of the $L = 1$ orbitally excited heavy baryons with light quarks in both the spin-flavor symmetric and the mixed representations are studied by the $1/N_c$ expansion method in the framework of the heavy quark effective theory. The mixing effect between the baryons in the two representations is also considered. The general pattern of the spectrum is predicted which will be verified by the experiments in the near future.

1 Introduction

Experimentally, a lot of data for orbitally excited heavy baryons have been accumulating[1]. The following charmed baryon states have been found, $\Lambda_c(2593)^+$ with $I(J^P) = 0(\frac{1}{2}^-)$ which is denoted as $\Lambda_{c1}(\frac{1}{2}^-)^+$, $\Lambda_c(2625)^+$ with $I(J^P) = 0(?)$ denoted as $\Lambda_{c1}(\frac{3}{2}^-)^+$, where ? is $\frac{3}{2}^-$ from the quark model, and its strange analogues $\Xi_{c1}(\frac{3}{2}^-)$.

Theoretical understanding of these baryons is necessary. The heavy quark effective theory (HQET) [2] provides a systematic way to investigate hadrons containing a single heavy quark. To obtain detailed prediction, however, some non-perturbative QCD methods have to be used, such as lattice simulation, $1/N_c$ expansion [3,4], chiral Lagrangian and QCD sum rules. In this talk we report the application of the $1/N_c$ expansion method[5,6].

Let us first make a brief review of the HQET [2]. It is an effective field theory of QCD for heavy hadrons. In the limit $m_Q/\Lambda_{QCD} \to \infty$, the heavy quark spin-flavor symmetry (HQS) is explicit. The 4-velocity v of the heavy quark becomes a good quantum number. Because m_Q is unusable, it can be

[a]Speaker

removed by redefining the heavy quark field:

$$h_v = e^{im_Q v\cdot x} Q \,. \tag{1}$$

The effective Lagrangian is then

$$\mathcal{L}_{\text{eff}} = \bar{h}_v v \cdot D h_v + O(1/m_Q) \,. \tag{2}$$

The hadron mass is expanded as

$$\bar{\Lambda}_H + O(1/m_Q) \tag{3}$$

in the effective theory, that is

$$M_H = m_Q + \bar{\Lambda}_H + O(1/m_Q) \,. \tag{4}$$

Second, let us come to the $1/N_c$ expansion [3]. This is a non-perturbative method of QCD. The idea is to extract out the non-perturbative information of $SU(N_c)$ gauge theory by taking $N_c \to \infty$. The N_c counting rules are given as follows. The interaction vertex is $g_s/\sqrt{N_c}$; the quark propagator keeps unchanged; and the gluon propagator is represented by double lines, one for quarks and the other for anti-quarks.

For the mesons, the non-perturbative properties can be observed from the analysis of the planar diagrams. The large N_c limit is quite successful. Because the meson decay amplitude $\sim 1/\sqrt{N_c}$, mesons and glue states are free and stable. This agrees qualitatively with color confinement. Another example is the explanation of the Zweig's rule.

For the baryons, the diagrammatic method does not work. The Hartree approximation can be adopted. The observation is that in the $N_c \to \infty$ limit, interaction between any pair of quarks is negligible ($\sim 1/N_c$); the total potential on an individual quark, which is ~ 1, is a sum of many small terms, therefore it can be regarded as the background potential or a c-number potential. For ground state baryons, the many-body wave function is written as

$$\Psi(x_1, ..., x_{N_c}, t) = \sum_1^{N_c} \phi(x, t) \,, \tag{5}$$

where $\phi(x,t)$ is one-body state. Interesting results for large N_c baryons can be obtained. The baryon-(anti-)baryon interaction is order N_c. And the baryon-meson interaction is order 1. However, the Hartree potential is not known, in which the two-body, three-body, and many-body interactions are the same important. One conjecture is that baryons are solitons of the mesonic theory, skyremions [7]!

Something more can be said about the large N_c baryons[4]. For the ground state baryons, there is a contracted SU($2N_f$) light quark spin-flavor symmetry (LQS) in the large N_c limit. This was first obtained from the chiral perturbation theory of baryon-pion interactions in deriving the consistent conditions for the coupling constants in the large N_c limit. It can be also understood in the Hartree picture. This makes a $1/N_c$ expansion based on the spin-flavor structure practically possible for the baryons. Many quantitative predictions and further extensions of the above result have been made [8]. In fact, this expansion is another scheme of the $1/N_c$ expansion. This can be simply seen from considering the masses of the non-strange baryons,

$$\begin{aligned} M_H &= N_c\Lambda_{\rm QCD} + O(1) + O(\frac{1}{N_c}) + ... \\ &= N_c\tilde{\Lambda}_{\rm QCD} + c_1\frac{S^2}{N_c} + c_2\frac{(S^2)^2}{N_c^2} + \dots , \end{aligned} \tag{6}$$

where the first line is the ordinary $1/N_c$ expansion, and the second one the expansion based on the spin-flavor structure. Of course, in the $N_c \to \infty$ limit, $M_H = \bar{\Lambda}_H = N_c\Lambda_{\rm QCD} = N_c\tilde{\Lambda}_{\rm QCD} = m_{proton}$ which is not so useful.

2 Excited Heavy Baryons in the $1/N_c$ Expansion

For the charmed baryons like $\Lambda_{c1}(\frac{1}{2}^-)^+$, $\Lambda_{c1}(\frac{3}{2}^-)^+$, $\Xi_{c1}(\frac{3}{2}^-)^+$, we analyze their masses $\bar{\Lambda}_H$ in the $1/N_c$ expansion. The classification of them is according to the angular momentum J, the isospin I and the total angular momentum of the light degrees of freedom J^l which becomes a good quantum number due to HQS. In this case, the excited hadron spectrum shows the degeneracy of pair of states which are related to each other by HQS. Constituently there are two ways for the $L = 1$ excitation. One is that the heavy quark is excited; the other is that one light quark is excited. Correspondingly under the LQS, the $N_c - 1$ light quarks are in the symmetric and mixed representation, respectively.

2.1 Symmetric Representation

In the symmetric representation, the picture for the light quarks is essentially the same as that of the ground state heavy baryons. The spin-flavor decomposition rule is $I = S^l$ for the non-strange baryons, where S^l is the total spin of the light quark system. Note that the light quark system as a whole has $L = 1$ orbital angular momentum. All possible states of excited heavy baryons are listed in Table 1.

In the Hartree–Fock picture of the baryons, the N_c counting rules require us to include many-body interactions in the analysis. However, a large part of

Table 1: Excited heavy baryon states of the symmetric representation of $N_c - 1$ light quarks.

(J, I)	(J^l, S^l)	$\bar{\Lambda}_H^0$
$(1/2, 0)$	$(1, 0)$	$N_c c_0 + 2c_1$
$(3/2, 0)$	$(1, 0)$	$N_c c_0 + 2c_1$
$(1/2, 1)$	$(0, 1)$	$N_c c_0 - 2c_1 + \frac{2c_2}{N_c}$
$(1/2, 1)$	$(1, 1)$	$N_c c_0 + \frac{2c_2}{N_c}$
$(3/2, 1)$	$(1, 1)$	$N_c c_0 + \frac{2c_2}{N_c}$
$(3/2, 1)$	$(2, 1)$	$N_c c_0 + 4c_1 + \frac{2c_2}{N_c}$
$(5/2, 1)$	$(2, 1)$	$N_c c_0 + 4c_1 + \frac{2c_2}{N_c}$

these interactions are spin-flavor irrelevant. Namely this part contributes in the order $N_c\Lambda_{\rm QCD}$ universally to all the baryons with different spin-flavor structure in Table 1. The mass splittings among the baryons can be obtained. For the purely light quark contribution to $\bar{\Lambda}_H$, the $1/N_c$ analysis goes the same as that to the ground state heavy baryons. There is LQS at the leading order of the $1/N_c$ expansion. The mass splitting due to the violation of LQS started from ${S^l}^2/N_c$. However, different from the ground state baryons, formally the orbital angular momentum of the heavy quark has more dominant contribution to $\bar{\Lambda}_H$ than $O(1/N_c)$. This is because of the orbital-light-quark-spin interactions. After summing up all the relevant many-body interactions, this order $O(1)$ contribution is $\vec{L}\cdot\vec{S^l} f(\frac{{S^l}^2}{N_c^2})$, where f is a general function which can be Taylor expanded. The mass $\bar{\Lambda}_H$ can be written simply as

$$\bar{\Lambda}_H^0 = N_c\tilde{c}_0 + \tilde{c}_1\vec{L}\cdot\vec{S^l} + O\left(\frac{1}{N_c}\right), \tag{7}$$

where coefficients $\tilde{c}_i \sim \Lambda_{\rm QCD}$ ($i = 0, 1$). There should be also a term proportional to L^2 in the above equation, which gives constant contribution to $\bar{\Lambda}_H^0$ for a given light quark representation, and therefore has been absorbed into the leading term. The term $\vec{L}\cdot\vec{S^l}$ can be rewritten as ${J^l}^2 - {S^l}^2$ with $\vec{J^l}$ being defined as $\vec{J^l} = \vec{L} + \vec{S^l}$. Therefore

$$\bar{\Lambda}_H^0 = N_c c_0 + c_1({J^l}^2 - {S^l}^2) + O\left(\frac{1}{N_c}\right), \tag{8}$$

where coefficients $c_i \sim \Lambda_{\rm QCD}$.

Table 2: Excited heavy baryon states of the mixed representation of $N_c - 1$ light quarks.

(J, I)	(J^l, S^l)	$\bar{\Lambda}^0_{H'}$
$(1/2, 0)$	$(0, 1)$	$-2c_{LS} + \frac{1}{18}c_T - \frac{1}{3}\bar{c}_1 + \frac{1}{2}\bar{c}_2$
$(1/2, 0)$	$(1, 1)$	$-c_{LS} + \frac{1}{18}c_T - \frac{1}{6}\bar{c}_1 - \frac{1}{4}\bar{c}_2$
$(3/2, 0)$	$(1, 1)$	$-c_{LS} + \frac{1}{18}c_T - \frac{1}{6}\bar{c}_1 - \frac{1}{4}\bar{c}_2$
$(3/2, 0)$	$(2, 1)$	$9c_{LS} + \frac{1}{18}c_T + \frac{1}{6}\bar{c}_1 + \frac{1}{20}\bar{c}_2$
$(5/2, 0)$	$(2, 1)$	$9c_{LS} + \frac{1}{18}c_T + \frac{1}{6}\bar{c}_1 + \frac{1}{20}\bar{c}_2$
$(1/2, 1)$	$(1, 0)$	$\frac{1}{18}c_T$
$(3/2, 1)$	$(1, 0)$	$\frac{1}{18}c_T$

2.2 *Mixed Representation*

In the mixed representation, all states are listed in Table 2. Again, $\bar{\Lambda}_{H'}$ is trivially $N_c\Lambda_{\rm QCD}$ at the leading order of the $1/N_c$ expansion. The spin-flavor dependence, however is more complicated. For the spectrum of excited light baryons, see ref. [8].

The many-body Hamiltonians related to the spin-flavor structure which involve orbital angular momentum L give $O(1)$ contribution. We use the following operators which were used in [8] to analyze $\bar{\Lambda}_H$,

$$\begin{aligned} H_{LS} &\propto \hat{a}^\dagger\, \vec{L}\cdot\vec{\sigma}\, \hat{a} \\ H_T &\propto \frac{1}{N_c} G^{ia} G_{ia} \\ H_1 &\propto \frac{1}{N_c} \hat{a}^\dagger\, L^i \otimes \tau^a\, \hat{a}\, G_{ia} \\ H_2 &\propto \frac{1}{N_c} \hat{a}^\dagger \{L^i, L^j\} \otimes \sigma_i \otimes \tau^a\, \hat{a}\, G_{ja}\,. \end{aligned} \tag{9}$$

The first one H_{LS} is one-body Hamiltonian, while the others are two-body Hamiltonians. G_{ia} are the generators of the spin-flavor symmetry group SU(4), given by

$$G^{ia} = \hat{a}^\dagger\, \sigma^i \otimes \tau^a\, \hat{a} \tag{10}$$

with σ^i and τ^a being the spin and isospin matrices, respectively. Such structure gives coherent addition over $N_c - 2$ core quarks. The first G^{ia} in H_T acts on the excited quark, the other G_{ia}'s on the $N_c - 2$ unexcited light quarks, namely the core quarks. In our case, all the operators must be understood as the ones acting on the light degrees of freedom. Note that the higher order many-body Hamiltonian which contains more factor of G_{ia} can be reduced to those given in Eq.(9).

The contributions to the baryon masses due to these Hamiltonians are obtained by calculating the baryonic matrix elements. The matrix elements of these operators between the states of light quarks which specify the states of excited heavy baryons are given as follows,

$$\begin{aligned}
&\langle I_c = \frac{1}{2};\ I\ I_3;\ S^{l\prime}\ S_3^{l\prime},\ l=1\ m' \mid H_T \mid I_c = \frac{1}{2};\ I\ I_3;\ S^l\ S_3^l,\ l=1\ m\rangle \\
&= 2c_T \delta_{S^{l\prime},S^l}\delta_{S_3^{l\prime},S_3^l}\delta_{m,m'}(-1)^{1-S^l-I}\left\{\begin{array}{ccc} S^l & \frac{1}{2} & \frac{1}{2} \\ 1 & \frac{1}{2} & \frac{1}{2}\end{array}\right\}\left\{\begin{array}{ccc} \frac{1}{2} & 1 & \frac{1}{2} \\ \frac{1}{2} & I & \frac{1}{2}\end{array}\right\},
\end{aligned} \tag{11}$$

$$\begin{aligned}
&\langle I_c = \frac{1}{2};\ I\ I_3;\ l=1,\ S^{l\prime},\ J^l\ J_3^l \mid H_{LS} \mid I_c = \frac{1}{2};\ I\ I_3;\ l=1,\ S^l,\ J^l\ J_3^l\rangle \\
&= c_{LS}(-1)^{S^l-S^{l\prime}}\sqrt{(2S^l+1)(2S^{l\prime}+1)}\sum_{j=\frac{1}{2},\frac{3}{2}}(2j+1)\{j(j+1)-2-3/4\} \\
&\quad \left\{\begin{array}{ccc} \frac{1}{2} & \frac{1}{2} & S^l \\ 1 & J^l & j\end{array}\right\}\left\{\begin{array}{ccc} \frac{1}{2} & \frac{1}{2} & S^{l\prime} \\ 1 & J^l & j\end{array}\right\},
\end{aligned} \tag{12}$$

$$\begin{aligned}
&\langle I_c = \frac{1}{2};\ I\ I_3;\ l=1,\ S^{l\prime},\ J^l\ J_3^l \mid H_1 \mid I_c = \frac{1}{2};\ I\ I_3;\ l=1,\ S^l,\ J^l\ J_3^l\rangle \\
&= 6\bar{c}_1(-1)^{I-J^l+S^l-S^{l\prime}-1}\sqrt{(2S^l+1)(2S^{l\prime}+1)}\left\{\begin{array}{ccc} \frac{1}{2} & 1 & \frac{1}{2} \\ \frac{1}{2} & I & \frac{1}{2}\end{array}\right\}\left\{\begin{array}{ccc} S^l & 1 & S^{l\prime} \\ \frac{1}{2} & \frac{1}{2} & \frac{1}{2}\end{array}\right\}\left\{\begin{array}{ccc} 1 & 1 & 1 \\ S^{l\prime} & J^l & S^l\end{array}\right\},
\end{aligned} \tag{13}$$

$$\begin{aligned}
&\langle I_c = \frac{1}{2};\ I\ I_3;\ l=1,\ S^{l\prime},\ J^l\ J_3^l \mid H_2 \mid I_c = \frac{1}{2};\ I\ I_3;\ l=1,\ S^l,\ J^l\ J_3^l\rangle \\
&= 3\bar{c}_2(-1)^{1+J^l+I+S^{l\prime}+2S^l}\sqrt{(2S^l+1)(2S^{l\prime}+1)}\left\{\begin{array}{ccc} \frac{1}{2} & 1 & \frac{1}{2} \\ \frac{1}{2} & I & \frac{1}{2}\end{array}\right\}\left\{\begin{array}{ccc} 2 & 1 & 1 \\ J^l & S^{l\prime} & S^l\end{array}\right\}\left\{\begin{array}{ccc} S^{l\prime} & S^l & 2 \\ \frac{1}{2} & \frac{1}{2} & 1 \\ \frac{1}{2} & \frac{1}{2} & 1\end{array}\right\}
\end{aligned} \tag{14}$$

where I_c is the isospin of the core quarks. In real world ($N_c = 3$), there is only one quark in the core so I_c always equals $\frac{1}{2}$. With these matrix elements, we can express the excited heavy baryon mass up to the zeroth order of $1/N_c$:

$$\bar{\Lambda}_{H'} = N_c\bar{c}_0 + \langle H_{LS}\rangle + \langle H_T\rangle + \sum_{i=1}^{2}\langle H_i\rangle\ , \tag{15}$$

where c_{LS}, c_T, and $\bar{c}_i$'s are the coefficients $\sim \Lambda_{\rm QCD}$.

2.3 Mixing

It is necessary to consider the mixing between the baryons with light quarks in the spin-flavor symmetric and mixed representations. When they have same good quantum numbers of $(J,\ I,\ J^l)$, there is no physical way to distinguish them. This consideration will give the physical spectrum. Because of the light quark spin-flavor symmetry at the leading order of $1/N_c$ expansion, the baryons with same $(J,\ I,\ J^l)$ quantum numbers but in different representations do not mix. (S^l is a good quantum number in $N_c \to \infty$.) [9]. The mixing occurs at the sub-leading order. The classification of baryons by the spin-flavor symmetry is therefore physical at the leading order. For the physical spectrum, the mixing results in a deviation from $\bar{\Lambda}^0_H$. By denoting the mixing mass as $\tilde{m}$ which is of $O(1)$, the mass matrix for the baryons with same (J, I, J^l) is written as

$$\begin{pmatrix} \bar{\Lambda}^0_H & \tilde{m} \\ \tilde{m} & \bar{\Lambda}^0_{H'} \end{pmatrix} . \tag{16}$$

The mass difference $\bar{\Lambda}^0_H - \bar{\Lambda}^0_{H'}$ is $O(1)$. Taking $\tilde{m} < \bar{\Lambda}^0_{H'} - \bar{\Lambda}^0_H$ for illustration, the physical masses are corrected to be

$$\begin{aligned} \bar{\Lambda}_H &\simeq \bar{\Lambda}^0_H - \frac{\tilde{m}^2}{\bar{\Lambda}^0_{H'} - \bar{\Lambda}^0_H} , \\ \bar{\Lambda}_{H'} &\simeq \bar{\Lambda}^0_{H'} + \frac{\tilde{m}^2}{\bar{\Lambda}^0_{H'} - \bar{\Lambda}^0_H} . \end{aligned} \tag{17}$$

The $1/N_c$ expansion of $\tilde{m}$ is parameterized as

$$\tilde{m} = \tilde{m}_0 + O(1/N_c) , \tag{18}$$

where $\tilde{m}_0$ is universal due to LQS. To the order of $O(1)$, the spectrum is,

$$\begin{aligned} \bar{\Lambda}_{(\frac{1}{2}(\frac{3}{2}),0,1)} &= N_c c_0 + 2c_1 - \frac{\tilde{m}_0^2}{k - c_{LS} - \frac{1}{6}\bar{c}_1 - \frac{1}{4}\bar{c}_2 - 2c_1} , \\ \bar{\Lambda}_{(\frac{1}{2},1,0)} &= N_c c_0 - 2c_1 , \\ \bar{\Lambda}_{(\frac{1}{2}(\frac{3}{2}),1,1)} &= N_c c_0 - \frac{\tilde{m}_0^2}{k} , \\ \bar{\Lambda}_{(\frac{3}{2}(\frac{5}{2}),1,2)} &= N_c c_0 + 4c_1 , \end{aligned} \tag{19}$$

where k is an $O(1)$ constant which is what remains after the $\bar{\Lambda}^0_{H'}$ and $\bar{\Lambda}^0_H$ cancellation. Note that the masses of the states $(\frac{1}{2}, 1, 0)$ and $(\frac{3}{2}(\frac{5}{2}), 1, 2)$ are not affected by the mixing, because there are no physical states with the same good quantum numbers in the mixed representation. From the above spectrum, we see that $c_1 > 0$. The states $(\frac{3}{2}(\frac{5}{2}), 1, 2)$ is always the highest states. They are heavier than the other states at least by $4c_1$ through requiring the states $(\frac{1}{2}(\frac{3}{2}), 0, 1)$ to be the lowest. If $2c_1 > \frac{\tilde{m}_0^2}{k}$, the requirement implies

$$\frac{\tilde{m}_0^2}{k - c_{LS} - \frac{1}{6}\bar{c}_1 - \frac{1}{4}\bar{c}_2 - 2c_1} > 4c_1 \, . \tag{20}$$

In this case, the spectrum pattern is

$$M(\frac{1}{2}(\frac{3}{2}), 0, 1) < M(\frac{1}{2}, 1, 0) < M(\frac{1}{2}(\frac{3}{2}), 1, 1) < M(\frac{3}{2}(\frac{5}{2}), 1, 2) \, . \tag{21}$$

On the other hand, if $2c_1 < \frac{\tilde{m}_0^2}{k}$, the requirement is

$$\tilde{m}_0^2 \left(\frac{1}{k - c_{LS} - \frac{1}{6}\bar{c}_1 - \frac{1}{4}\bar{c}_2 - 2c_1} - \frac{1}{k} \right) > 2c_1 \, , \tag{22}$$

which gives the spectrum

$$M(\frac{1}{2}(\frac{3}{2}), 0, 1) < M(\frac{1}{2}(\frac{3}{2}), 1, 1) < M(\frac{1}{2}, 1, 0) < M(\frac{3}{2}(\frac{5}{2}), 1, 2) \, . \tag{23}$$

The experimentally found baryons $\Lambda_{c1}(\frac{1}{2})$ and $\Lambda_{c1}(\frac{3}{2})$ correspond to the $(\frac{1}{2}(\frac{3}{2}), 0, 1)$ states. More data are needed to fix the unknown parameters c_i's, $\bar{c}_i$'s, k and c_{LS}. In the near future, experiments will check the above predicted spectrum. Hopefully one of the above mass patterns will be picked out. It will be a check for the validity of our method, if the parameters are in the reasonable range (Λ_{QCD}) and meanwhile satisfy the relations given above.

3 Summary

In summary, we have reported the $1/N_c$ expansion method in studying the spectra of the $L = 1$ orbitally excited heavy baryons within the framework of HQET. The analysis is very simple for the baryons with light quarks being in the spin-flavor symmetric representation, compared to that for the heavy baryons with light quarks in the mixed representation. The simplicity is an unique feature. It can be seen from the point that the light quark system is in the ground state and it is the heavy quark that is orbitally excited. However the mixing effect due to the baryon states in the mixed representation corrects the spectrum pattern in the sub-leading order of $1/N_c$ expansion. The effect is important to get the realistic spectra at this order. The general pattern of the baryon spectrum has been given, which will be verified by the experiments in the near future. The $1/m_Q$ and SU(3) corrections have been considered in ref. [5]. Certain mass relations for the baryons $\Lambda_{c1}^{(*)}$, $\Sigma_{c1}^{(*)}$, $\Xi_{c1}^{(')(*)}$, and $\Omega_{c1}^{(*)}$ have been derived. The same analysis can be applied to the bottom baryons.

Acknowledgments

This work was supported in part by the National Natural Science Foundation of China and BK21 Program of the Ministry of Education of Korea.

References

1. D.E. Groom *et al.* (Particle Data Group), *Eur. Phys. J. C* **15**, 1 (2000).
2. For reviews, see N. Isgur and M.B. Wise, in *B Decays*, ed. B. Stone, 2nd edition (World Scientific, Singapore, 1994), P. 231; M. Neubert, *Phys. Rep.* **245**, 259 (1994); I. Bigi, M. Shifman, and N. Uraltsev, *Ann. Rev. Nucl. Part. Sci.* **47**, 591 (1997).
3. G. 't Hooft, *Nucl. Phys.* B **72**, 461 (1974); E. Witten, *Nucl. Phys.* B **160**, 57 (1979).
4. J.-L. Gervais and B. Sakita, *Phys. Rev. Lett.* **52**, 87 (1984); R.F. Dashen and A.V. Manohar, *Phys. Lett.* B **315**, 425 (1993); E. Jenkins, *Phys. Lett.* B **315**, 431 (1993); C. Carone, H. Georgi, and S. Osofsky, *Phys. Lett.* B **322**, 227 (1994); M.A. Luty and J. March-Russell, *Nucl. Phys.* B **426**, 71 (1994).
5. J.-P. Lee, C. Liu and H.S. Song, hep-ph/0006267.
6. J.-P. Lee, C. Liu and H.S. Song, *Phys. Rev.* D **59**, 034002 (1999).
7. E. Witten, *Nucl. Phys.* B **223**, 433 (1983).
8. J. Dai *et al.*, *Phys. Rev.* D **53**, 273 (1996); E. Jenkins, *Phys. Rev.* D **54**, 4515 (1996); *Phys. Rev.* D **55**, R10 (1997); C. Liu, *Phys. Rev.* D **57**, 1991 (1998); C. Carone *et al.*, *Phys. Rev.* D **50**, 5793 (1994); J.L. Goity, *Phys. Lett.* B **414**, 140 (1997).
9. J.G. Körner, *Nucl. Phys. B (Proc. Suppl.)* **50**, 140 (1996).

LATTICE STUDY OF SCALAR AND TENSOR GLUEBALLS

CHUAN LIU
Department of Physics
Peking University
Beijing, 100871, P. R. China
E-mail: liuchuan@mail.phy.pku.edu.cn

Glueball spectrum is studied using an improved gluonic action on asymmetric lattices in the pure $SU(3)$ gauge theory. The smallest spatial lattice spacing is about $0.08fm$ which makes the extrapolation to the continuum limit more reliable. In particular, attention is paid to the scalar glueball mass which is known to have problems in the extrapolation. Converting our lattice results to physical units using the scale set by the static quark potential, we obtain the following results for the glueball masses: $M_G(0^{++}) = 1730(90)MeV$ for the scalar glueball mass and $M_G(2^{++}) = 2400(95)MeV$ for the tensor glueball.

1 Introduction

QCD is believed to be the theory which describes strong interactions among quarks and gluons. A direct consequence of this is the existence of excitations of gluonic degrees of freedom, i.e. glueballs. However, due to their non-perturbative nature, the spectrum of glueballs can only be investigated reliably with non-purterbative methods like lattice QCD [1,2,3,4]. Recently, it was advocated that such a calculation can be performed on a relatively coarse lattice using an improved gluonic action on asymmetric lattices [5,6,7]. In this paper, we present our results on scalar and tensor glueball spectrum calculation. The spatial lattice spacing in our simulations ranges from $0.08fm$ to $0.25fm$ which enables us to extrapolate more reliably to the continuum limit. The improved gluonic action we used is the tadpole improved gluonic action on asymmetric lattices as described in [6,7]. It is given by:

$$S = -\beta\sum_{i>j}\left[\frac{5}{9}\frac{TrP_{ij}}{\xi u_s^4} - \frac{1}{36}\frac{TrR_{ij}}{\xi u_s^6} - \frac{1}{36}\frac{TrR_{ji}}{\xi u_s^6}\right] - \beta\sum_{i}\left[\frac{4}{9}\frac{\xi TrP_{0i}}{u_s^2} - \frac{1}{36}\frac{\xi TrR_{i0}}{u_s^4}\right] . \quad (1)$$

In the above expression, β is related to the bare gauge coupling, $\xi = a_s/a_t$ is the (bare) aspect ratio of the asymmetric lattice with a_s and a_t being the lattice spacing in spatial and temporal direction respectively. The parameter u_s is the tadpole improvement parameter to be determined self-consistently from the spatial plaquettes in the simulation. P_{ij} and P_{0i} are the spatial and temporal plaquette variables. R_{ij} designates the 2×1 Wilson loop (2 in direction i and 1 in direction j). The bare aspect ratio is taken to be 3 in our simulation. Using the non-perturbatively determined tadpole improvement [8] parameter u_s, the renormalization effects of the aspect ratio, which is ignored in this calculation, is small [6,7].

This paper is organized as follows. In the next section, we will explain our calculation of the Wilson loops, glueball correlation functions which give us the mass values of the glueballs. Finite volume errors are discussed and more importantly, finite lattice spacing errors are analyzed. Special attention is paid to the scalar sector where the extrapolation used to be troublesome at coarse lattices. In the last section, we will have some discussion of out result and conclude.

2 Monte Carlo Simulations

Glueball spectrum calculations in pure gauge theory basically involve three stages. The first stage, gauge field configurations are generated using some algorithm. We have utilized a Hybrid Monte Carlo algorithm to update gauge field configurations. Several lattice sizes have been simulated and the detailed information can be found in Table.1. For each lattice with fixed bare parame-

Table 1: Simulation parameters and the corresponding lattice spacing in physical units obtained from Wilson loop measurements.

Lattices	β	λ_W	n_W	r_0/a_s
$8^3 \times 24$	2.4	$0.20 \sim 0.40$	$2 \sim 4$	1.98(2)
$8^3 \times 24$	2.6	$0.20 \sim 0.40$	$2 \sim 4$	2.48(2)
$8^3 \times 24$	3.0	$0.20 \sim 0.40$	$4 \sim 6$	4.11(4)
$8^3 \times 24$	3.2	$0.25 \sim 0.50$	$4 \sim 6$	5.89(8)
$10^3 \times 30$	3.2	$0.25 \sim 0.50$	$4 \sim 6$	5.89(8)

ters, order of 10^3 configurations have been accumulated. Each gauge field configuration is separated from the previous one by several Hybrid Monte Carlo trajectories, typically $5 \sim 10$, to make sure that they are sufficiently decorrelated. Further binning of the data has been performed and no noticeable remaining autocorrelation has been observed.

The second stage of the calculation is to perform measurements of physical observables using the accumulated gauge field configurations. In fact, two independent measurements have to be done. One is to set the scale in physical units, i.e. to determine the lattice spacing a_s in physical units. The second is to measure glueball mass values in lattice units. This is done by measuring various glueball correlation functions. With the scale being set in the first step, the mass values of the glueballs can then be converted into physical units.

In the final stage of the calculation, glueball mass values obtained from finite lattices have to be extrapolated to the continuum limit. This means that finite volume effects and finite lattice spacing errors have to be eliminated. Typically, finite volume errors are found to be small in these calculations[6,7]. It is the finite lattice spacing errors that are more troublesome, especially for the scalar glueball mass. It was observed that, the scalar glueball sector exhibits a dip in the extrapolation, making the extrapolation less dependable compared with other channels[6]. To remedy this situation, a simulation at a smaller values of a_s has been performed. In our calculation, we also performed a simulation at a smaller lattice spacing where $a_s \sim 0.08 fm$. These two simulations now makes the extrapolation in the scalar sector more dependable and less sensitive to the form of the functions used in the extrapolation.

2.1 Setting the scale

In our simulation, the scale is set by measuring Wilson loops from which the static quark anti-quark potential $V(R)$ is obtained. Using the static potential between quarks, we are able to determine the lattice spacing in physical units by measuring r_0, a pure gluonic scale determined from the static potential [9]. The definition of the scale r_0 is given by: $R^2 dV(R)/dR]_{R=r_0} = 1.65$. In physical units, r_0 is roughly $0.5 fm$. For a recent determination of r_0, please consult Ref. [10].

In order to measure the Wilson loops accurately, it is the standard procedure to perform single link smearing [2,6] on the spatial gauge links of the configurations. In this procedure, each spatial gauge link is replaced by a linear combination of the original link and its spatial staples. Each spatial gauge staple is weighted by a parameter λ_W relative to the original gauge link. The final result is then projected back into $SU(3)$. This smearing scheme can be performed iteratively on the spatial links of gauge fields for as many as n_W times. The effect of smearing is to projects out higher excitation contaminations from the Wilson loop measurements. It is also useful to form thermally averaged temporal links. Then, Wilson loops are constructed using these smeared links. For a Wilson loop of size $R \times T$, it is fitted against:

$$W(R,T) \stackrel{T\to\infty}{\sim} Z(R)e^{-V(R)T} \quad . \tag{2}$$

Therefore, by extracting the effective mass plateau at large temporal separation, the static quark potential $V(R)$ is obtained. We have measured the Wilson loops along different lattice axis. It is seen that the measured data points for the quark anti-quark potential along different lattice axis lie on a universal line which is an indication that the improved action restores the rotational symmetry quite well. The smearing parameter λ_W used in this process are also listed in Table. 1. The static quark potential is fitted according to a Coulomb term plus a linear confining potential which is known to work well at these lattice spacings[7]. The potential is parameterized as:

$$V(R) = V_0 + e_c/R + \sigma R \ . \tag{3}$$

From this and the definition of r_0, it follows that:

$$r_0/a_s = \sqrt{\frac{1.65 + e_c}{\sigma a_s^2}} \ . \tag{4}$$

To convert the measured result to r_0/a_s, we have also used the bare value of ξ. The result of the spatial lattice spacing in physical units is also included in Table. 1. The errors for the ratio r_0/a_s are obtained by blocking the whole data set into smaller blocks and obtain the error from different blocks.

2.2 Glueball mass measurement

To obtain glueball mass values, it is necessary to constructed glueball operators in various symmetry sectors of interests. On a lattice, the full rotational symmetry is broken down to only cubic symmetry, which is a finite point group with 5 irreducible representations [a]. They are labeled as : A_1, A_2, E, T_1 and T_2. The first two irreducible representations are one-dimensional. The third is two-dimensional while the last two are three-dimensional. In practice, one is interested in the scalar, tensor and vector glueballs in the continuum limit. The correspondence is the following: scalar glueball is in the A_1 representation of the cubic group; tensor glueball is in representation $E + T_2$, which forms a 5-dimensional representation; vector glueballs are in the representation T_1. Of course, this correspondence is not one to one but infinite to one. Therefore, what we can measure is the lowest excitation in the corresponding representation of the cubic group[11,3].

Glueball correlation functions are notoriously difficult to measure. They die out so quickly that it is very difficult to get a clear signal. To enhance

[a] Since we only consider the glueball states with zero momentum, it suffices to study the cubic group. For glueball states with non-vanishing momenta, the whole space group has to be considered.

the signal of glueball correlation functions, smearing and fuzzying have to be performed on spatial links of the gauge fields [2,6,7]. These techniques greatly enhance the overlap of the glueball states and thus provide possibility of measuring the mass values. In our simulations, after performing single link smearing and double link fuzzying on spatial links, we first construct raw operators at a given time slice t: $\{\mathcal{R}_t^{(i)} = \sum_{\mathbf{x}} \mathcal{W}_{\mathbf{x},t}^{(i)}\}$, where $\mathcal{W}_{t,\mathbf{x}}^{(i)}$ are closed Wilson loops originates from a given lattice point $x = (t, \mathbf{x})$.Then, elements from the cubic group is applied to these raw operators and the resulting set of loops now forms a basis for a representation of the cubic group. Suitable linear combinations of these operators are constructed to form a basis for a particular irreducible representation of interest [3]. We denote these operators as $\{\mathcal{O}_\alpha^{(R)}(t)\}$ where R labels a specific irreducible representation and α labels different operators at a given time slice t. In order to maximize the overlap with one glueball state, we construct a glueball operator $\mathcal{G}^{(R)}(t) = \sum_\alpha v_\alpha^{(R)} \bar{\mathcal{O}}_\alpha^{(R)}(t)$, where $\bar{\mathcal{O}}_\alpha^{(R)}(t) = \mathcal{O}_\alpha^{(R)}(t) - \langle 0 \mid \mathcal{O}_\alpha^{(R)}(t) \mid 0 \rangle$. The coefficients $v_\alpha^{(R)}$ are to be determined from a variational calculation. To do this, we construct the correlation matrix:

$$\mathbf{C}_{\alpha\beta}(t) = \sum_\tau \langle 0 \mid \bar{\mathcal{O}}_\alpha^{(R)}(t+\tau)\bar{\mathcal{O}}_\alpha^{(R)}(\tau) \mid 0 \rangle \ . \tag{5}$$

The coefficients $v_\alpha^{(R)}$ are chosen such that they minimize the effective mass

$$m_{eff}(t_C) = -\frac{1}{t_C} \log \left[\frac{\sum_{\alpha\beta} v_\alpha^{(R)} v_\beta^{(R)} \mathbf{C}_{\alpha\beta}(t_C)}{\sum_{\alpha\beta} v_\alpha^{(R)} v_\beta^{(R)} \mathbf{C}_{\alpha\beta}(0)} \right] , \tag{6}$$

where t_C is time separation for the optimization. In our simulation $t_C = 1$ is taken. If we denote the optimal values of $v_\alpha^{(R)}$ by a column vector $\mathbf{v}^{(R)}$, this minimization is equivalent to the following eigenvalue problem:

$$\mathbf{C}(t_C) \cdot \mathbf{v}^{(R)} = e^{-t_C m_{eff}(t_C)} \mathbf{C}(0) \cdot \mathbf{v}^{(R)} \ . \tag{7}$$

The eigenvector $\mathbf{v}_0^{(R)}$ with the lowest effective mass then yields the coefficients $v_{0\alpha}^{(R)}$ for the operator $\mathcal{G}_0^{(R)}(t)$ which best overlaps the lowest lying glueball in the channel with symmetry R. Higher-mass eigenvectors of this equation will then overlap predominantly with excited glueball states of a given symmetry channel.

With these techniques, the glueball mass values are obtained in lattice units and the final results are listed in Table.2. The errors are obtained by binning the total data sets into several blocks and doing jackknife on the blocks.

Table 2: Glueball mass estimates for the symmetry channel A_1^{++}, E^{++} and T_2^{++} at various lattice spacings. The entries corresponding to the highest *beta* value are the values after the infinite volume extrapolation. The last row tabulated the continuum extrapolated result of the glueball mass values in units of $1/r_0$.

β	$a_t M_{A_1^{++}}$	$a_t M_{E^{++}}$	$a_t M_{T_2^{++}}$
2.4	0.552(8)	0.980(10)	1.002(9)
2.6	0.482(12)	0.760(16)	0.798(15)
3.0	0.322(8)	0.460(13)	0.470(13)
3.2	0.233(7)	0.323(12)	0.340(10)
∞	4.23(22)	5.77(34)	5.92(32)

2.3 Extrapolation to the continuum limit

As has been mentioned, finite volume errors are eliminated by performing simulations at the same lattice spacing but different physical volumes. This also helps to purge away the possible toleron states whose energy are sensitive to the size of the volume. A simulation at a larger volume is done for the smallest lattice spacing in our calculation. We found that the mass of the scalar glueball remains unchanged when the size of the volume is increased. The mass of the tensor glueball is affected, which is consistent with the known result that tensor glueballs have a rather large size and therefore feel the finiteness of the volume more heavily. The infinite volume is obtained by extrapolating the finite volume results using the relation [12]:

$$a_t M^{(R)}(L_s) = a_t M^{(R)}(\infty)\left(1 - \lambda^{(R)} \exp(-\sqrt{3}z/2)/z\right) , \qquad (8)$$

where $z = M^{(A_1^{++})} L_s$. Using the results for the mass of the E^{++} and T_2 glueballs on $8^3 24$ and $10^3 30$ lattices for the same value of β, the final result for the mass of these glueball states are obtained. Glueball mass values for other symmetry sectors are not so sensitive to the finite volume effects. Therefore, in Table.2, only the extrapolated values for the smallest physical volume are tabulated. Other entries are obtained from $8^3 24$ lattice results.

As for the finite lattice spacing errors, special attention is paid to the scalar glueball sector where the continuum limit extrapolation was known to have problems. Due to the simulation points at small lattice spacings, around $0.1 fm$ and below, the ambiguity in this extrapolation is greatly reduced. We have tried to extrapolate the result using different formula suggested in Ref. [7], the extrapolated results are all consistent within statistical errors. For definiteness, we take the simple form:

$$r_0 M_G(a_s) = r_0 M_G(0) + c_1(a_s/r_0)^2 + c_2(a_s/r_0)^4 \ , \tag{9}$$

and the result is illustrated in Fig.2. The final extrapolated results for the glue-

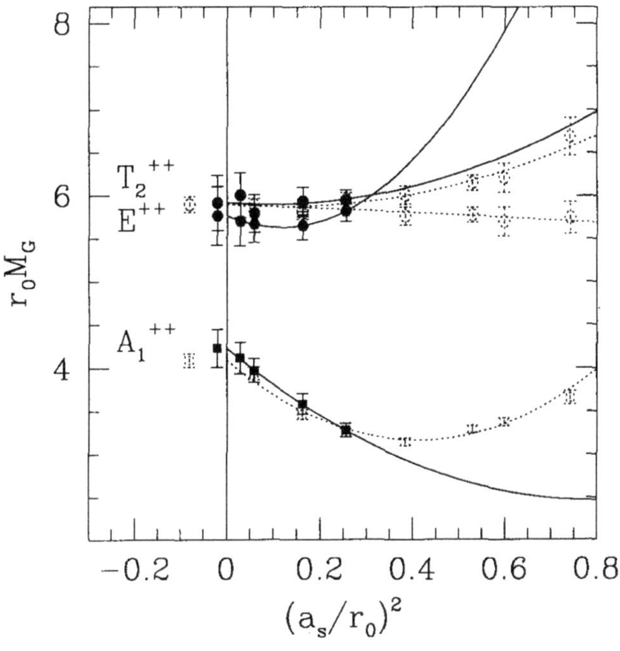

Figure 1: The continuum limit extrapolation of glueball mass values in scalar and tensor channels. The solid symbols are results from this calculation with the corresponding continuum limit extrapolation represented by the solid lines. For comparison, the corresponding results from are also shown with open symbols and dashed lines.

ball mass values are also listed in Table.2. The data points from our simulation results are shown with solid symbols and the corresponding extrapolations are plotted as solid lines. It is also noticed that the extrapolated mass values for E^{++} and T_2^{++} channels coincide within statistical errors, indicating that in the continuum limit, they form the tensor representation of the rotational group. For comparison, simulation results from [7] are also shown with open symbols and the corresponding extrapolation are represented by the dashed lines. We also tried to extrapolate linearly in $(a_s/r_0)^2$ using three data points with smallest lattice spacing. The results are statistically consistent with the results using the extrapolation (9) within errors. It is seen that, due to data points at lattice spacings around $0.1 fm$ and below, the uncertainties in the extrapolation for the glueball mass values are reduced.

Finally, to convert our simulation results on glueball masses into physical units, we use the result $r_0^{-1} = 410 MeV$. The errors for the hadronic scale r_0 is neglected. For the scalar glueball we obtain $M_G(0^{++}) = 1730(90) MeV$. For the tensor glueball mass in the continuum, we combine the results for the T_2^{++} and E^{++} channels and obtain $M_G(2^{++}) = 2400(95) MeV$ for the tensor glueball mass.

3 Conclusions

We have studied the scalar and tensor glueball spectrum at zero momentum in the pure $SU(3)$ gauge theory using Monte Carlo simulations on asymmetric lattices with the lattice spacing in the spatial directions ranging from $0.08 fm$ to $0.25 fm$. This helps to make extrapolations to the continuum limit with more confidence for the scalar and tensor glueball states. The mass values of the glueballs are converted to physical units in terms of the hadronic scale r_0. We obtain the mass for the scalar glueball and tensor glueball to be: $m_G(0^{++}) = 1730(90) MeV$ and $m_G(2^{++}) = 2400(95) MeV$.

References

1. G. Bali *et al.*, Phys. Lett. B **309**, 378 (1993).
2. C. Michael and M. Teper, Nucl. Phys. B **314**, 347 (1989).
3. B. Berg and A. Billoire, Nucl. Phys. B **221**, 109 (1983).
4. B. Berg and A. Billoire, Nucl. Phys. B **226**, 405 (1983).
5. M. Alford *et al.*, Phys. Lett. B **361**, 87 (1995).
6. C. Morningstar and M. Peardon, Phys. Rev. D **56**, 4043 (1997).
7. C. Morningstar and M. Peardon, Phys. Rev. D **60**, 034509 (1999).
8. G. P. Lepage and P. B. Mackenzie, Phys. Rev. D **48**, 2250 (1993).
9. R. Sommer, Nucl. Phys. B **411**, 839 (1994).
10. M. Guagnelli, R. Sommer, and H. Wittig, Nucl. Phys. B **535**, 389 (1998).
11. R. C. Johnson, Phys. Lett. B **114**, 147 (1982).
12. M. Lüscher, Commun. Math. Phys. **104**, 177 (1986).

PRELIMINARY RESULTS FOR LIGHT HADRON MASSES FROM LATTICE QCD ON ASYMMETRIC LATTICES

C. LIU
Department of Physics
Peking University
Beijing, 100871, P. R. China
E-mail: liuchuan@mail.phy.pku.edu.cn

J.P. MA
Institute of Theoretical Physics
Academia Sinica
P.O. Box 2735, Beijing, 100080, P. R. China
E-mail: majp@itp.ac.cn

Preliminary results for light hadron masses are presented from lattice QCD calculations on asymmetric lattices with improved actions. The used lattices have a fixed spatial lattice spacing a_s at 0.4fm, and a temporal lattice spacing a_t from 0.14 to 0.08fm. The discretization errors of a_s is reduced through improving actions at certain level. We find that masses of mesons have a weak a_t dependence, while the a_t dependence for baryons is more significant. Extrapolating to $a_t = 0$ limit, we obtain the mass of the nucleon, which has a 9% difference compared with the experimental value.

Recently it has been shown that light hadron masses can be calculated by simulations of lattice QCD with improved actions on small lattices and results are comparable with experiment [1]. In these works, improved actions on symmetric lattice were used. Because actions are improved to reduce the effect of the finite lattice spacing, especially the effect of the lattice spacing in the time direction, the actions in the free case describe particles in an unexpected way. For example, a free quark propagator has more than one pole in the momentum space. This means that the propagator describes more than one particle, one of them is physically expected, others are unphysical which appear in high energy regimes. Effects of unphysical poles have not been found in previous simulations [1]. However it is somewhat disturbing in the sense that we are not sure how these poles affect physical results.

It is pointed out that one can use asymmetric lattices to avoid the unphysical poles [2,3], where the lattice has a small lattice spacing a_t in the time direction and rather large lattice spacing a_s in spatial directions. Actions are then improved only to reduce the effect of a_s while taking a_t small enough so that the effect of a_t is negligible or under control. Simulations with this type of improved actions have been performed to calculate masses of glueballs [3,4], hybrid mesons [5], $\cdots$ etc.

We are currently simulating lattice QCD on asymmetric lattices to calculate masses of light hadrons. In this report, we present our preliminary results. The used gluonic action is:

$$S_G = \beta \left\{ \frac{5}{3}\frac{P_{sp}}{\xi u_s^4} + \frac{4}{3}\frac{\xi P_{tp}}{u_s^2} - \frac{1}{12}\frac{P_{sr}}{\xi u_s^6} - \frac{1}{12}\frac{\xi P_{str}}{u_s^4} \right\} \tag{1}$$

where P_{sp} is the sum of all spatial plaquetts, P_{st} is for temporal plaquetts, P_{sr} is for the product of gauge link around 2×1 spatial rectangular loops, and P_{str} is for 2×1 rectangular loops with 2 spatial links and one temporal. ξ is the aspect ratio:

$$\xi = \frac{a_s}{a_t}. \tag{2}$$

The tadpole improvement is implemented only for spatial links with the parameter u_s. The quark action used in the simulations is

$$S_q = \sum_x \bar{\psi}(x) \Big\{ m_0 + \frac{1}{2}[\xi\gamma_4 \nabla_4 + \sum_i \gamma_i \nabla_i (1 - \frac{1}{6}\Delta_i)$$
$$-\xi\Delta_4 - \frac{1}{\xi}\sum_i (\Delta_i - \frac{1}{12}\Delta_i^2) + \sum_i \sigma_{4i}F_{4i} + \frac{1}{\xi}\sum_{i>j}\sigma_{ij}F_{ij}] \Big\} \psi(x),$$
$$\nabla_\mu \psi(x) = u_\mu(x)\psi(x+\hat{\mu}) - u_\mu^\dagger(x-\hat{\mu})\psi(x-\hat{\mu}),$$
$$\Delta_\mu \psi(x) = u_\mu(x)\psi(x+\hat{\mu}) + u_\mu^\dagger(x-\hat{\mu})\psi(x-\hat{\mu}) - 2\psi(x), \tag{3}$$

where m_0 is the bare mass of the quark. This action is obtained by taking the SW action [6] on asymmetric lattice and then lattice derivatives in spatial directions are improved with terms at higher orders in a_s. At the classical level the action is accurate at $O(a_s^4)$. In the above action we take the coefficient c_{sw} for the clover term to be unity. The tadpole improvement is also implemented by dividing each spatial link $u_i(x)$ by u_s. It should be noted that one can put a coefficient instead of 1 in front of each term in the action, and tune these coefficients nonperturbatively to obtain optimal results. In this work only tree-level values are taken.

We use the quenched approximation and the gluonic action to generate gauge field configurations. The results, which will be given in this report, are from the configurations with simulation parameters shown in Table 1.

The lattices given in Table 1 have the same lattice spacing a_s in spatial directions, while they have different lattice spacing a_t in the time direction. This allows us to study how large the effect due to a_t is. We will also perform simulations with different a_s and with different volumes to determine the effects

Table 1: Simulation parameters

β	Lattice	No.	ξ	u_s^4
1.9	$6^3 \times 18$	200	3	0.340
1.9	$6^3 \times 24$	200	4	0.332
1.9	$6^3 \times 30$	200	5	0.328

of a_s and the finite-volume. The whole calculation is carried out on several PC's.

To see the effect of the improvement, we check dispersion relations with π and with ρ. On asymmetric lattice the dispersion relation in unit of lattice spacing reads:

$$E^2(|\mathbf{p}|) = E^2(0) + \frac{1}{\xi^2}c^2|\mathbf{p}|^2. \tag{4}$$

We calculate hadron propagators with first three low-lying momenta and determine corresponding energies. Fitting these energies with the relation we can obtain c^2. For the lattice with $\xi = 3$ we have

$$\begin{aligned} c^2 &= 1.04(11), \text{from } \pi, \\ c^2 &= 0.828(81), \text{from } \rho, \end{aligned} \tag{5}$$

where $\xi = 3$ is used. It should be pointed out that one should use the renormalized aspect ratio instead of the bare one. The renormalization has a small effect on this lattice, its effect is at several percent level. The c^2 given above is close to 1, indicating that the rotation invariance may be restored, or the effect of the finite lattice spacing may be small. Similar results are also obtained from lattices with $\xi = 4, 5$. We also simulated lattice QCD with unimproved actions on a symmetric lattice with a lattice spacing a, which is roughly the same as the a_s of the lattice with $\xi = 3$, and checked the dispersion relation. We obtained:

$$\begin{aligned} c^2 &= 0.408(80), \text{from } \pi, \\ c^2 &= 0.263(52), \text{from } \rho. \end{aligned} \tag{6}$$

It is seen from Eq.(6) that the value of c^2 is far from 1, it indicates that the rotation invariance is badly violated and the effect of the finite lattice spacing is large. Comparing Eq.(5) with Eq.(6) we see that the effect of the finite lattice spacing is really reduced by using improved actions. However, the c^2 determined from different hadrons is different. We tried to tune the coefficient

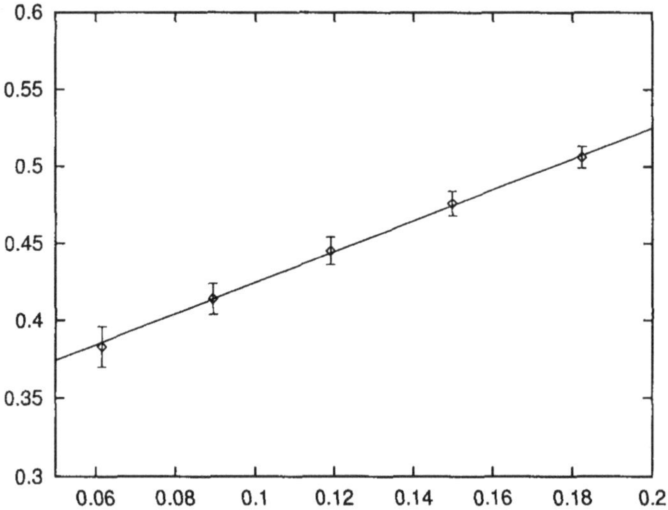

Figure 1: Chiral extrapolation for m_ρ versus m_π^2.

c_{sw} to obtain the same c^2 without success, and it seems that it is not possible to have the same c^2 only by tuning a single coefficient.

We calculated hadron masses with local operators, and extrapolate them to chiral limit. The extrapolation is done by fitting pseudoscalar meson, vector-meson and baryon with the chiral formula:

$$\begin{aligned} m_\pi &= u_0 + u_1 m_0, \\ m_{P,V} &= v_0 + v_1 m_\pi^2, \\ m_B &= w_0 + w_1 m_\pi^2 + w_2 m_\pi^3. \end{aligned} \tag{7}$$

We observe that the dependence of m_π on m_0 is well described by the linear relation and for baryon we observe clear signal indicating the cubic term in m_π. We show Fig.1, Fig.2 and Fig.3 to illustrate our chiral extrapolation for the lattice with $\xi = 5$. In the chiral limit($m_\pi = 0$) hadron masses are obtained which are summarized in Table 2.

In Table 2 the lattice spacing is determined by m_ρ, the mass parameter m_0 for the strange quark is determined by the experimental value $m_K^2/m_\rho^2 = 0.412$. The calculation for hadrons containing a s-quark with the $\xi = 4$ lattice is not completed, and we have finished for this lattice calculations of hadron masses only with three different quark masses. The calculations with other quark

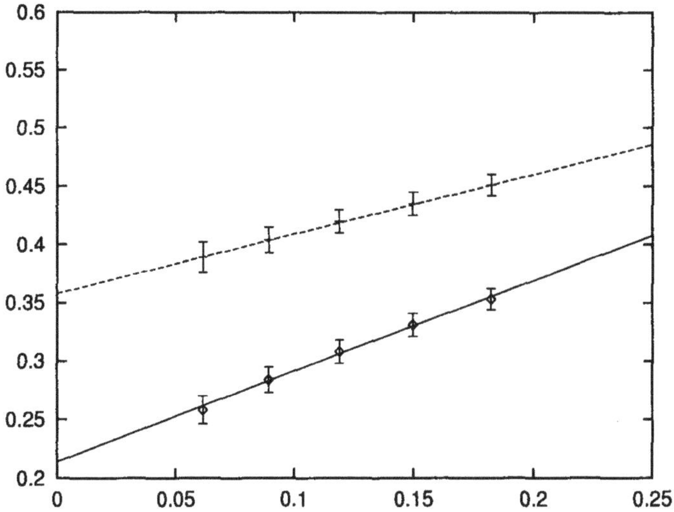

Figure 2: Chiral extrapolation for m_K and m_{K^*} versus m_π^2.

masses are still on the way. From the above results we realize that the a_t-dependence of m_K/m_ρ, m_{K^*}/m_ρ and m_ϕ/m_ρ is rather weak. The dependence is overwhelmed by statistical errors and ratios are comparable with experiment. In contrast, we observe a stronger a_t-dependence in ratios for baryon. We find that this dependence for m_N/m_ρ can well be fitted with

$$m_N/m_\rho = 1.328(16) + 3.328(156) \cdot a_t, \qquad \chi^2 = 0.03. \tag{8}$$

which is shown in Fig.4. We also tried to fit with a quadratic relation, but the fitting quality is worse than the one given above. With this extrapolation, we predicted within our approximations the ratio for m_N to be 1.328(16), which has a 9% difference compared with the experimental value. This difference may be explained with approximations used here, and it also may be caused by the effect of a_s.

We are currently trying to complete the results in Table 2 and do simulations on lattices with smaller a_s to check a_s-dependence of hadron masses.

Acknowledgment: This work is supported by National Science Foundation of P.R. China and by the Hundred Young Scientist Program of Sinica Academia of P.R.China.

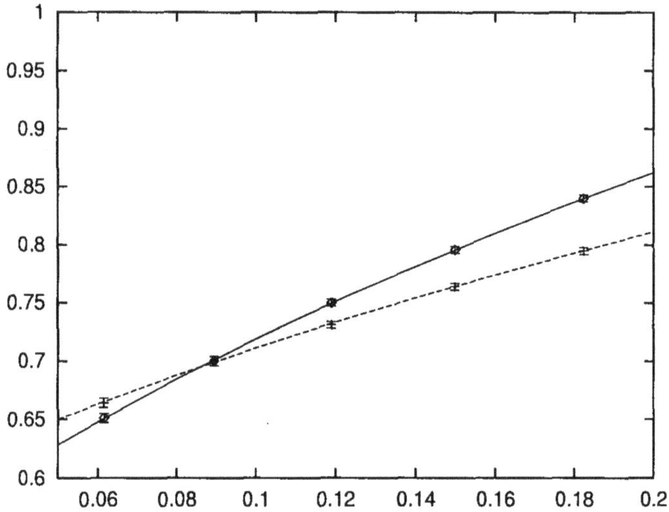

Figure 3: Chiral extrapolation for m_N and m_Λ versus m_π^2.

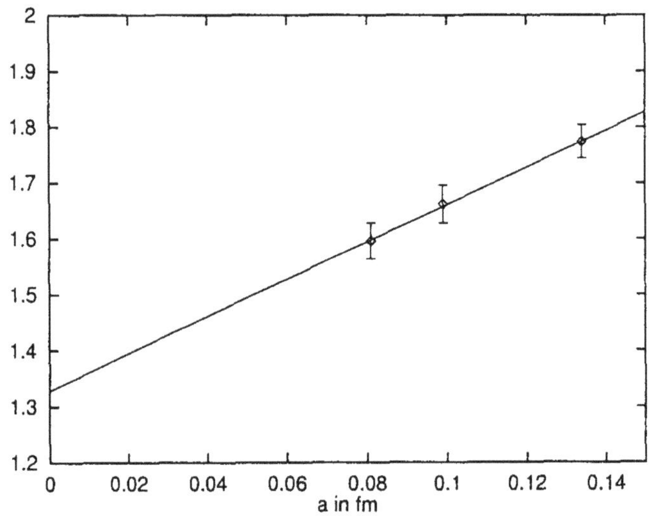

Figure 4: Extrapolation to the $a_t = 0$ limit for m_N/m_ρ.

Table 2: Preliminary results of the simulation.

a_t^{-1}	1.428GeV	1.940GeV	2.372GeV	exp.
m_ρ	0.536(7)	0.396(7)	0.324(4)	768.5MeV
m_K/m_ρ	0.681(26)		0.660(23)	0.64
m_{K^*}/m_ρ	1.134(30)		1.105(29)	1.16
m_ϕ/m_ρ	1.291(52)		1.216(51)	1.32
m_N/m_ρ	1.774(30)	1.662(34)	1.596(23)	1.22
m_Λ/m_ρ	1.946(31)		1.781(26)	1.45

References

1. M. Alford, T. Klassen and G.P. Lepage, Nucl. Phys. B47(Proc. Suppl), (1996) 370; H.R. Fiebig and R.M. Woloshyn, Phys. Lett. B385 (1996) 273; J.P. Ma and B.H.J. McKellar, Phys. Rev. D57 (1998) 6723.
2. M. Alford, T. Klassen and G.P. Lepage, Nucl. Phys. B496 (1997) 377.
3. C.J. Morningstar and M. Peardon, Phys. Rev. D56 (1997) 4043.
4. C. Liu, in these proceedings; see also hep-lat/0004018.
5. T. Manke, et.al., (CP-PACS Collaboration), Phys. Rev. Lett. Vol. 82 (1999) 4396.
6. B. Sheikoleslami and R. Wohlert, Nucl. Phys. B259 (1985) 572.

3 Numerical Algorithms

In the Chen family temple.

A STUDY OF LATTICE SU(2) GAUGE CONFIGURATIONS BY COOLING

YING CHEN, JI MIN WU, BING HE
Institute of High Energy Physics, Chinese Academy of Sciences,
P.O.Box 918-4, Beijing 100039, P. R. China
E-mail: wujm@alpha02.ihep.ac.cn

We study lattice SU(2) configurations by fast cooling with Wilson action and an improved action. The classical lattice gauge configurations are obtained. Its some properties are presented. We also show its some properties in momentum space.

In the last several years, many authors[1] studied gauge configurations on lattice by cooling and get many interesting results. Here, we still want to study it in this approach to understand more.[2]

We study SU(2) gauge field on lattice by cooling with Wilson action and tadpole-improved action. Wilson action is:

$$S = \frac{1}{g^2} \sum_{n,\mu<\nu} Tr(U_\mu(n)U_\nu(n+\mu)U_\mu^+(n+\nu)U_\nu^+(n) + h.c. - 2). \tag{1}$$

The local action associated with link $U_\mu(n)$ is

$$\begin{aligned} S(U_\mu) &= Tr(U_\mu \cdot staples + h.c.) \\ &\equiv Tr(U_\mu V_\mu + h.c.) \end{aligned} \tag{2}$$

where staple is defined as:

$$V_\mu(n) \equiv \sum_{\nu\neq\mu} U_\nu(n+\mu)U_\mu^+(n+\nu)U_\nu^+(n) \tag{3}$$

We work with fast cooling approach which lower the local action with such substitution:

$$U_\mu(n) \to U_\mu^{'}(n) = V_\mu^+(n), \tag{4}$$

and then, local action and total action decrease:

$$\begin{aligned} S(U_\mu) \to S(U_\mu^{'}) &< S(U_\mu) \\ S \to S' &< S. \end{aligned}$$

So-called 'one cooling step' is to sweep all links once (one by one).

In the path integral language, in cooling approach, the quantum field path approaches to the classical path which is determined by minimum action principle, i.e. we get the lattice classical gauge configuration.

Under such cooling, physical quantities varies as:

* gauge fields,

$$A_\mu(n) \to A'_\mu(n) = A_\mu(n) + \frac{1}{6}a^2 D_\nu F_{\nu\mu} + O(a^4) \tag{5}$$

* field tensor,

$$F_{\mu\nu}(n) \to F'_{\mu\nu}(n) = F_{\mu\nu}(n) + \frac{1}{6}a^2[D_\mu D_\alpha F_{\alpha\nu} - D_\nu D_\alpha D_{\alpha\mu}] + O(a^4) \tag{6}$$

* energy density,

$$S(n) \propto Tr\ F_{\mu\nu}F_{\mu\nu}(n) \to Tr\ F'_{\mu\nu}\ F'_{\mu\nu}(n)$$
$$Tr\ F'_{\mu\nu}F'_{\mu\nu}(n) = Tr\ F_{\mu\nu}F_{\mu\nu}(n) - \frac{2}{3}a^2\ Tr(D_\mu F_{\mu\nu})^2 + O(a^4) \tag{7}$$

* topological charge density,

$$Q(n) \propto Tr\ F_{\mu\nu}\ \tilde{F}_{\mu\nu}(n) \to Tr\ F'_{\mu\nu}\tilde{F}'_{\mu\nu}(n)$$
$$Tr\ F'_{\mu\nu}\tilde{F}'_{\mu\nu}(n) = Tr\ F_{\mu\nu}\ \tilde{F}_{\mu\nu}(n) - \frac{2}{3}\ a^2 Tr(D_\mu F_{\mu\nu} D_\alpha \tilde{F}_{\alpha\nu})^2 + O(a^4) \tag{8}$$

On the other hand, it is will known that the lattice classical equation of motion for Wilson action is,

$G_\mu(n) \equiv \sum_{\nu\neq\mu} \Big($ 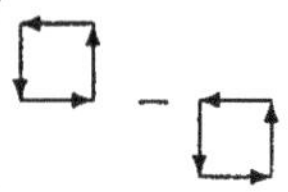 $- h.c. \Big) = 0.$ (9)

We parametrize SU(2) link as $U_\mu(n) = A_0 I + i\vec{A}\cdot\vec{\sigma}$ and cool the SU(2) configurations on a $8^3 \times 24$ lattice. $G_\mu(n)$ is also a 2×2 matrix $a_0 + i\vec{a}\cdot\vec{\sigma}$. Fig.1 shows $G_\mu(n)$ for an arbitrary link $U_\mu(n)$. One can find this matrix equation is satisfied with cooling steps. Therefore, we get lattice classical gauge configuration by cooling on lattice.

Fig.2 shows an arbitrary link matrix evoluated with cooling steps. It shows the link variables approaches to its fixed point.

$$U_\mu(n) \to U'_\mu(n) = U_\mu(n),$$
$$A_\mu(n) \to A'_\mu(n) = A_\mu(n)\ .$$

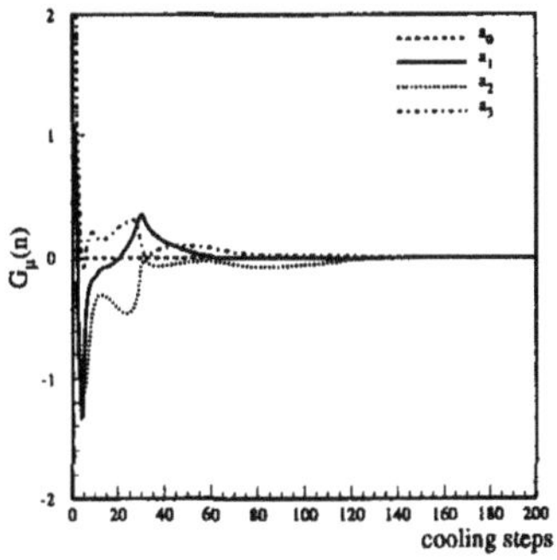

Figure 1: $G_\mu(n)$ for an arbitrary link $U_\mu(n)$.

Comparing with eq.(5), we find the classical equation of motion in continuum is satisfied for each site to $O(a^4)$,

$$D_\nu F_{\nu\mu}(n) = 0. \tag{10}$$

It also shows, we get lattice classical gauge configuration by cooling.

Now, let's measure the topological charge on the cooled configurations. In continuum, the topological charge is defined as

$$Q = \frac{1}{16\pi^2}\int d^4x\ Tr\ F_{\mu\nu}(x)\tilde{F}_{\mu\nu}(x). \tag{11}$$

On lattice, it defined as

$$Q_L = \sum_n\ Q_L(n)$$

$$Q_L(n) = \frac{1}{16\pi^2}\sum_{\pm\mu\nu\rho\sigma}\tilde{\epsilon}_{\mu\nu\rho\sigma}\ \ Tr(\Box_{\mu\nu}(n)\cdot\Box_{\rho\sigma}(n)). \tag{12}$$

Fig.3 shows the distribution of topological charge for Wilson action and improved action (500 configurations, after 25 cooling steps). The latter is :

$$S_1 = \beta\sum_{x,s>s'}\left[\frac{5}{3}\frac{P_{ss'}}{\xi u_s^4} - \frac{1}{12}\frac{R_{ss'}}{\xi u_s^6} - \frac{1}{12}\frac{R_{s's}}{\xi u_s^6}\right]$$
$$+\beta\sum_{x,s}\left[\frac{4}{3}\frac{P_{st}\xi}{u_s^2u_t^2} - \frac{1}{12}\frac{R_{st}\xi}{u_s^4u_t^2}\right], \tag{13}$$

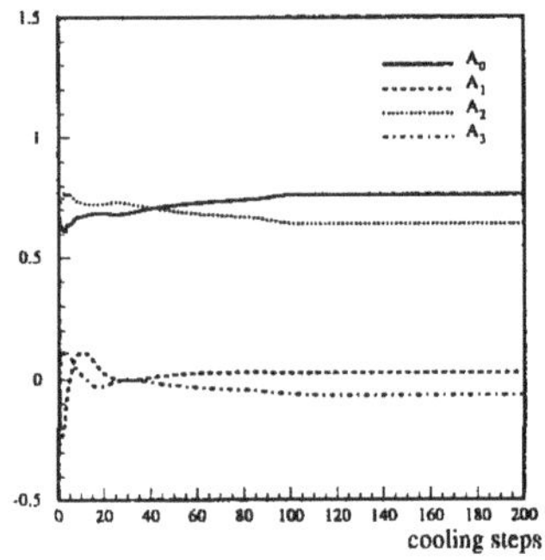

Figure 2: An arbitrary link matrix evoluated with cooling steps.

One can see the topological charge distributes nearly being an integer. It is the multiply of 0.86 in Wilson case. But, it is the multiply of 0.97 in an improved action case. Obviously, the renomalization effect is deduced for the improved action.

In continuum, it is known, the equation of motion for classical gauge field is $D_\nu F_{\nu\mu} = 0$. One knows its solutions:

(1) instanton and multi-instanton solution[3] for self-dual field, $F_{\mu\nu} = \pm\ \tilde{F}_{\mu\nu}$;

(2) saddle-point solution, sphaleron,[4] in non-self-dual field case.

But,lattice space is T^4 topology (periodical condition), which is different from continuum R^4 topology. One have to find the classical solution on lattice carefully. Much work remains to be done.

Now, let's investigate some field quantities is momentum space. It will provide us more information.

(1) Gauge field. It is defined as

$$A_\mu(n) = \frac{1}{2iga}(U_\mu(n) - U_\mu^+(n)) + O(a^2). \tag{14}$$

By taking Fourier transformation, we get momentum component of $A_\mu(n)$,

$$A_\mu(q) = \sum_x e^{iqx} A_\mu(x) = \sum_n e^{iqa\cdot n} A_\mu(n) \tag{15}$$

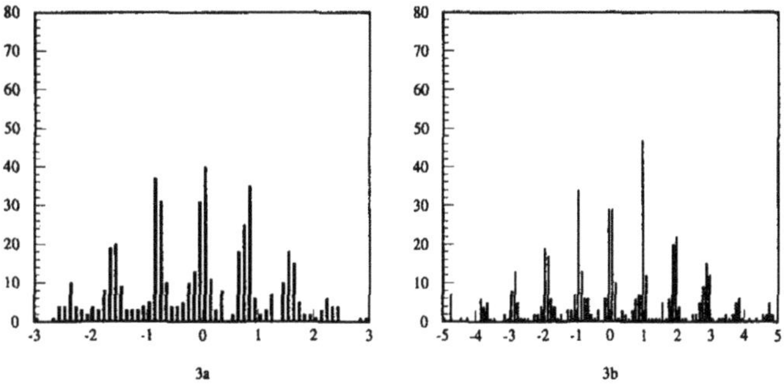

Figure 3: The distribution of topological charge for Wilson action (3a) and improved action (3b).

where q is the momentum,

$$q_\mu a = \frac{2\pi}{L_\mu} n_\mu \ , \qquad n_\mu = 0, 1, 2, \cdots L_\mu - 1 \ ,$$

$$q \ a = (q_1, q_2, q_3, q_4).$$

We take its modul

$$| A_\mu(q) |^2 = | Re \ A_\mu(q) |^2 + | Im \ A_\mu(q) |^2 . \tag{16}$$

Fig.4 shows $| A_\mu(q) |^2$ for an arbitrary link. Its each component are nearly stable with the cooling steps.

Although $A_\mu(x)$ is not gauge invariant. Its statistical average should be zero. Here is only an individual and arbitrary link variable shown in Fig.4. We just have a looking to see its behavior under cooling.

(2) Strength tension. It's defined on lattice as

$$F_{\mu\nu}(n) = \frac{1}{4} \ Im$$

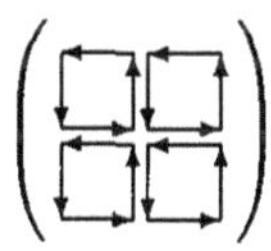

. (17)

Similarly, taking Fourier transformation of $F_{\mu\nu}(n)$, Fig.5 shows its modul

$$| F_{\mu\nu}(q) |^2 = | Re \ F_{\mu\nu}(q) |^2 + | Im \ F_{\mu\nu}(q) |^2 . \tag{18}$$

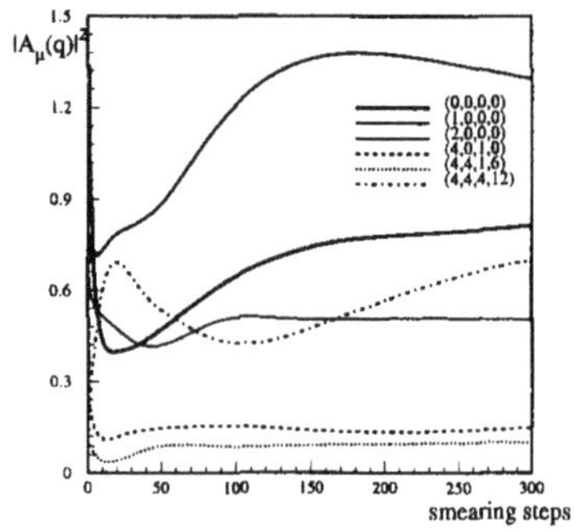

Figure 4: The modul $\mid A_\mu(q) \mid^2$ for an arbitrary link.

One can find the damp of $\mid F_{\mu\nu}(q) \mid$ is very fast and there is no sign that only the high momentum component are suppressed.

(3) Topological charge density. Its definition is

$$Q_L(n) = -\frac{1}{512\pi^2}(\frac{5}{3}Q_P(n) - \frac{1}{6}Q_R(n)) \tag{19}$$

where

$$Q_P(n) = \sum_{\mu\nu\rho\sigma=\pm 1}^{\pm 4} \tilde{\epsilon}_{\mu\nu\rho\sigma}\ Tr(P_{\mu\nu}(n)P_{\rho\sigma}(n))$$

$$Q_R(n) = \sum_{\mu\nu\rho\sigma=\pm 1}^{\pm 4} \tilde{\epsilon}_{\mu\nu\rho\sigma}\ Tr[(R^{1\times 2}_{\mu\nu}(n) + R^{2\times 1}_{\rho\sigma}(n)) \cdot (R^{1\times 2}_{\rho\sigma}(n) + R^{2\times 1}_{\rho\sigma}(n))]$$

and $P_{\mu\nu}(n)$ is the plaquett in $\mu\nu$ plant, $R^{1\times 2}_{\mu\nu}$ is (1×2) rectangle in $\mu\nu$ plant.

Fig.6 shows its modul

$$\mid Q_L(q) \mid^2 = \mid Re\ Q_L(q) \mid^2 + \mid Im\ Q_L(q) \mid^2 . \tag{20}$$

One can find most of modes damp to almost zero or tend to be stable. Only the zero momentum component and some other lower modes have some metastable plateaus with cooling steps.

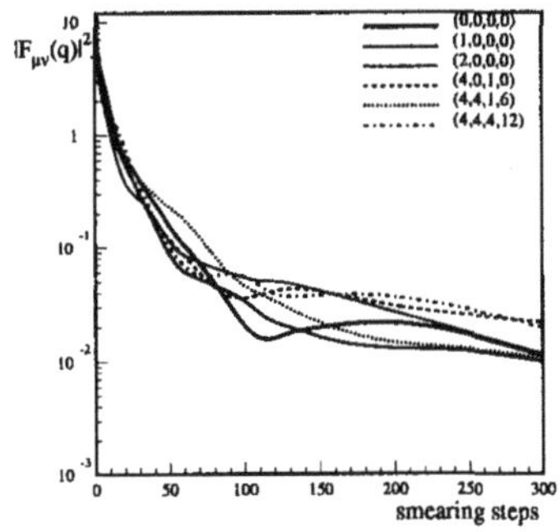

Figure 5: The modul $\mid F_{\mu\nu}(q) \mid^2$ evoluated with cooling steps

Summary One can get lattice classical gauge configuration by cooling procedure. After enough steps of cooling, a stable configuration appears. It corresponds to the fixed point of the smoothing transformation. This classical configuration is still a complicated one to solve. Recently, many results from phenomenology and lattice study[5] suggest that instatons probably really play an improtant role in QCD, especially in light quark physics. Therefore, it is very interesting that one can study the contriburion from quantum configurations and from classical configurations respectively to the same physical quantity to get more physical picture.

In the momentum space, we find the momentum component of gauge field $\mid A_\mu(q) \mid^2$ didn't suppressed by cooling steps. But, momentum component of strength tensore $\mid F_{\mu\nu}(q) \mid^2$ is suppressed. Some Fourier components of topological charge density $\mid Q_L(q) \mid^2$ survives the cooling procedure. This gives rise to the local structure of the smoothed configuration. This shows to us that not the high momentum component of all the the physical quantities damp by cooling. One have to show thier behavior one by one.

Acknowledgement

This work is partially surpported by the National Nature Science Foundations of China under the Grant No.19677205. The authors thank the National Research Center for Intelligent Computing System of China for the use of the Dawning 2000I parallel computer by Grant No. 99128.

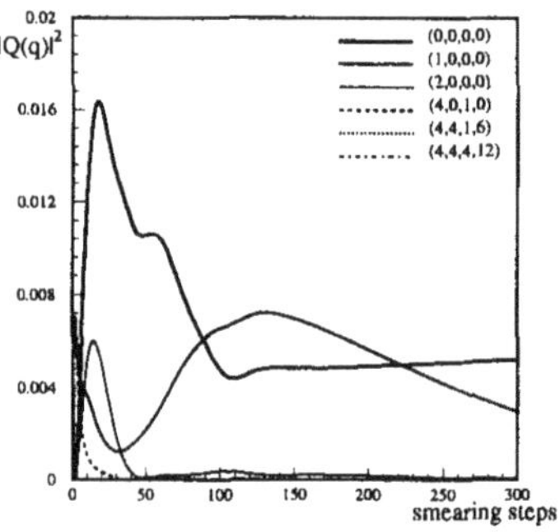

Figure 6: The modul | $Q_L(q)$ |^2 evoluated with cooling steps

References

1. B.Berg, Phys. Lett. B104(1981)475; M. -C. Chu, J. M. Grandy, S. Huang, and J. W. Negele, Phys. Rev.Lett.70(1993) 225; Phys.Rev. D48(1993)3340; Phys. Rev. D49(1994)6039; J.Hoek, M.Teper and J.Waterhouse Nucl. Phys. B285(1987)589; B.Alles, M.D'Elia and A.Di Giacomo, Nucl.Phys. B494(1997)281; M.G.Perez et. el. Nucl.Phys. B413(1994)535; D.A.Smith et. al. Phys. Rev.D58(1998)014505;
2. Ying Chen, Bing He and Ji Min Wu, Phys. Lett. B486(2000)448; Chinese Phys. Lett. 17(2000)640.
3. A.A.Belavin,A.M.Polyakov,A.A.Schwartz and Y.S.Tyupkin Phys. Lett. B59(1975)85;
4. F.R.Klinkhamer and M.Manton, Phys. Rev. D30(1984)2212;
5. T. Schäfer and E. V. Shuryak, Rev. Mod. Phys. 70(1998)323.

GREEN'S FUNCTION MONTE CARLO APPROACH TO SU(3) YANG-MILLS THEORY IN (3+1)D

C. J. HAMER, M. SAMARAS AND R.J. BURSILL
School of Physics, University of New South Wales,
Sydney, 2006, Australia
E-mail: C.Hamer@unsw.edu.au

A 'forward walking' Green's Function Monte Carlo algorithm is used to obtain expectation values for SU(3) lattice Yang-Mills theory in (3+1) dimensions. The ground state energy and Wilson loops are calculated, and the finite-size scaling behaviour is explored. Crude estimates of the string tension are derived, which agree with previous results at intermediate couplings; but more accurate results for larger loops will be required to establish scaling behaviour at weak coupling.

1 Introduction

Classical Monte Carlo simulations provide a very powerful and accurate method for the study of Euclidean lattice gauge theories. In the Hamiltonian formulation, on the other hand, the corresponding quantum Monte Carlo methods have been somewhat neglected. Here we present a study of SU(3) Yang-Mills theory in (3+1) dimensions, using the Green's Function Monte Carlo approach.[1]

Heys and Stump[2] and Chin et al.[3] pioneered the use of "Green's Function Monte Carlo" (GFMC) or "Diffusion Monte Carlo" techniques in Hamiltonian LGT, in conjunction with a weak-coupling representation involving continuous gauge field link variables. This was successfully adapted to non-Abelian Yang-Mills theories,[4,5] with no minus sign problem arising. In this representation, however, one is simulating the wave function in gauge field configuration space by a discrete ensemble or density of random walkers: it is not possible to determine the derivatives of the gauge fields for each configuration, or to enforce Gauss's law explicitly. Hence the ensemble always relaxes back to the ground state sector. In order to compute the string tension or mass gap, one must measure an appropriate correlation function, and estimate the mass gap as the inverse of the correlation length. We have introduced the 'forward-walking' technique, well-known in many-body theory,[1,6] to measure the expectation values and correlation functions. The technique has been demonstrated for the cases of the transverse Ising model in (1+1)D,[7] and the U(1) LGT in (2+1)D.[8]

Here we apply the technique for the first time to a non-Abelian model, namely SU(3) Yang-Mills theory in (3+1)D. The ground state energy and Wilson loop values are calculated, and approximate values are extracted for the string tension in the weak-coupling regime. Comparisons are made with earlier calculations, where they are available.[9]

2 Method

2.1 Lattice Hamiltonian

The Green's Function Monte Carlo formalism has been adapted to SU(2) Yang-Mills theory by Chin, van Roosmalen, Umland and Koonin,[4] and sketched for the SU(3) case by Chin, Long and Robson.[5]

The SU(3) lattice Hamiltonian is given by [5]

$$H = \frac{g^2}{2a}\{\sum_l E_l^a E_l^a - \frac{\lambda}{3}\sum_p Tr(U_p + U_p^\dagger)\}, \tag{1}$$

where E_l^a is a component of the electric field at link l, $\lambda = 6/g^4$, the index a runs over the 8 generators of SU(3), and U_p denotes the product of four link operators around an elementary plaquette. We will work with the dimensionless operator

$$H = \frac{1}{2}\sum_l E_l^a E_l^a - \frac{\lambda}{6}\sum_p Tr(U_p + U_p^\dagger). \tag{2}$$

The link variables are elements of the group SU(3) in the fundamental representation

$$U = \exp(-i\frac{1}{2}\lambda^a A^a). \tag{3}$$

2.2 Green's Function Monte Carlo method

The Green's Function Monte Carlo method employs the operator $\exp(-\tau(H - E))$, i.e. the time evolution operator in imaginary time, as a *projector* onto the ground state $|\psi_0\rangle$:

$$|\psi_0\rangle \propto \lim_{\Delta\tau\to 0, N\Delta\tau\to\infty} e^{-N\Delta\tau(H-E)}|\Phi\rangle \tag{4}$$

where $|\Phi\rangle$ is any suitable trial state. To procure some variational guidance, one performs a "similarity transformation" with the trial wave function Φ, and evolves the *product* $\Phi|\psi_0\rangle$ in imaginary time. The heart of the procedure is the calculation of the matrix element correrspondimg to a single small time step $\Delta\tau$. Chin et al. [4] show that

$$\begin{aligned}\langle \mathbf{x}'|\Phi e^{-\Delta\tau(H-E)}\Phi^{-1}|\mathbf{x}\rangle &= \prod_l \langle U_l'|N\{\exp(-\frac{1}{2}\Delta\tau E_l^a E_l^a)\exp[\Delta\tau E_l^a(E_l^a \ln\Phi)]\}|U_l\rangle \\ &\quad \exp\{\Delta\tau[E - \Phi^{-1}H\Phi(\mathbf{x})]\} + O(\Delta\tau^2) \\ &\equiv p(\mathbf{x}', \mathbf{x})w(\mathbf{x}) + O(\Delta\tau^2),\end{aligned} \tag{5}$$

where $\mathbf{x} = \{U_l\}$ denotes an entire lattice configuration of link fields.

The product $\Phi|\psi\rangle$ is simulated by the density of an ensemble of random walkers. At the kth. step, the 'weight' of each walker at $\mathbf{x}_k$ is multiplied by $w(\mathbf{x}_k)$. The effect of $p(\mathbf{x_{k+1}}, \mathbf{x_k})$ is to alter each link variable U_l in $\{\mathbf{x_k}\}$ to U_l' by a Gaussian random walk plus a "drift step" guided by the trial wave function:

$$U' = \Delta U U_d U, \tag{6}$$

where $U_d = \exp[i\frac{1}{2}\lambda^a(i\Delta\tau E^a \ln\Phi)]$ is the drift step, and ΔU is an SU(3) group element randomly chosen from a Gaussian distribution around the identity, with variance $\langle\Delta s^2\rangle$, where

$$\langle\Delta s^2\rangle \approx \sum_a A^a A^a = 8\Delta\tau, \tag{7}$$

for small A^a.

The simulation is carried out for a large number of iterations $\Delta\tau$, until an equilibrium distribution $\Phi|\psi_0\rangle$ is reached. The energy E in (5) is adjusted after each iteration so as to maintain the total ensemble weight constant. The average value of E can then be taken as an estimate of E_0, the ground-state energy.

As time evolves, the weights of some walkers grow larger, while others grow smaller, which would produce an increased statistical error. To avoid this, a "branching" process is employed, whereby a walker with weight larger than some threshold is split into two independent walkers, while others with weights lower than another threshold are amalgamated.

2.3 Trial Wave Function

The trial wave function is chosen to be the one-parameter form [5]

$$\Phi = \exp[\alpha \sum_p Tr(U_p + U_p^\dagger)]. \tag{8}$$

Then the drift step for each link is [4]

$$U_d = \exp[-i\frac{\lambda^a}{2}A_l^a], \tag{9}$$

$$A_l^a = -i\Delta\tau\frac{\alpha}{2}\sum_{p\in l} Tr[\lambda^a U_l..U_4^\dagger - h.c.]. \tag{10}$$

Finally, the trial energy factor is

$$\Phi^{-1}H\Phi = \sum_l \{\tfrac{\alpha^2}{8}(\sum_{p\in l} Tr[\lambda^a U_l..U_4^\dagger - h.c.])^2 + (\tfrac{2\alpha}{3} - \tfrac{\lambda}{24})\sum_{p\in l} Tr(U_p + U_p^\dagger)\} \tag{11}$$

2.4 Forward Walking estimates

The "forward walking" technique is used to estimate expectation values.[1] Its application to the U(1) lattice gauge theory in (2+1)D was discussed by Hamer et al.[8] It is implemented for an operator Q (assumed diagonal, for simplicity) by[6] recording the value $Q(\mathbf{x}_i)$ for each "ancestor" walker at the beginning of a measurement; propagating the ensemble as normal for J iterations, keeping a record of the "ancestor" of each walker in the current population; and taking the weighted average of the $Q(\mathbf{x}_i)$ with respect to the weights of the descendants of $\mathbf{x}_i$ after the J iterations, using sufficient iterations J that the estimate reaches a 'plateau'.

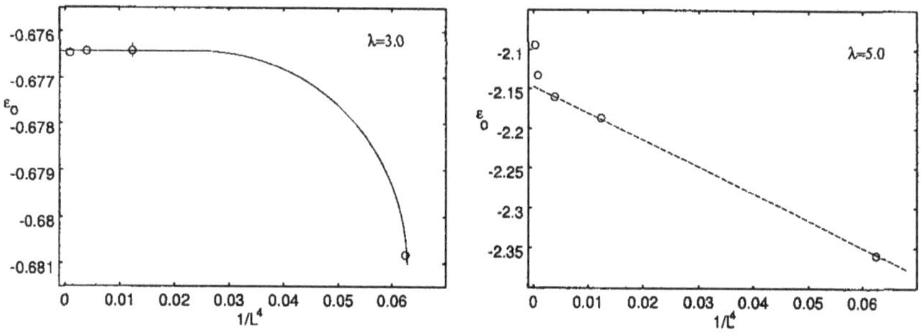

Figure 1: Ground-state energy per site graphed against $1/L^4$, where L is the lattice size: a) at coupling $\lambda = 3.0$, b) at coupling $\lambda = 5.0$. The lines are merely to guide the eye.

3 Results

Simulations were carried out for $L \times L \times L$ lattices up to L=8 sites, using runs of typically 4000 iterations and an ensemble size of 250 to 1000 depending (inversely) on lattice size. Time steps $\Delta\tau$ of 0.01 and 0.05 "seconds" were used, with each iteration consisting of 5 sweeps and 1 sweep through the lattice, respectively, followed by a branching process. The first 400 iterations were discarded to allow for equilibration. The data were block averaged over blocks of up to 256 iterations, to minimize the effect of correlations on the error estimates.

The results taken at $\Delta\tau = 0.01$ and $\Delta\tau = 0.05$ were extrapolated linearly to $\Delta\tau = 0$. The variational parameter c was given values as used by Chin et al.,[5] obtained from a variational Monte Carlo calculation. We checked that these were approximately the optimum values for small lattices.

Forward-walking measurements were taken over J iterations, where J ranged from 20 to 100, depending on the coupling λ. Ten separate measurements were taken over this time interval, in order to check whether the value measured by forward-walking had reached equilibrium. A new measurement was started soon after the previous one had finished.

3.1 Ground-state Energy

The dependence of the ground-state energy per site on lattice size is illustrated in Figure 1, at two fixed couplings $\lambda = 3.0$ and $\lambda = 5.0$. In the "strong-coupling" case, $\lambda = 3.0$, it can be seen that the results converge exponentially fast in L, whereas in the "weak-coupling" regime, $\lambda = 5.0$, the convergence is more like $1/L^4$ at these lattice sizes. This behaviour merits some further explanation.

A similar phenomenon occurs in the case of the U(1) theory in (2+1)D.[10,8] In the strong-coupling regime, where the mass gap is large, the usual exponential convergence occurs. In the weak-coupling regime, however, where the mass gap M is very small, the finite-size scaling behaviour for small lattice sizes is that of a massless theory, and it is only at much larger lattice sizes $L \approx 1/M$ that a crossover to exponential convergence occurs. An "effective Lagrangian" corresponding to free, massless gluons (non-interacting QCD) should describe the finite-size behaviour in the present case, in line with the idea of asymptotic freedom. By analogy with the (2+1)D case, we expect a $1/L^4$ dependence for the corrections to the ground-state energy per site. We hope to pursue this analysis further at a later date.

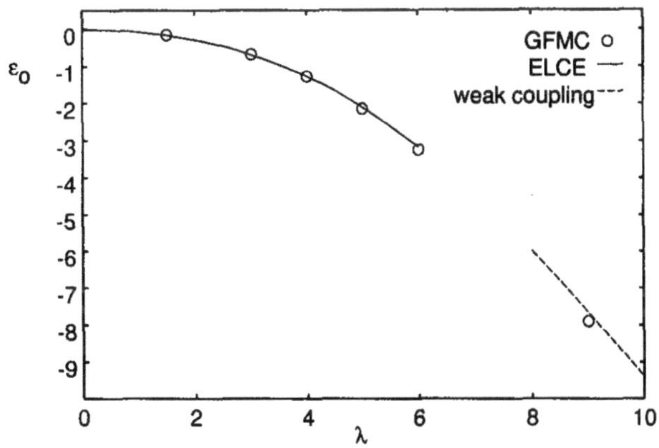

Figure 2: The bulk ground-state energy per site graphed against coupling λ. The points are our Monte Carlo estimates; the solid line represents earlier ELCE estimates[9]; and the dashed line represents the asymptotic weak-coupling behaviour.

An anomalous feature in Figure 1b) is that the $L = 8$ point lies well out of line with the others. This occurs at other couplings also. We suspect that the results for $L = 8$ are not reliable, and that the trial wave function will have to be further improved to give reliable results for such large lattices.

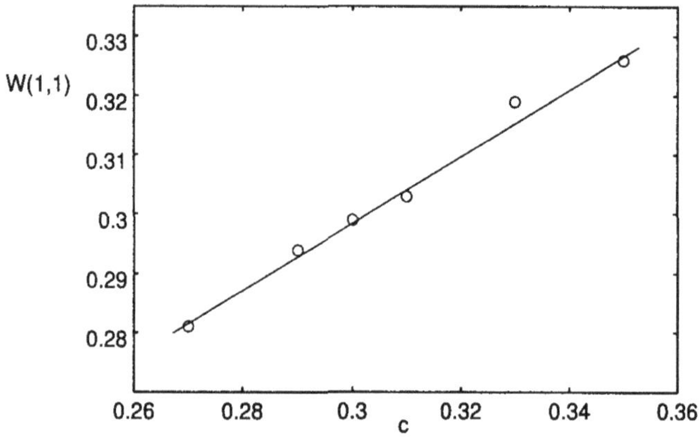

Figure 3: Estimated value for the mean plaquette $W(1,1)$ as a function of the variational parameter c, for $L = 6, \lambda = 5.0$.

We have made estimates of the bulk limit, extrapolating mainly from the smaller L values where possible. The estimates for the bulk ground-state energy per site are graphed as a function of coupling in Figure 2, where they are compared with previous estimates [9] obtained by an 'Exact Linked Cluster Expansion' (ELCE) procedure, and with the asymptotic weak-coupling series.[11] The Monte Carlo results agree very well with the ELCE estimates, and appear to match nicely onto the expected weak-coupling behaviour for $\lambda \geq 6$.

3.2 Wilson Loops

The forward-walking method was used to estimate values for the $m \times n$ Wilson loops, $W(m,n)$. A graph of the 'mean plaquette' $W(1,1)$ versus the variational parameter c is shown in Figure 3. A problem is immediately apparent. The estimate for $W(1,1)$ is not independent of c, in fact it depends linearly on c over this range, and the size of the variation is such that the probable systematic error due to the choice of c is an order of magnitude larger than the random statistical error in the results. Thus it would be advantageous in future studies to put more effort into improving the trial wave function, rather than merely improving the statistics.

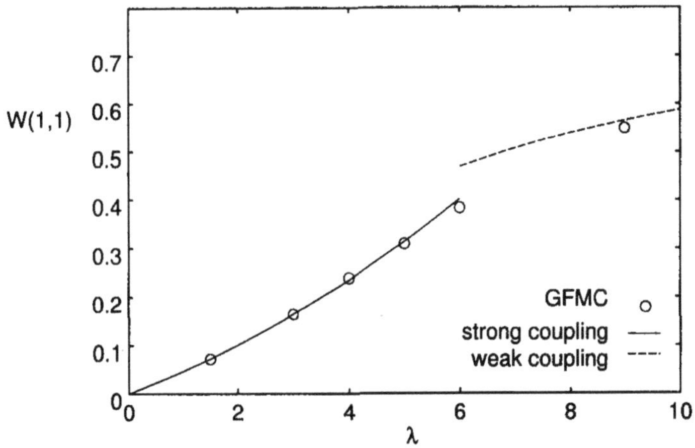

Figure 4: The mean plaquette $W(1,1)$ for the bulk system graphed against coupling λ. The solid line represents the strong-coupling series expansion[9], and the dashed line the asymptotic weak-coupling behaviour.

The finite-size behaviour for the Wilson loops is similar to that of the ground-state energy. The estimates for the mean plaquette in the bulk limit are graphed as a function of coupling λ in Figure 4, and compared with series estimates at strong and weak coupling.[9,11] The agreement is quite good.

3.3 String Tension

Having obtained estimates for the Wilson loop values on the bulk lattice, one can extract estimates for the 'spacelike' string tension using the Creutz ratios:

$$Ka^2 \simeq R_n = -\ln\left[\frac{W(n,n)W(n-1,n-1)}{W(n,n-1)^2}\right]. \tag{12}$$

The results are shown in Figure 5. Also shown in Figure 5 are some previous estimates derived from the 'axial' string tension, obtained[9] using an 'Exact Linked Cluster Expansion' (ELCE) method. The axial string tension aT is calculated as an energy per link, and must be converted to a dimensionless, 'spacelike' tension by dividing by the 'speed of light'[12] c. We have also used the weak-coupling relationship between the scales of Euclidean and Hamiltonian lattice Yang-Mills theory calculated by Hasenfratz et al. [12] to plot the results against the Euclidean coupling $\beta = 6/g_E^2$.

It can be seen that the present GFMC results are in rough agreement with the axial string tension results in the region $4 \leq \beta \leq 5$, which is also the region

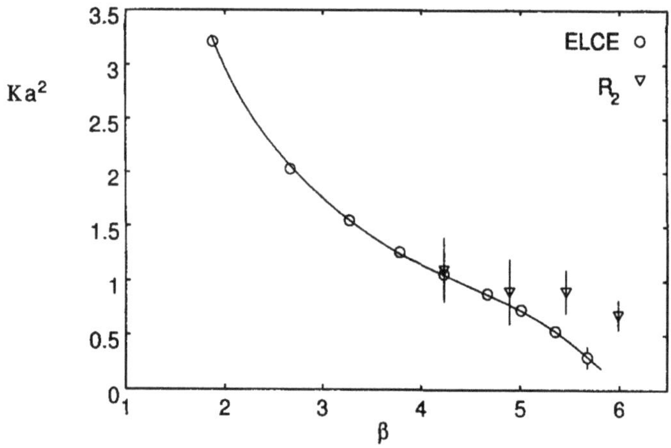

Figure 5: The string tension Ka^2 graphed against coupling β. The circles are obtained from ELCE estimates of the axial string tension[9]; the triangles are Monte Carlo estimates of R_2.

where the 'roughening' transition occurs in the string tension.[9] For $\beta > 5$, however, the Creutz ratio R_2 runs above the ELCE estimate, and shows no sign of the expected crossover to an exponentially decreasing scaling behaviour at $\beta \simeq 6$. We presume that this is a finite-size effect, and that the Creutz ratios R_n for larger n will show a substantial decrease in the 'weak-coupling' regime $\beta \geq 6$. That is certainly the pattern seen in the Euclidean calculations, or in the $\mathrm{U}(1)_{2+1}$ model.[8] Unfortunately, however, our present results for the larger Wilson loops are not of sufficient accuracy to allow worthwhile estimates of R_n for $n \geq 2$.

4 Conclusions

Some significant problems with the GFMC method have emerged from this study. The 'forward-walking' technique was introduced specifically to avoid any variational bias from the trial wave function [1,6]. As it turns out, however, the results for the Wilson loops show a substantial dependence on the trial wave function parameter c. The systematic error due to this dependence is an order of magnitude larger than the statistical error, so it would pay to put more effort in future studies into improving the trial wave function, rather than simply increasing the statistics. Furthermore, the effective ensemble size decreases during each measurement as the descendants of each 'ancestor' state die out, and this produces a substantial loss in statistical accuracy at weak coupling, as well.

It would be preferable if one were able to do away entirely with all the paraphernalia of trial wave function, weights, branching algorithms, etc, and just rely on some sort of Metropolis-style accept/reject algorithm to produce a correct distribution of walkers. Within a quantum Hamiltonian framework, a way is known to do this, namely the Path Integral Monte Carlo (PIMC) approach.[13] We conclude that the PIMC approach may be better suited than GFMC to the study of large and complicated lattice Hamiltonian systems.

Acknowledgments

This work is supported by the Australian Research Council. Calculations were performed on the SGI Power Challenge Facility at the New South Wales Centre for Parallel Computing and the Fujitsu VPP300 vector machine at the Australian National University Supercomputing Facility: we are grateful for the use of these facilities. Part of the work was carried out while CJH was on study leave at the Institute for Theoretical Physics, University of California at Santa Barbara, and at the Los Alamos National Laboratory: he would like to thank Dr. R. Singh and Dr. J Gubernatis for their hospitality. This research was supported in part by the National Science Foundation under grant no. PHY94-07194.

References

1. D. M. Ceperley and M. H. Kalos, in *Monte Carlo Methods in Statistical Mechanics*, ed. K. Binder (Springer-Verlag, New York, 1979).
2. D. W. Heys and D. R. Stump, Phys. Rev. **D 28**, 2067 (1983).
3. S. A. Chin, J. W. Negele and S. E. Koonin, Ann. Phys. (N.Y.) **157**, 140 (1984).
4. S. A. Chin, O. S. van Roosmalen, E. A. Umland and S. E. Koonin, Phys. Rev. **D 31**, 3201 (1985).
5. S.A. Chin, C. Long, and D. Robson, Phys. Rev. **D 37**, 3001 (1988);
6. M. H. Kalos, J. Comp. Phys. **1**, 257 (1966).
7. M. Samaras and C. J. Hamer, Aust. J. Phys. **52**, 637 (1999)
8. C.J. Hamer, R.J. Bursill and M. Samaras, submitted to Phys. Rev.**D**
9. C.J. Hamer, A.C. Irving and T.E. Preece, Nucl. Phys. **B 270**, 553 (1986)
10. C.J. Hamer and Zheng Weihong, Phys. Rev. **D 48**, 4435 (1993)
11. T. Höfsass and R. Horsley, Phys. Letts. **123B**, 65 (1983)
12. A. Hasenfratz and P. Hasenfratz, Nucl. Phys. **B 193**, 210 (1981)
13. D.M. Ceperley, Rev. Mod. Phys. **67**, 279 (1995)

TUTORIAL ON QUASI-SPARSE EIGENVECTOR DIAGONALIZATION

DEAN LEE
University of Massachusetts, Amherst, MA 01003 USA
dlee@physics.umass.edu

We review several topics related to the diagonalization of quantum field Hamiltonians using the quasi-sparse eigenvector (QSE) method.

1 Introduction

Quasi-sparse eigenvector (QSE) diagonalization is a new computational method which finds the low-lying eigenvalues and eigenvectors for a general quantum field Hamiltonian [1]. It is able to handle the exponential increase in the size of Fock space for large systems by exploiting the sparsity of the Hamiltonian. QSE diagonalization can even be applied directly to infinite-dimensional systems. The method is most effective when the splitting between low-lying eigenvalues is not too small compared to the size of the off-diagonal Hamiltonian matrix entries. In such cases the low-lying eigenvectors are quasi-sparse, meaning that the vector is dominated by a small fraction of its largest components. The QSE algorithm can then be applied to the Hamiltonian H using the following steps:

1. Select a subset of orthonormal basis vectors $\{e_{i_1}, \cdots, e_{i_n}\}$ and call the corresponding subspace S.
2. Diagonalize H restricted to S and find one eigenvector v.
3. Sort the basis components $\langle e_{i_j}|v\rangle$ according to their magnitude and remove the least important basis vectors.
4. Replace the discarded basis vectors by new basis vectors. These are selected at random from a pool of candidate basis vectors which are connected to the old basis vectors through non-vanishing matrix elements of H.
5. Redefine S as the subspace spanned by the updated set of basis vectors and repeat steps 2 through 5.

If the subset of basis vectors is sufficiently large, the exact eigenvectors will be stable fixed points of the update process.

The purpose of this short tutorial is to provide additional information for those interested in writing or using codes that implement QSE methods. We discuss the calculation of Hamiltonian matrix elements in the momentum Fock space representation and the generation and selection of new basis states. We also include a simple program which applies the QSE method to find the ground state of an input matrix. Readers interested in an introductory overview of QSE diagonalization are encouraged to consult the references in [1].

2 Fock states and the Hamiltonian matrix

It is most effective to describe the details of a QSE calculation in the context of a specific example. In this discussion we consider ϕ^4 theory in $1+1$ dimensions, the same example as in [1]. Some of the techniques described here have been specifically optimized for scalar field theories. Other systems require a somewhat different set of tools and will be discussed in the future.

The Hamiltonian density for ϕ^4 theory in $1+1$ dimensions has the form

$$\mathcal{H} = \tfrac{1}{2}\left(\tfrac{\partial\phi}{\partial t}\right)^2 + \tfrac{1}{2}\left(\tfrac{\partial\phi}{\partial x}\right)^2 + \tfrac{\mu^2}{2}\phi^2 + \tfrac{\lambda}{4!}{:}\phi^4{:},$$

where the normal ordering is with respect to the mass μ. We consider the system in a periodic box of length $2L$. We then expand in momentum modes and reinterpret the problem as an equivalent Schrödinger equation [2]. We will use a momentum cutoff scheme where the modes q_n are restricted in momentum so that $|n| \leq N_{max}$. The resulting Hamiltonian is

$$H = -\tfrac{1}{2}\sum_n \tfrac{\partial}{\partial q_{-n}}\tfrac{\partial}{\partial q_n} + \tfrac{1}{2}\sum_n\left(\omega_n^2(\mu) - \tfrac{\lambda b(\mu)}{4L}\right) q_{-n}q_n \\ + \tfrac{\lambda}{4!2L}\sum_{n_1+n_2+n_3+n_4=0} q_{n_1}q_{n_2}q_{n_3}q_{n_4}, \tag{1}$$

where

$$\omega_n(\mu) = \sqrt{\tfrac{n^2\pi^2}{L^2} + \mu^2} \tag{2}$$

and $b(\mu)$ is the coefficient for the mass counterterm

$$b(\mu) = \sum_n \tfrac{1}{2\omega_n(\mu)}. \tag{3}$$

It is convenient to split the Hamiltonian into free and interacting parts with respect to an arbitrary mass μ':

$$H_{free} = -\tfrac{1}{2}\sum_n \frac{\partial}{\partial q_{-n}}\frac{\partial}{\partial q_n} + \tfrac{1}{2}\sum_n \omega_n^2(\mu')\, q_{-n}q_n, \tag{4}$$

$$H = H_{free} + \tfrac{1}{2}\sum_n \left(\mu^2 - \mu'^2 - \tfrac{\lambda b(\mu)}{4L}\right) q_{-n}q_n \tag{5}$$
$$+ \frac{\lambda}{4!2L} \sum_{n_1+n_2+n_3+n_4=0} q_{n_1}q_{n_2}q_{n_3}q_{n_4}.$$

μ' will be used to define the basis states of our Fock space. Since H is independent μ', it is useful to perform calculations for different μ' to obtain an estimate of the error. It is also helpful to find the range of values for μ' which maximizes the quasi-sparsity of the eigenvectors. This can be used to improve the accuracy of the QSE calculation.

Let us define ladder operators with respect to μ',

$$a_n(\mu') = \frac{1}{\sqrt{2\omega_n(\mu')}}\left[q_n\omega_n(\mu') + \frac{\partial}{\partial q_{-n}}\right] \tag{6}$$

$$a_n^\dagger(\mu') = \frac{1}{\sqrt{2\omega_n(\mu')}}\left[q_{-n}\omega_n(\mu') - \frac{\partial}{\partial q_n}\right]. \tag{7}$$

These satisfy the usual commutation relations,

$$\left[a_n, a_m^\dagger\right] = \delta_{nm} \tag{8}$$

$$\left[a_n, a_m\right] = \left[a_n^\dagger, a_m^\dagger\right] = 0. \tag{9}$$

The Hamiltonian can now be rewritten as

$$H = \sum_n \omega_n(\mu')a_n^\dagger a_n + \tfrac{1}{4}(\mu^2 - \mu'^2 - \tfrac{\lambda b(\mu)}{4L})\sum_n \frac{(a_{-n}+a_n^\dagger)(a_n+a_{-n}^\dagger)}{\omega_n(\mu')} \tag{10}$$
$$+ \frac{\lambda}{192L}\sum_{n_1+n_2+n_3+n_4=0}\left[\frac{(a_{n_1}+a_{-n_1}^\dagger)}{\sqrt{\omega_{n_1}(\mu')}}\frac{(a_{n_2}+a_{-n_2}^\dagger)}{\sqrt{\omega_{n_2}(\mu')}}\frac{(a_{n_3}+a_{-n_3}^\dagger)}{\sqrt{\omega_{n_3}(\mu')}}\frac{(a_{n_4}+a_{-n_4}^\dagger)}{\sqrt{\omega_{n_4}(\mu')}}\right].$$

In (10) we have omitted constants contributing only to the vacuum energy.

We can represent any momentum-space Fock state as a string of occupation numbers,

$$\left|o_{-N_{\max}}, \cdots, o_{N_{\max}}\right\rangle, \tag{11}$$

where

$$a_n^\dagger a_n \left|o_{-N_{\max}}, \cdots, o_{N_{\max}}\right\rangle = o_n \left|o_{-N_{\max}}, \cdots, o_{N_{\max}}\right\rangle. \tag{12}$$

From the usual ladder operator relations, it is straightforward to calculate the matrix element of H between two arbitrary Fock states. In [1] auxiliary cutoffs were used to render the dimension of Fock space finite. This is in fact an unnecessary restriction and QSE diagonalization encounters no difficulties dealing with the full infinite-dimensional space.

3 New basis vectors

Aside from calculating matrix elements, the only other fundamental operation involving basis vectors is the generation of new basis vectors. As mentioned before, the new states should be connected to an old basis vector through non-vanishing matrix elements of H. Let us refer to the old basis vector as $|e\rangle$. For this example there are two types of terms in our interaction Hamiltonian, a quartic interaction

$$\sum_{n_1,n_2,n_3} \left(a_{n_1} + a^{\dagger}_{-n_1}\right) \left(a_{n_2} + a^{\dagger}_{-n_2}\right) \left(a_{n_3} + a^{\dagger}_{-n_3}\right) \left(a_{-n_1-n_2-n_3} + a^{\dagger}_{n_1+n_2+n_3}\right), \tag{13}$$

and a quadratic interaction

$$\sum_{n} \left(a_{-n} + a^{\dagger}_{n}\right) \left(a_{n} + a^{\dagger}_{-n}\right). \tag{14}$$

To produce a new vector from $|e\rangle$ we simply choose one of the possible operator monomials

$$\begin{aligned} &a_{n_1} a_{n_2} a_{n_3} a_{-n_1-n_2-n_3},\ a^{\dagger}_{-n_1} a_{n_2} a_{n_3} a_{-n_1-n_2-n_3}, \cdots, \\ &a_{-n} a_{n},\ a^{\dagger}_{n} a^{\dagger}_{-n}, \cdots \end{aligned} \tag{15}$$

and let it act upon $|e\rangle$. Our experience is that the interactions involving the small momentum modes are generally more important than those for the large momentum modes, a signal that the ultraviolet divergences have been properly renormalized. For this reason it is best to arrange the selection probabilities such that the smaller values of $|n_1|$, $|n_2|$, $|n_3|$ and $|n|$ are chosen more often.

We note that the new basis vector selection will occasionally fail. Either the vector is zero due to an annihilation operator acting on an unoccupied state or the vector is already in our subset of basis vectors and therefore not new. In either case we simply select a new basis vector again. If the selection process fails repeatedly then a different old basis vector $|e'\rangle$ is used.

4 Sample code

We have considered two basic operations, the calculation of matrix elements and the generation of new basis states. The subroutines which perform these tasks will depend on the form of the quantum field Hamiltonian. The main structure of the QSE algorithm, however, is independent of the details of the Hamiltonian. We demonstrate the essential features of the algorithm with the following short MATLAB program. In this example we find the ground state of a finite matrix. To keep the example as simple as possible, we avoid calculating matrix elements and generating new basis states by assuming that the entire Hamiltonian matrix can be stored in memory.

```
niter = 30;
nactive = 100;
nretain = 80;
H = loadsparse('samplematrix');
vecs = [1:nactive];
for iter = 1:niter
    [v,d] = eigs(H(vecs,vecs),'SR',1);
    weights1 = v.^2;
    [sortelements, sortorder] = sort(-weights1);
    vecs = vecs(sortorder(1:nretain));
    wtsum1 = cumsum(weights1)/sum(weights1);
    while (length(vecs) <nactive)
        over1 = find(rand*wtsum1(nretain) <wtsum1);
        seedvec = vecs(over1(1));
        candidates = find(H(:,seedvec));
        candidates = candidates(candidates ~= seedvec);
        weights2 = abs(H(candidates,seedvec));
        wtsum2 = cumsum(weights2)/sum(weights2);
        over2 = find(rand <wtsum2);
        newvec = candidates(over2(1));
        vecs = unique([vecs newvec]);
    end
end
save results.mat vecs v d
```

`niter` is the number of iterations of the algorithm. `nactive` is the number of basis states used to form the subspace S, and `nretain` is the number of basis states kept after removing the tail of the eigenvector distribution. `loadsparse` is a program which reads in a datafile called `samplematrix`. This data is stored in the matrix variable H. The remaining code can be broken down in terms of the five basic steps of the QSE algorithm.

Select a subset of orthonormal basis vectors $\{e_{i_1}, \cdots, e_{i_n}\}$ *and call the corresponding subspace* S.

```
vecs = [1:nactive];
```

The variable vecs is a row vector which labels the basis vectors in the subspace S. In our example vecs is initialized to be the first nactive basis vectors. We are assuming that the ground state of the system has a non-zero overlap with at least one of the first nactive basis vectors. A more general initialization would be to choose a random subset of basis vectors.

Diagonalize H *restricted to* S *and find one eigenvector* v.

```
[v,d] = eigs(H(vecs,vecs),'SR',1);
```

The function eigs is a sparse matrix Arnoldi diagonalization routine. It is being called with the parameters 'SR',1 which tells eigs to find the eigenvalue and eigenvector with the smallest real part. The Hamiltonian submatrix corresponding with basis vectors in vecs is diagonalized. The eigenvector is stored in the column vector v and the eigenvalue is stored as the variable d.

Sort the basis components $\langle e_{i_j}|v\rangle$ *according to their magnitude and remove the least important basis vectors.*

```
weights1 = v.^2;
[sortelements, sortorder] = sort(-weights1);
vecs = vecs(sortorder(1:nretain));
```

weights1 is a column vector which contains the squares of the components of the eigenvector v. sortelements is a list of the elements of -weights1 from smallest to greatest. sortorder is the corresponding list of locations for these elements. If for example weights1 = [.2 .5 .3], then sortorder = [2 3 1]. The list sortelements is not used. Only the nretain most important basis vectors are kept in vecs.

Replace the discarded basis vectors by new basis vectors. These are selected at random from a pool of candidate basis vectors which are connected to the old basis vectors through non-vanishing matrix elements of H.

```
wtsum1 = cumsum(weights1)/sum(weights1);
while (length(vecs) <nactive)
    over1 = find(rand*wtsum1(nretain) <wtsum1);
    seedvec = vecs(over1(1));
    candidates = find(H(:,seedvec));
    candidates = candidates(candidates ~= seedvec);
    weights2 = abs(H(candidates,seedvec));
```

```
        wtsum2 = cumsum(weights2)/sum(weights2);
        over2 = find(rand <wtsum2);
        newvec = candidates(over2(1));
        vecs = unique([vecs newvec]);
    end
```

cumsum is a function which takes the input list and creates another list consisting of partial sums. wtsum1 is therefore an ascending string of numbers between 0 and 1 such that the difference between an entry and its preceding entry is proportional to the corresponding value of weights1. over1 is a location list of elements of wtsum1 which are greater than the product of a generated random number between 0 and 1 and wtsum1(nretain). over1(1) is the first entry of over1. We note that over1(1) must be less than or equal to nretain. seedvec is the basis vector which corresponds with over1(1). candidates is a column vector consisting of all basis vectors which have a non-zero matrix element with seedvec. Since the task is to find new basis vectors, seedvec is removed from candidates.

weights2 is a column vector containing the absolute values of the matrix transition elements from seedvec. wtsum2 is an ascending string between 0 and 1 such that the difference between an entry and its preceding entry is proportional to the value of weights2. over2 is a location list of elements of wtsum2 greater than a generated random number. newvec is the basis vector which corresponds with over2(1). newvec is appended to the set vecs. The function unique sorts the input list and removes any repetitions. This loop is repeated until vecs has nactive elements.

Redefine S as the subspace spanned by the updated set of basis vectors and repeat steps 2 through 5.

```
    for iter = 1:niter
        ...
    end
    save results.mat vecs v d
```

The algorithm is iterated niter times and the final results are saved in the file results.mat. More examples of QSE diagonalization codes for various field theory systems written in MATLAB, Fortran, and/or C++ will be made available in the future.

5 Summary

In this short tutorial we have listed some supplemental material for those interested in writing or using codes that implement QSE methods. We have discussed the calculation of Hamiltonian matrix elements and the selection of new basis states. We have also presented a simple MATLAB program which applies the QSE method to find the ground state of a matrix.

The discussion here was limited to the basic QSE method. As presented in the Guangzhou workshop there have also been interesting new developments in QSE diagonalization regarding the technique of stochastic error correction. This technique will be presented in a forthcoming work [4].

Acknowledgments

I thank my collaborators on the papers cited here and the organizers and participants of the International Workshop on Nonperturbative Methods and Lattice QCD in Guangzhou. Financial support was provided by the National Science Foundation.

References

1. D. J. Lee, N. Salwen, D. D. Lee, hep-th/0002251; D. Lee, hep-th/0003296.
2. P. Marrero, E. Roura, D. Lee, Phys. Lett. B471 (1999) 45.
3. N. Salwen, D. Lee, Phys. Rev. D62 (2000) 025006.
4. D. Lee, N. Salwen, M. Windoloski, work in progress.

MONTE CARLO HAMILTONIAN: GENERALIZATION TO QUANTUM FIELD THEORY

XIANG-QIAN LUO
Department of Physics, Zhongshan University Guangzhou 510275, China

HAMZA JIRARI, HELMUT KRÖGER
Département de Physique, Université Laval, Québec, Québec G1K 7P4, Canada

KEVIN J.M. MORIARTY
Department of Mathematics, Statistics and Computer Science, Dalhousie University, Halifax, Nova Scotia B3H 3J5, Canada

Monte Carlo techniques with importance sampling have been extensively applied to lattice gauge theory in the Lagrangian formulation. Unfortunately, it is extremely difficult to compute the excited states using the conventional Monte Carlo algorithm. Our recently developed approach: the Monte Carlo Hamiltonian method, has been designed to overcome the difficulties of the conventional approach. In this paper, we extend the method to many body systems and quantum field theory. The Klein-Gordon field theory is used as a testing ground.

1 Introduction

Quantum theory is one of the most important achievements in modern science, but a lot of non-perturbative aspects are still poorly understood. There are two standard formulations in quantum theory: Hamiltonian and Lagrangian. A comparison of the conventional approaches is given in Tab.1.

The Lagrangian formulation is very suitable for applying the Monte Carlo (MC) method to systems with many degrees of freedom, and in the last two decades, it has been widely applied to lattice gauge theory[1,2,3]. In the standard Lagrangian MC method, however, it is extremely difficult to compute the spectrum and wave function beyond the ground state. On the other hand, the standard Hamiltonian formulation is capable of doing it. If fact, some efforts have been made to solve the lattice QCD Schrödinger equation[4]. But due to the complexity of non-Abelian gauge theory and high degrees of freedom, only the ground state[5] and the lowest lying glueball states[6,7,8] have been investigated. There have also been some attempts to perform numerical simulations in the Hamiltonian formulation, e.g. the quantum MC technique[9], MC method using a guided random walk[10,11], and the MC Green's function method[12,13]. For an review of MC methods applied to solve the quantum many-body problem see Ref.[14].

Table 1. Comparison of the conventional methods in the standard formulations.

Formulation	Hamiltonian	Lagrangian
Approach	Schrödinger Eq. $H\|E_n\rangle = E_n\|E_n\rangle$	Path Integral $\langle O\rangle = \frac{\int[d\varphi]O[\varphi]\exp(-S[\varphi]/\hbar)}{\int[d\varphi]\exp(-S[\varphi]/\hbar)}$
Algorithm	Series expansion, variational, Runge-Kutta ...	MC simulation
Advantage	Both the ground state, and the excited states can be computed.	It generates the most important configs. for the measurements.
Disadvantage	Analytical methods are too tedious for many body systems; Runge-Kutta works only in 1-D.	It is difficult to study the excited states, and finite density QCD.

A natural question is whether in Lagrangian MC simulations one can construct an effective Hamiltonian? If yes, the excited states can also be computed. We have recently proposed a new approach[15] (we called it Monte Carlo Hamiltonian method) to investigate this problem. A lot of models[15,16,17,18,19,20] in quantum mechanics (QM) have been used to test the method, as briefly reviewed in Sect. 2. In this paper, we will extend the method to many body systems and quantum field theory (QFT).

2 Algorithm

2.1 *Effective Hamiltonian*

The basic idea of the MC Hamiltonian method for one body quantum mechanics has been described in details in Ref.[15]. The (imaginary time) transition amplitude between an initial state at position x_i, and time t_i, and final state at x_f and t_f is related to the Hamiltonian H by

$$\begin{aligned} M_{fi} &= \langle x_f, t_f|x_i, t_i\rangle = \langle x_f|e^{-H(t_f-t_i)/\hbar}|x_i\rangle \\ &= \sum_{n=1}^{\infty}\langle x_f|E_n\rangle e^{-E_nT/\hbar}\langle E_n|x_i\rangle, \end{aligned} \tag{1}$$

where $T = t_f - t_i$. According to Feynman's path integral formulation of QM, the transition amplitude is also related to the path integral:

$$M_{fi} = \int [dx] \exp(-S[x]/\hbar)|_{x_i,t_i}^{x_f,t_f}, \tag{2}$$

where $S = S_0 + S_V$ is the action for a given path, $S_0 = \int_{t_i}^{t_f} dt\ m\dot{x}^2/2$, and $S_V = \int_{t_i}^{t_f} dt\ V(x)$.

The results above are standard ones in Lagrangian formulations. What is new in our method[15] is to construct an effective Hamiltonian H_{eff} (finite $N \times N$ matrix) through

$$M_{fi} = \langle x_f|e^{-H_{\text{eff}}T/\hbar}|x_i\rangle = \sum_{n=1}^{N} \langle x_f|E_n^{\text{eff}}\rangle e^{-E_n^{\text{eff}}T/\hbar} \langle E_n^{\text{eff}}|x_i\rangle. \tag{3}$$

The eigenvalues E_n^{eff} and wave function $|E_n^{\text{eff}}\rangle$ can be obtained, by diagonalizing M using a unitary transformation $M = U^\dagger DU$, where $D = diag(e^{-E_1^{\text{eff}}T/\hbar}, ..., e^{-E_N^{\text{eff}}T/\hbar})$. Once the spectrum and wave functions are available, all physical information can also be obtained.

Since the theory described by H, which has an infinite basis in Hilbert space, is now approximated by a theory described by a finite matrix H_{eff}, which has a finite basis, the physics of H and H_{eff} might be quite different at high energy. Therefore we expect that we can only reproduce the low energy physics of the system. This is good enough for our purpose. In Refs.[15,16,17,18,19,20], we investigated many 1-D, 2-D and 3-D QM models (Tab.2) using this MC Hamiltonian algorithm. We computed the spectrum, wave functions and some thermodynamical observables. The results are in very good agreement with those from analytical and/or Runge-Kutta methods.

2.2 Basis in Hilbert Space

To get the correct scale for the spectrum, the position state $|x_n\rangle$ at initial time t_i or final time t_f (Bargman states or box states) should be properly normalized. We denote a normalized basis of Hilbert states as $|e_n\rangle$, $n = 1, ..., N$. In position space, it can be expressed as

$$e_n(x) = \begin{cases} 1/\sqrt{\Delta x_n}, & x \in [x_n, x_{n+1}] \\ 0, & x \notin [x_n, x_{n+1}] \end{cases} \tag{4}$$

where $\Delta x_n = x_{n+1} - x_n$.

There is some arbitrariness in choosing a basis for the initial and final states. The simplest choice is a basis with $\Delta x_n = \text{const.}$, which is called the

Table 2. QM systems, investigated by the MC Hamiltonian method with a regular basis.

System	Potential
QM in 1-D	$V(x)=0$
	$V(x)=\frac{1}{2}m\omega^2x^2$
	$V(x)=-V_0\mathrm{sech}^2(x)$
	$V(x)=\frac{1}{2}x^2+\frac{1}{4}x^4$
	$V(x)=\frac{1}{2}\lvert x\rvert$
	$V(x)=\{\begin{matrix}\infty, & x<0\\ Fx, & x\geq 0\end{matrix}$
QM in 2-D	$V(x,y)=\frac{1}{2}m\omega^2x^2+\frac{1}{2}m\omega^2y^2$
	$V(x,y)=\frac{1}{2}m\omega^2x^2+\frac{1}{2}m\omega^2y^2+\lambda xy$
QM in 3-D	$V(x,y,z)=\frac{1}{2}m\omega^2x^2+\frac{1}{2}m\omega^2y^2+\frac{1}{2}m\omega^2z^2$

"regular basis". In Refs.[15,16,17,18], the regular basis was used. For many body systems or QFT, the regular basis will encounter problems. For example, in a system with a 1-D chain of oscillators (see later), if the number of oscillators is 30, the minimum non-trivial regular basis is $N=2^{30}=1073741824\approx 10^9$, which is prohibitively large for numerical calculations of the matrix elements.

Guided by the idea of importance sampling, in Refs.[19,20], we proposed to select a basis from the Boltzmann weight proportional to the transition amplitude between $x'_i=0$ at $t'_i=0$ and $x'_f=x_n$ at some t'_f. In a free particle or harmonic oscillator case, the distribution is just a Gaussian

$$P_{basis}[x_n]=\frac{1}{\sqrt{2\pi}\sigma}\exp\left(-\frac{x_n^2}{2\sigma^2}\right), \tag{5}$$

where $\sigma=\sqrt{\hbar T'/m}$ for the free case and $\sigma=\sqrt{\hbar\sinh(\omega T')/(m\omega)}$ for the harmonic oscillator. We call such a basis the "stochastic basis", to be used as states at initial time t_i or final time t_f. For more complicated models, one may still use the Gaussian distribution, but with an adjustable parameter σ.

2.3 Matrix elements

As explained above, the calculation of the transition matrix elements is an essential ingredient of our method. The matrix element in the normalized basis is related to $\langle x_{n'},t_f|x_n,t_i\rangle$ by

$$M_{n'n}=\langle e_{n'},t_f|e_n,t_i\rangle=\int_{x'_n}^{x_{n'+1}}dx'\int_{x_n}^{x_{n+1}}dx''\,\frac{\langle x',t_f|x'',t_i\rangle}{\sqrt{\Delta x_{n'}\Delta x_n}}$$

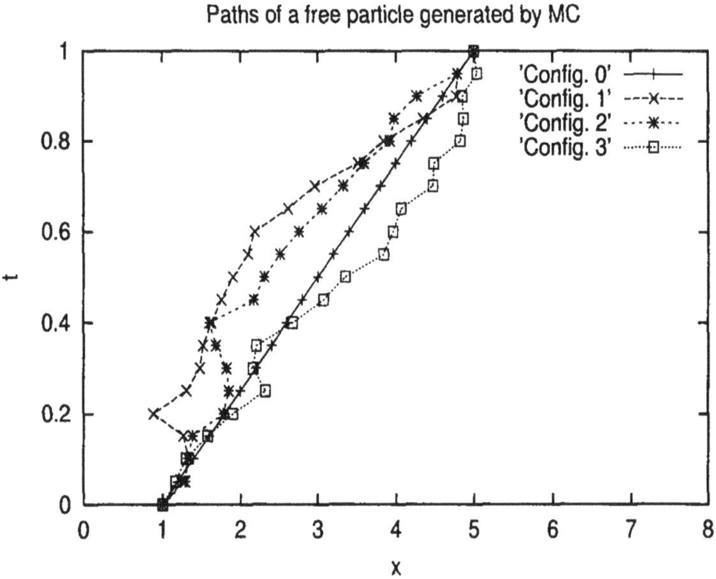

Figure 1. Config. 0 stands for the classical path, Config. 1, Config. 2 and Config. 3 stand the paths generated by MC. Here $N_t = 20$ (there are 19 time slices between t_i and t_f).

$$\approx \sqrt{\Delta x_{n'} \Delta x_n} \langle x_{n'}, t_f | x_n, t_i \rangle, \tag{6}$$

where for the regular basis, $\Delta x_n =$ const., and for a stochastic basis, $\Delta x_n = 1/(P_{basis}[x_n]N)$. $\langle x_{n'}, t_f | x_n, t_i \rangle$ can be calculated by MC as follows.
(a) Discretize the continuous time with time interval $\Delta t = T/N_t$.
(b) Generate free configurations $[x]$ between $t \in (t_i, t_f)$ obeying the Boltzmann distribution

$$P_0[x] = \frac{\exp(-S_0[x]/\hbar)}{\int [dx] \exp(-S_0[x]/\hbar)} \Big|_{x_n, t_i}^{x_{n'}, t_f}. \tag{7}$$

Figure 1 shows a sample of free configurations generated by MC method with importance sampling.
(c) Measure

$$\langle O_V \rangle = \int [dx] \ \exp(-S_V(x)/\hbar) \Big|_{x_n, t_i}^{x_{n'}, t_f} \ P_0[x]. \tag{8}$$

The path integral in Eq. (6) is then

$$\langle x_{n'}, t_f | x_n, t_i \rangle = \langle O_V \rangle \sqrt{\frac{m}{2\pi\hbar T}} \exp\left[-\frac{m}{2\hbar T}(x_{n'} - x_n)^2 \right]. \tag{9}$$

3 Quantum Field Theory

3.1 The Klein-Gordon model

The main purpose of the algorithm is to study many body systems and QFT. As an example, we consider a free scalar field theory: the Klein-Gordon model in 1+1 dimensions, which has the Hamiltonian

$$H = \int dx \frac{1}{2} \left(\pi^2 + \nu^2 \left(\frac{\partial \varphi}{\partial x} \right)^2 + \Omega_0^2 \varphi^2 \right). \tag{10}$$

Discretizing the theory on a finite string with $N_{\rm osc}$ sites and lattice spacing a, and making the replacement $x \to x_j = ja$, $\pi(x) \to a^{-1/2}\Pi(x_j)$, $\varphi(x) \to a^{-1/2}\phi(x_j)$, $\int \to a\sum_{j=1}^{N_{\rm osc}}$ and $\nu \to a\Omega$, the Hamiltonian becomes

$$H = \sum_{j=1}^{N_{\rm osc}} \frac{1}{2} \left(\Pi^2(x_j) + \Omega^2(\phi(x_j) - \phi(x_{j+1}))^2 + \Omega_0^2 \phi^2(x_j) \right), \tag{11}$$

which describes a chain of $N_{\rm osc}$ coupled oscillators[21] with $m = 1$ in 1 spatial dimension. The field $\phi(x_j)$ and its conjugate momentum $\Pi(x_j) = \dot{\phi}(x_j)$ at the j-th site correspond respectively to the displacement and velocity of the j-th oscillator.

In QFT, the initial and final states are defined by

$$|\Phi_i, t_i\rangle = |\phi(x_1, t_i), ..., \phi(x_{N_{\rm osc}}, t_i)\rangle,$$

$$|\Phi_f, t_f\rangle = |\phi(x_1, t_f), ..., \phi(x_{N_{\rm osc}}, t_f)\rangle. \tag{12}$$

The transition amplitude between them is

$$\begin{aligned} M_{fi} &= \langle \Phi_f, t_f | \Phi_i, t_i \rangle = \langle \Phi_f | e^{-H(t_f - t_i)/\hbar} | \Phi_i \rangle \\ &= \int [d\phi] \exp(-S[\phi]/\hbar) \Big|_{\phi_i, t_i}^{\phi_f, t_f} \\ &= \lim_{N_t \to \infty} \int \prod_{k=1}^{N_t - 1} \prod_{j=1}^{N_{\rm osc}} \left(\sqrt{\frac{1}{2\pi\hbar\Delta t}} d\phi(x_j, t_k) \right) \exp(-S[\phi]/\hbar), \end{aligned} \tag{13}$$

where the Euclidean action is

$$S = \frac{\Delta t}{2} \sum_{k=1}^{N_t-1} \sum_{j=1}^{N_{osc}} \left[\frac{(\phi(x_j,t_{k+1}) - \phi(x_j,t_k))^2}{(\Delta t)^2} \right.$$

$$\left. + \Omega^2 \left(\phi(x_j,t_k) - \phi(x_{j+1},t_k)\right)^2 + \Omega_0^2 \phi^2(x_j,t_k) \right]. \quad (14)$$

For a finite N_{osc} and non-zero Ω, one has to introduce a boundary condition for $\phi(x_{N_{osc}+1})$. We implement the periodic boundary condition $\phi(x_{N_{osc}+1}) = \phi(x_1)$. The spectrum is analytically known[21]:

$$E_n = \sum_{n_l} \left(n_l + \frac{1}{2}\right) \hbar\omega_l, \quad (15)$$

where $\omega_l = \sqrt{\Omega_0^2 + 4\Omega^2 \sin^2(p_l \Delta x/2)}$, and $n_1, ..., n_{N_{osc}} = 0, 1, \dots$. We can also compute the transition amplitude analytically, using the discrete Fourier transformation

$$\phi(x_j,t) = \sum_l \frac{\Delta p}{\sqrt{2\pi}} \exp(ip_l x_j)\tilde{\phi}(p_l,t),$$

$$\tilde{\phi}(p_l,t) = \sum_j \frac{\Delta x}{\sqrt{2\pi}} \exp(-ip_l x_j)\phi(x_j,t), \quad (16)$$

where $\Delta x = a$, $\Delta p = 2\pi/(N_{osc}\Delta x)$, $x_j = [-(N_{osc}-1)/2 + (j-1)]\Delta x$, $p_l = [-(N_{osc}-1)/2 + (l-1)]\Delta p$, and $j, l = 1, ..., N_{osc}$. The result is

$$M_{fi} = \prod_{l=1}^{N_{osc}} \sqrt{\frac{\omega_l}{2\pi\hbar \sinh(\omega_l T)}} \exp\left[-\frac{\omega_l \Delta p/\Delta x}{2\hbar \sinh(\omega_l T)} \right.$$

$$\left. \times \left(\left(|\tilde{\phi}(p_l,t_f)|^2 + |\tilde{\phi}(p_l,t_i)|^2 \right) \cosh(\omega_l T) - 2\mathrm{Re}\left(\tilde{\phi}(p_l,t_f)\tilde{\phi}^\star(p_l,t_i) \right) \right) \right]. \quad (17)$$

3.2 *Stochastic basis*

In QFT, a stochastic basis has to be used for the initial and final states, because in the thermodynamical limit, $N_{osc} \to \infty$. Although the variables ϕ are coupled in position space, we may still use a stochastic basis $[\Phi_n]$, $n = 1, ..., N$ according to the distribution with an adjustable parameter σ

$$P_{basis}[\Phi_n] = \prod_{j=1}^{N_{osc}} \frac{1}{\sqrt{2\pi}\sigma} \exp\left(-\frac{\phi_n^2(x_j)}{2\sigma^2} \right). \quad (18)$$

The matrix element between the normalized initial and final states is

$$M_{n'n} = \langle e_{n'}, t_f | e_n, t_i \rangle \approx \sqrt{\Delta\Phi_{n'}\Delta\Phi_n}\langle \Phi_{n'}, t_f | \Phi_n, t_i \rangle, \tag{19}$$

where the infinitesimal volumes for the initial and final states are

$$\Delta\Phi_n = d\phi_n(x_1)...d\phi_n(x_{N_{osc}}) = \frac{1}{P_{basis}[\Phi_n]N},$$
$$\Delta\Phi_{n'} = d\phi_{n'}(x_1)...d\phi_{n'}(x_{N_{osc}}) = \frac{1}{P_{basis}[\Phi_{n'}]N}. \tag{20}$$

4 Results

We consider $N_{osc} = 9$, $a = 1$, $\Omega = 1$, $\Omega_0 = 2$, $m = 1$, $\hbar = 1$, and $T = 2$. For the adjustable parameter σ in the stochastic basis, we choose $\sigma = \sqrt{\hbar \sinh(\omega T')/(m\omega)}$ with $T' = T$ for simplicity. After the stochastic basis with $N = 1000$ is generated, we obtain the matrix elements $M_{n'n}$ using Eqs. (17), (19) and (20). (In the future work, we will calculate $\langle \Phi_{n'}, t_f | \Phi_n, t_i \rangle$ directly by MC). Then we compute the eigenvalues and eigenvectors using the method described in Sect. 2.1.

Table 3 gives a comparison between the spectrum from the effective Hamiltonian with the stochastic basis and the analytical formular Eq. (15) for the first 20 states. They agree very well. This means that an arbitrary choice of Gaussian distribution for the stochastic basis is good enough. (Of course, one should study systematically the dependence of the results on σ).

We have also computed thermodynamical quantities such as the partition function Z, average energy $\overline{E}$ and specific heat C. The analytical results are

$$Z(\beta) = \mathrm{Tr}\left(\exp\left(-\beta H\right)\right) = \prod_{l=1}^{N_{osc}} \frac{1}{2\sinh\left(\beta\hbar\omega_l/2\right)},$$
$$\overline{E}(\beta) = \frac{1}{Z}\mathrm{Tr}\left(H\exp\left(-\beta H\right)\right) = -\frac{\partial \log Z}{\partial\beta} = \sum_{l=1}^{N_{osc}} \frac{\hbar\omega_l}{2}\coth\left(\beta\hbar\omega_l/2\right),$$
$$C(\beta) = k_B\frac{\partial\overline{E}}{\partial T} = -k_B\beta^2\frac{\partial\overline{E}}{\partial\beta} = k_B\sum_{l=1}^{N_{osc}}\left(\frac{\beta\hbar\omega_l/2}{2\sinh\left(\beta\hbar\omega_l/2\right)}\right)^2, \tag{21}$$

where $\beta = T/\hbar$, the temperature $\mathcal{T} = 1/(\beta k_B)$, and k_B is the Boltzmann constant. Since we have approximated H by $H_{\rm eff}$, we can express those thermodynamical observables via the eigenvalues of the effective Hamiltonian

$$Z^{\rm eff}(\beta) = \sum_{n=1}^{N} e^{-\beta E_n^{\rm eff}},$$

Table 3. Comparison of the spectrum of the Klein-Gordon model on the lattice, between the MC Hamiltonian method with a *stochastic* basis and the analytic ones.

n	E_n^{eff}	E_n^{exact}
1	10.904663192168	10.944060480668
2	12.956830557334	12.944060480668
3	12.985023578737	13.057803869484
4	13.044311582647	13.057803869484
5	13.299967341242	13.321601993380
6	13.345480638394	13.321601993380
7	13.552195133687	13.589811791733
8	13.585794986361	13.589811791733
9	13.680136748933	13.751084748745
10	13.744919087477	13.751084748745
11	14.984737011385	14.944060480668
12	15.012353803145	15.057803869484
13	15.057295761044	15.057803869484
14	15.108904652020	15.171547258300
15	15.125356713561	15.171547258300
16	15.187413290039	15.171547258300
17	15.308536490102	15.321601993380
18	15.396255686587	15.321601993380
19	15.420708031412	15.435345382196
20	15.432823810789	15.435345382196

$$\overline{E}^{\text{eff}}(\beta) = \sum_{n=1}^{N} \frac{E_n^{\text{eff}} e^{-\beta E_n^{\text{eff}}}}{Z^{\text{eff}}(\beta)},$$

$$C^{\text{eff}}(\beta) = k_B \beta^2 \left(\sum_{n=1}^{N} \frac{(E_n^{\text{eff}})^2 e^{-\beta E_n^{\text{eff}}}}{Z^{\text{eff}}(\beta)} - \left(\overline{E}^{\text{eff}}(\beta)\right)^2 \right). \tag{22}$$

Since this is a static system, the eigenvalues should in principle not vary with β, which is also confirmed numerically within statistical errors when β and N are not too small. Using this assumption and the spectrum at $\beta = 2$, we obtain the thermodynamical quantities for other β. The results for the free energy, average energy and specific heat as a function of β are shown in Figs. 2, 3 and 4. Again, the results from the MC Hamiltonian are in good agreement with the analytical ones when $\beta > 1$.

Preliminary results for larger N_{osc} indicate that it is not necessary to

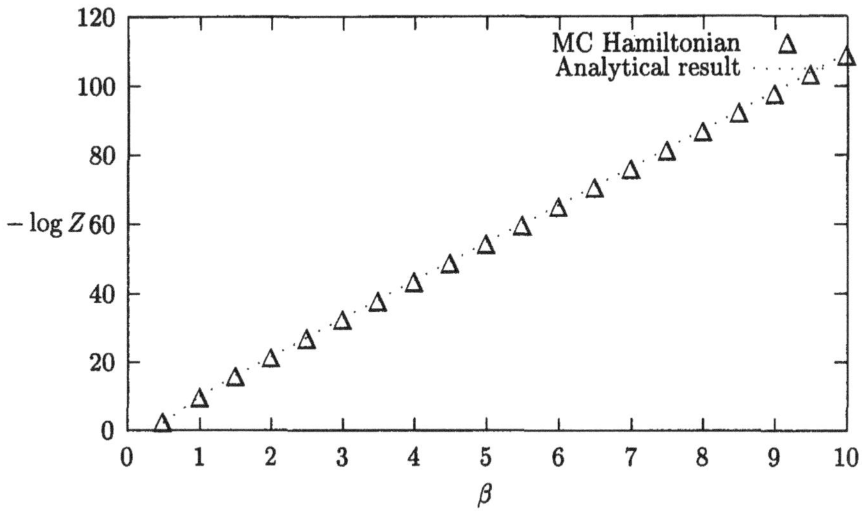

Figure 2. Free energy of the Klein-Gordon model on a 1+1 dimensional lattice.

increase N accordingly. This property is very important for a feasible application of the algorithm to many body systems and QFT.

5 Summary

In this paper, we have extended the effective Hamiltonian method with a stochastic basis to QFT, and taken the Klein-Gordon model as an example. The results are very encouraging. We believe that the application of the algorithm to more complicated systems will be very interesting for non-perturbative investigation beyond the ground state.

Acknowledgments

We are grateful to the attendees of the workshop for useful discussions. X.Q.L. is supported by the National Science Fund for Distinguished Young Scholars (19825117), National Natural Science Foundation (10010310687), Guangdong Provincial Natural Science Foundation (990212), and the Ministry of Education of China. H.K. has been supported by NSERC Canada. We would also thank E.B. Gregory, C.Q. Huang, J.Q. Jiang, J.J. Liu, and H. Xu for collaborations.

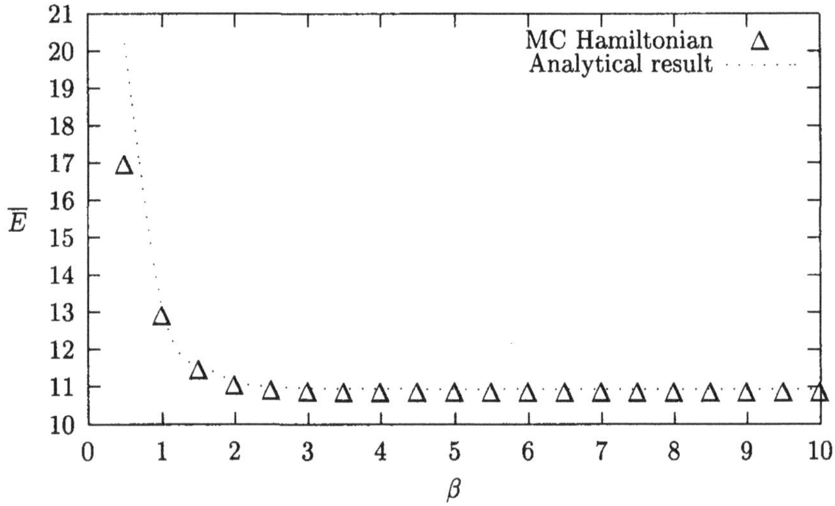

Figure 3. Average energy of the Klein-Gordon model on a 1+1 dimensional lattice.

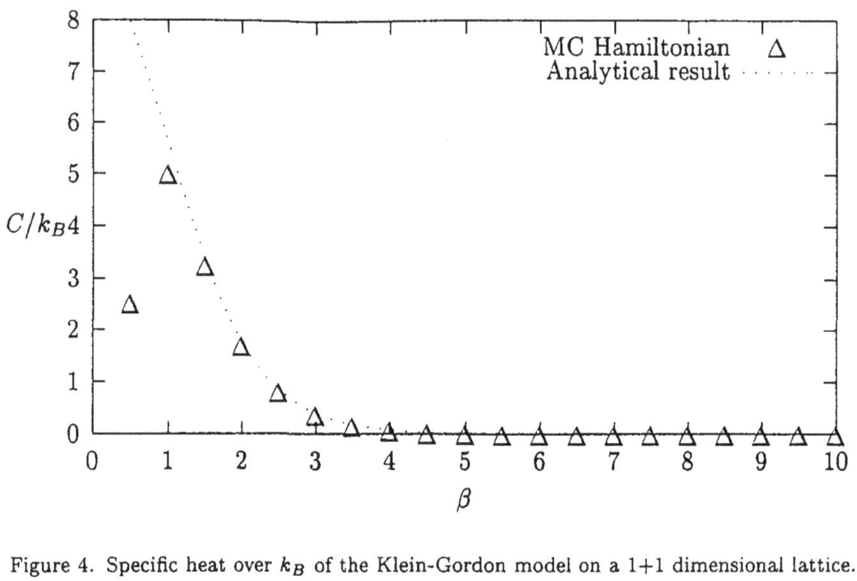

Figure 4. Specific heat over k_B of the Klein-Gordon model on a 1+1 dimensional lattice.

References

1. M. Creutz, *Quarks, Gluons and Lattices*, Cambridge University Press, Cambridge (1983).
2. H. Rothe, *Lattice Gauge Theory: an Introduction*, World Scientific, Singapore (1992).
3. I. Montvay and G. Münster, *Quantum Fields on a Lattice*, Cambridge University Press, Cambridge (1994).
4. S. Guo and X.Q. Luo, hep-lat/9706017, and references therein.
5. Q. Chen, X.Q. Luo, and S. Guo, *Phys. Lett.* **B341** (1995) 349.
6. Q. Chen, X.Q. Luo, S. Guo, X. Fang, *Phys. Lett.* **B348** (1995) 560.
7. X.Q. Luo, and Q. Chen, *Mod. Phys. Lett.* **A11** (1996) 2435.
8. X.Q. Luo, Q. Chen, S. Guo, X. Fang, and J. Liu, *Nucl. Phys.* **B (Proc. Suppl.)53** (1997) 243.
9. M. Honma, T. Mizusaki and T. Otsuka, *Phys. Rev. Lett.* **75** (1995) 1284; M. Honma, T. Mizusaki and T. Otsuka, nucl-th/9609043.
10. S.A. Chin, O.S. van Roosmalen E.A. Umland, and S.E. Koonin, *Phys. Rev.* **D31** (1985) 3201; S.A. Chin, C. Long and D. Robson, *Phys. Rev.* **D37** (1988) 3001; ibid 3006.
11. T. Barnes and D. Kotchan, *Phys. Rev.* **D35** (1987) 1947.
12. M. Beccaria, *Phys. Rev.* **D62** (2000) 034510.
13. C. Hamer, M. Smaras, and R. Bursill, these proceedings, hep-lat/0007031.
14. W. von der Linden, *Phys. Repts.* **220** (1992) 53.
15. H. Jirari, H. Kröger, X.Q. Luo, and K. Moriarty, *Phys. Lett.* **A258** (1999) 6.
16. X.Q. Luo, C. Huang, J. Jiang, H. Jirari, H. Kröger, and K. Moriarty, *Physica* **A281** (2000) 201.
17. X.Q. Luo, J. Jiang, C. Huang, H. Jirari, H. Kröger, and K. Moriarty, *Nucl. Phys.* **B(Proc. Suppl.)83-84** (2000) 810.
18. J. Jiang, C. Huang, X.Q. Luo, H. Jirari, H. Kröger, K. Moriarty, *Commun. Theor. Phys.* **34** (2000) 723.
19. H. Jirari, H. Kröger, C. Huang, J. Jiang, X.Q. Luo, and K. Moriarty, *Nucl. Phys.* **B(Proc. Suppl.)83-84** (2000) 953.
20. C. Huang, J. Jiang, X.Q. Luo, H. Jirari, H. Kröger, and K. Moriarty, quant-ph/9912051.
21. E. Henley and W. Thirring, *Elementary Quantum Field Theory*, McGraw-Hill (1962).

ODD-FLAVOR SIMULATIONS BY HYBRID MONTE CARLO [a]

TETSUYA TAKAISHI
Hiroshima University of Economics, Hiroshima, 731-0192, JAPAN.
E-mail:tt-taka@hue.ac.jp

PHILIPPE DE FORCRAND
CERN, Theory Division, CH-1211 Geneva 23, Switzerland
ETH-Hönggerberg, CH-8093 Zürich, Switzerland
E-mail:forcrand@itp.phys.ethz.ch

The standard hybrid Monte Carlo algorithm is known to simulate even flavors QCD only. Simulations of odd flavors QCD, however, can be also performed in the framework of the hybrid Monte Carlo algorithm where the inverse of the fermion matrix is approximated by a polynomial. In this exploratory study we perform three flavors QCD simulations. We make a comparison of the hybrid Monte Carlo algorithm and the R-algorithm which also simulates odd flavors systems but has step-size errors. We find that results from our hybrid Monte Carlo algorithm are in agreement with those from the R-algorithm obtained at very small step-size.

1 Introduction

Recent lattice QCD simulations include effects of dynamical fermions. The standard established algorithm to simulate dynamical QCD is the Hybrid Monte Carlo (HMC) algorithm [1] but it is limited to simulations of an even number of degenerate flavors. It would be desirable to simulate lattice QCD with three flavors since there exist three light quarks, i.e. u,d,s quarks, in the real world. Simulations with an odd number of flavors have been performed using the R-algorithm [2]. This algorithm, however, is not exact: it causes systematic errors of order $\Delta\tau^2$, where $\Delta\tau$ is the step-size of the Molecular Dynamics evolution. A careful extrapolation to zero step-size is therefore needed to obtain exact results. Nevertheless, it is common practice to forego this extrapolation and to perform simulations with a single step-size chosen small enough that the expected systematic errors are smaller than the statistical ones. We want to point out that there is an alternative to the R-algorithm, which gives arbitrarily accurate results without any extrapolation[3].

Some time ago, a local algorithm, the so-called "Multiboson algorithm", was proposed by Lüscher [4], in which the inverse of the fermion matrix is approximated by a suitable Chebyshev polynomial. Originally he proposed it for two flavors QCD. Boriçi and de Forcrand [5] noticed that the determinant of a

[a]Talk presented by T.Takaishi

fermion matrix can be written in a manifestly positive way using a polynomial approximation, so that one can simulate odd flavors QCD with the multiboson method. Indeed, using this method, one flavor QCD was simulated successfully [6]. The same polynomial approximation can be applied for the HMC algorithm [7], which means that the HMC algorithm also has the possibility of simulating odd flavors. Here we give the formulation for simulating odd flavors with the HMC algorithm and perform three flavors simulations. Then we compare our results with those of the R-algorithm.

2 Formulation

2.1 n_f=*2*

The lattice QCD partition function with n_f degenerate quark flavors is given by

$$Z = \int dU \det(D)^{n_f} \exp(-S_{gauge}), \tag{1}$$

where D is the fermion matrix and in this study we use Wilson fermions. For $n_f = 2$, the partition function is

$$Z = \int dU \det(D)^2 \exp(-S_{gauge}). \tag{2}$$

In the formulation of the HMC algorithm, $\det(D)^2$ is treated as

$$\det D^2 \sim \int d\phi^\dagger d\phi \exp(-\phi^\dagger D^{\dagger -1} D^\dagger \phi), \tag{3}$$

where the γ_5 hermiticity of the fermion matrix D, i.e. $D = \gamma_5 D^\dagger \gamma_5$, is used.

Introducing momenta P conjugate to the link variables U, the partition function is rewritten as

$$Z = \int dU dP \exp(-H), \tag{4}$$

where the Hamiltonian H is defined by

$$H = \frac{1}{2}P^2 + S_{gauge} + \phi^\dagger D^{\dagger -1} D^{-1} \phi. \tag{5}$$

This Hamiltonian is used for the Molecular Dynamics (MD) simulation of the standard HMC algorithm. Eq.(5) has a computational difficulty in MD simulations since one must solve $x = D^{-1}\phi$ type equations which in general take a large amount of computational time for a large fermion matrix and/or for a small quark mass.

Following Lüscher[4], the inverse of D can be approximated by a polynomial:

$$1/D \approx P_n(D) \equiv \prod_{k=1}^{n}(D - Z_k) \tag{6}$$

where Z_k are roots of the polynomial $P_n(D)$:

$$Z_k = 1 - exp(i\frac{2\pi k}{n+1}). \tag{7}$$

The rate of convergence of the approximation depends on the quark mass (See Sec.3).

Replacing D^{-1} in eq.(5) by $P_n(D)$ we obtain an approximate Hamiltonian,

$$H_n = \frac{1}{2}P^2 + S_{gauge} + \phi^\dagger P_n(D)^\dagger P_n(D)\phi. \tag{8}$$

An advantage of using H_n is that no solver calculation is required in the MD evolution. Instead, one needs n multiplications by the matrix D. Originally H_n was introduced to reduce computational work. Indeed, it was shown that H_n can provide some gain over the standard HMC algorithm[7].

H_n does introduce some systematic errors from the polynomial approximation. For the n_f=even case, however, these errors are easily corrected at the Metropolis step by using the exact Hamiltonian of eq.(5)[7].

The domain of convergence of $P_n(D)$ is bounded by a circle centered at (1,0) which goes through the origin. If all eigenvalues of D fall inside this domain, $P_n(D)$ converges exponentially. Otherwise, $P_n(D)$ does not converge, which may happen for some exceptional configurations. Our algorithm rejects these configurations at the Metropolis step. This domain of convergence can be changed by adopting another approximating polynomial. However, the origin must be excluded. Together with connectedness and conjugate symmetry of the spectrum, this implies that the real negative axis is always excluded from the domain of convergence for any polynomial. Configurations with real negative Dirac eigenvalues will be rejected by our polynomial algorithm.

2.2 $n_f = 1$

In this case, we have to consider $\det D$. $\det D$ can not be expressed in a manifestly positive manner using the same treatment of eq.(3). Thus the standard HMC algorithm can not handle $n_f = 1$ or n_f=odd simulations.

The multiboson algorithm was originally developed for a simulation of n_f=2 QCD [4]. After invention of the multiboson algorithm, Boriçi and de Forcrand [5] noticed that a single $\det D$ can be treated in a manifestly positive way and an $n_f = 1$ multiboson simulation was performed to study thermodynamics of $n_f = 1$ QCD [6].

As before, the inverse of the fermion matrix D, using a polynomial of degree $2n$, is approximated as [4,5]

$$1/D \approx \prod_{k=1}^{2n} (D - Z_k), \tag{9}$$

where $Z_k = 1 - \exp(i\ 2\pi k/(2n+1))$. Noticing that the Z_k's come in complex conjugate pairs, eq.(9) is rewritten as

$$1/D \approx \prod_{k=1}^{n} (D - \bar{Z}_k)(D - Z_k). \tag{10}$$

Using the γ_5 hermiticity of the fermion matrix, we find that $\det(D - \bar{Z}_k) = \det(D - Z_k)^\dagger$. Thus the determinant of D is written as

$$\det(D) \sim \det(T_n^\dagger(D) T_n(D))^{-1}, \tag{11}$$

where $T_n(D) \equiv \prod_{k=1}^{n}(D - Z_k)$. Using an integral form of the determinant, we obtain

$$\det(D) \sim \int d\phi^\dagger d\phi \exp(-\phi^\dagger T_n^\dagger(D) T_n(D)\phi). \tag{12}$$

The term $\phi^\dagger T_n^\dagger(D) T_n(D)\phi$ is manifestly positive. Then we may define the Hamiltonian of $n_f = 1$ QCD as

$$H = \frac{1}{2}P^2 + S_{gauge} + \phi^\dagger T_n^\dagger(D) T_n(D)\phi. \tag{13}$$

With this Hamiltonian there is no difficulty to perform HMC algorithm. The domain of convergence of the approximation eq.(12) is the same as for $n_f = 2$. Exceptional configurations for which eigenvalues fall outside this domain will be rejected at the Metropolis step.

2.3 $n_f = 2 + 1$

The partition function of $n_f = 2 + 1$ QCD is given by

$$Z = \int dU \det \tilde{D}^2 \det D \exp(-S_{gauge}), \tag{14}$$

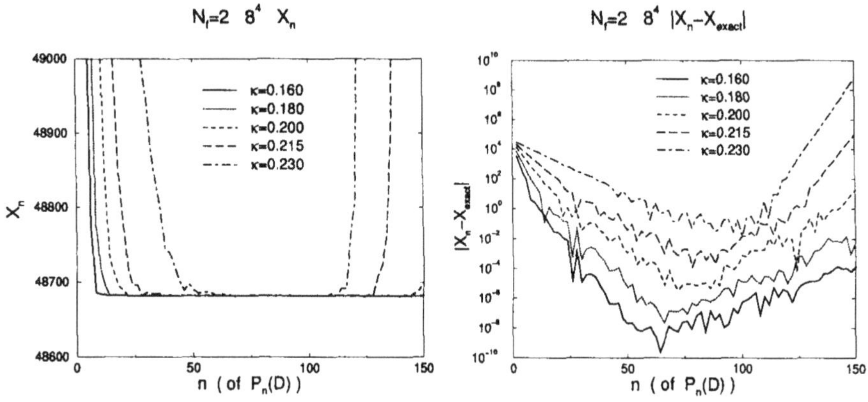

Figure 1: (left): X_n versus degree n. (right):$|X_n - X_{exact}|$ versus degree n.

where the notations $\tilde{D}$ and D are introduced to distinguish the two different quark masses. Using eq.(3) for $\det \tilde{D}^2$ and eq.(12) for $\det D$,

$$\det \tilde{D}^2 \det D \sim \int d\tilde{\phi}^\dagger d\tilde{\phi} d\phi^\dagger d\phi \exp(-\tilde{\phi}^\dagger \tilde{D}^{\dagger -1} \tilde{D}^{-1} \tilde{\phi} - \phi^\dagger T_n^\dagger(D) T_n(D) \phi). \quad (15)$$

We define $n_f = 2 + 1$ Hamiltonian by

$$H = \frac{1}{2} P^2 + S_g + \tilde{\phi}^\dagger \tilde{D}^{\dagger -1} \tilde{D}^{-1} \tilde{\phi} + \phi^\dagger T_n^\dagger(D) T_n(D) \phi. \quad (16)$$

Two remarks are in order: (i) the degree n of the approximating polynomial may be different during the Molecular Dynamics trajectory and for the Metropolis acceptance test; the former can be made arbitrarily small and tuned for maximum efficiency, while the latter should be taken very large to enforce the correct measure; (ii) the two bosonic fields ϕ and $\tilde{\phi}$ could be replaced by a single one, with action $\phi^\dagger T_n^\dagger(D) \tilde{D}^{\dagger -1} \tilde{D}^{-1} T_n(D) \phi$. For simplicity, in this exploratory study we use two distinct bosonic fields and a single degree for the approximating polynomial.

3 Convergence

3.1 $n_f = 2$

In order to see the rate of convergence of $P_n(D)$, we calculate the quantity $X_n = \phi^\dagger P_n^\dagger(D) P_n(D) \phi$. In the limit $n \to \infty$, X_n goes to $X_{exact} \equiv$

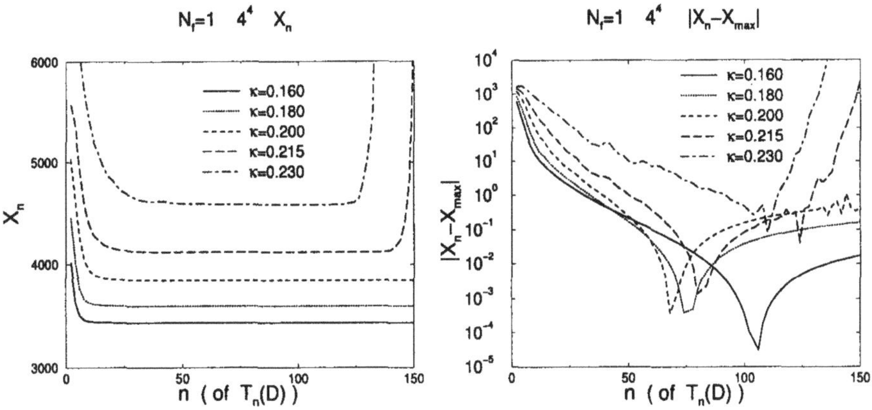

Figure 2: (left): X_n versus degree n. (right):$|X_n - X_{max}|$ versus degree n.

$\phi^\dagger {D^\dagger}^{-1} D^{-1}\phi$. First, we choose $X_{exact} = \eta^\dagger\eta$ where η is a random gaussian vector. Then the vector ϕ is set to $\phi \equiv D\eta$. The accuracy of X_n is measured by the difference between X_n and X_{exact}. We use a random gauge configuration for this analysis. Figure 1:(left) shows X_n versus the degree n for different quark masses. Here the same η is used for each calculation of X_n. X_n converges to one value as n increases, but at high degree n, X_n diverges, which can be understood due to the rounding errors of our computer, where calculations are performed with 64-bit accuracy.

Figure 1:(right) shows the accuracy of X_n by $|X_n - X_{exact}|$. Exponential convergence is seen for each quark mass, but the rate of convergence is slow for small quark masses.

3.2 $n_f = 1$

We do the same analysis as for $n_f = 2$, but for $n_f = 1$, the value of X_{exact} is not known. So we calculate the quantity $X_n = \phi^\dagger T_n^\dagger(D) T_n(D)\phi$, where the vector ϕ is a gaussian random vector, and we use a random gauge configuration. We assume that X_n goes to a certain value in the limit of $n \to \infty$.

Figure 2:(left) shows X_n as a function of degree n. X_n seems to converges to a certain value when the degree n increases. At high degree n, X_n diverges as in the case of $n_f = 2$.

To see the rate of convergence, we calculate $|X_n - X_{max}|$ where X_{max} is defined by $X_{max} = X_m$, $m >> n$. Due to the rounding errors, we can not

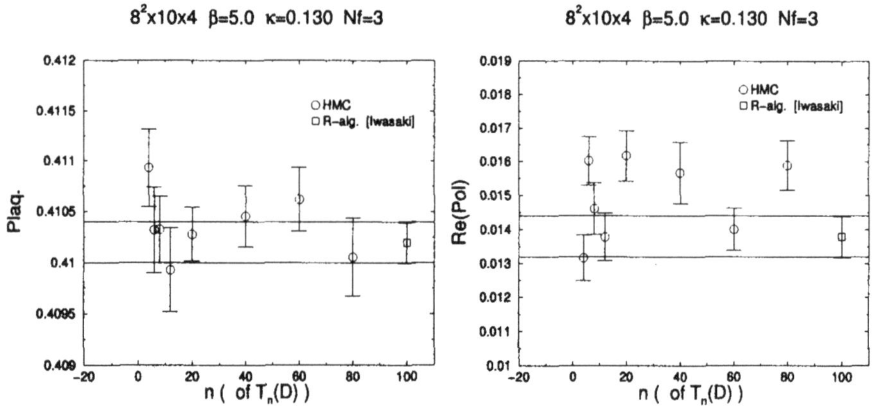

Figure 3: (left): Plaquette of $n_f = 3$ flavor QCD on an $8^2 \times 10 \times 4$ lattice at $\beta = 5.0$ and at $\kappa = 0.130$ as a function of degree n. (right): Real part of Polyakov loop.

take very large m. We take a maximum number m where the rounding errors still do not appear. Figure 2:(right) shows $|X_n - X_{max}|$ as a function of degree n. The dips seen in the figure are just due to the fact that at those points $X_n = X_{max} \equiv X_m$. The convergence seems to be exponential, but the rate of convergence is slow for small quark masses as in the $n_f = 2$ case.

4 Simulations

We perform simulations of three flavors QCD on an $8^2 \times 10 \times 4$ lattice at $\beta = 5.0$ with $\kappa = 0.130$ and 0.160. We measure the plaquette and Polyakov loop varying the degree n and compare them with those from the R-algorithm obtained with a step-size $\Delta\tau = 0.01$[8]. Figures 3 and 4:(left) show the plaquette as a function of n at $\kappa = 0.130$ and 0.160, respectively. Except for very small n, the results from the HMC algorithm agree with those from the R-algorithm within statistical errors. Results of the Polyakov loop are shown in figures 3 and 4:(right). Except for a small discrepancy seen in figure 3, the results from the HMC algorithm are in agreement with those from the R-algorithm. Note that convergence is not monotonic in n.

5 Conclusions

We formulated an odd-flavor HMC algorithm using a polynomial approximation. Simulations of three flavors QCD were performed. We found that the plaquette values are consistent with those from the R-algorithm at very small

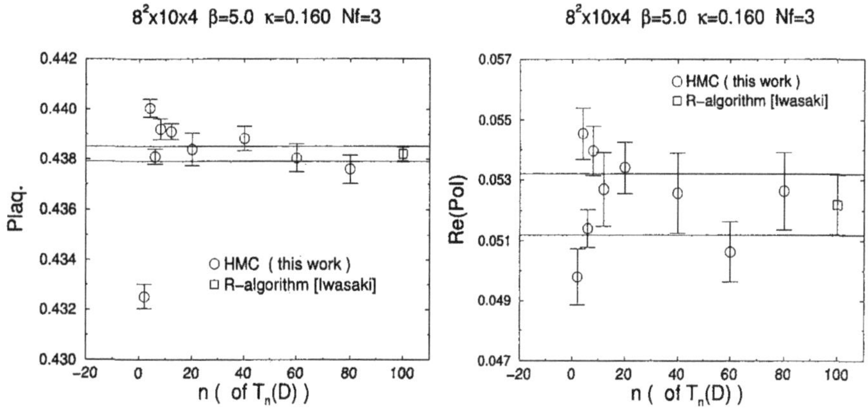

Figure 4: (left): Plaquette of $n_f = 3$ flavor QCD on an $8^2 \times 10 \times 4$ lattice at $\beta = 5.0$ and at $\kappa = 0.160$ as a function of degree n. (right): Real part of Polyakov loop.

step-size. In principle the HMC algorithm is able to simulate any flavors of QCD, with arbitrary accuracy and without extrapolation [as long as all Dirac eigenvalues are not real negative]. However the rounding errors should be under control when we use a large lattice or/and small quark masses where one may need a polynomial of high degree n to achieve sufficient approximation.

Acknowledgements

This work is partially supported by the Grant-in-Aid for Scientific Research by Monbusho, Japan(No.11740159).

References

1. S.Duane, A.Kennedy, B.Pendleton and D.Roweth, *Phys. Lett.* B **195**, 216 (1987)
2. S. Gottlieb, W.Liu, D.Toussaint, R.L.Renken and R.L.Sugar, *Phys. Rev.* D **35**, 2531 (1987).
3. For yet another approach, see T.Lippert, hep-lat/0007029, Lect.Notes Comput.Sci. and Eng. 15 (2000) 166-177.
4. M. Lüscher, *Nucl. Phys.* B **418**, 637 (1994);
   B.Bunk, K.Jansen, B.Jegerlehner, M.Lüscher, H.Simma and R.Sommer, *Nucl. Phys.* B(Proc.Suppl.) **42**, 49 (1995)
5. A. Boriçi and Ph. de Forcrand, *Nucl. Phys.* B **454**, 645 (1995).

6. C. Alexandrou, A.Boriçi, A.Feo, Ph. de Forcrand, A.Galli, F.Jegerlehner and T.Takaishi, *Phys. Rev.* D **60**, 034504 (1999).
7. Ph. de Forcrand and T. Takaishi *Nucl. Phys.* B(Proc.Suppl.) **53**, 968 (1997)
8. Y. Iwasaki, K.Kanaya, S.Kaya, S.Sakai and T.Yoshie, *Phys. Rev.* D **54**, 7010 (1996).

4 QCD at Finite Density and Temperature

Participants enjoying Seven Star Crag Park in Zhaoqing.

DIQUARK CONDENSATION IN TWO COLOUR QCD

R. ALOISIO AND A. GALANTE*
Dipartimento di Fisica dell'Università di L'Aquila, 67100 L'Aquila, Italy
E-mail: aloisio,galante@lngs.infn.it

V. AZCOITI
Departamento de Física Teórica, Facultad de Ciencias, Universidad de Zaragoza, 50009 Zaragoza, Spain
E-mail: azcoiti@azcoiti.unizar.es

G. DI CARLO
Istituto Nazionale di Fisica Nucleare, Laboratori Nazionali di Frascati, P.O.B. 13 - 00044 Frascati, Italy
E-mail: gdicarlo@lnf.infn.it

A. F. GRILLO
Istituto Nazionale di Fisica Nucleare, Laboratori Nazionali del Gran Sasso, 67010 Assergi (L'Aquila), Italy
E-mail: grillo@lngs.infn.it

Unquenched lattice SU(2) is studied at nonzero chemical potential in the strong coupling limit. The topic of diquark condensation is addressed analyzing the probability distribution function of the diquark condensate. We present results at zero external source without using any potentially dangerous extrapolation procedure. We find strong evidences for a (high density) second order phase transition where a diquark condensate appears, and show quantitative agreement of lattice calculations with low-energy effective Lagrangian calculations.

1 Introduction

Recently the standard scenario for the phase diagram of QCD in the chemical potential-temperature plane has changed; besides the standard hadronic and quark-gluon plasma phases, the existence of a new state of matter has been claimed by several groups[1]. This new phase is characteristic of the high density-low temperature regime; the asymptotic freedom of QCD and the known instability of large Fermi spheres in presence of (whatever weak) attractive forces results in a pairing of quarks of higher momenta (in analogy with the Cooper pairing in solid state systems at low temperature). A condensation of quark pairs should be the distinctive signal of the new phase and

*TALK PRESENTED BY A. GALANTE

has been indeed predicted using simplified phenomenological models of QCD.

Unfortunately the lattice approach, the most powerful tool to perform first principles non perturbative studies, is affected in the case of finite density QCD by the well known sign problem that has prevented until now any step towards the understanding of this new physics. This is not the case for the SU(2) theory where the fermions are in the pseudo-real representation and finite density numerical simulations are feasible.

In the following we present a detailed study of diquark condensation for the unquenched two colours model. This work follows a previous one where we considered chiral and diquark susceptibilities and found strong evidence for a phase transition separating the ordinary low density phase from a high density one where chiral symmetry is restored and the baryon number symmetry is broken[2]. In the present work we mainly focus on an approach, based on the analysis of the probability distribution function (p.d.f.) of the order parameter, to extract the value of the order parameter directly at zero external source without using any potentially dangerous extrapolation procedure, implicit in the standard approach.

The study of diquark condensation in two colours QCD is an interesting topic by itself and, notwithstanding the relevant differences between SU(2) and SU(3) at non zero baryon density, we can hope to use some of the results to get insights about the three colours case. Even if the phase diagram of SU(3) and SU(2) are expected to share similarities, we have to be aware of the differences between the two models.

- The quark-quark condensate $\langle qq \rangle$ is coloured for the $N_c = 3$ theory and colourless for the $N_c = 2$ one. This implies that in the SU(3) case the diquark condensation has to be interpreted as an Higgs-like mechanism leading to the breaking of the local colour symmetry, while in the SU(2) case we have a standard spontaneous symmetry breaking (SSB) of the $U_V(1)$ symmetry associated with conservation of baryon number.

- The zero temperature critical chemical potential, related to the mass of the lightest baryonic state, is different in the two cases: it is 1/3 of the nucleon mass for SU(3) and 1/2 of the pion mass for SU(2) (i.e. zero at vanishing quark mass). For $N_c = 2$, in order to have a phase diagram similar to the SU(3) one, it is appropriate to consider a non zero bare quark mass.

- The lightest baryonic state is a fermion for the tree colours model and a boson for the two colours one.

Even if our results can not be directly extended to the SU(3) case we can use lattice calculations to do a highly non trivial, quantitative check of the continuum predictions for the SU(2) theory. In particular we have considered results coming from low energy effective Lagrangian calculations[3].

In the next section we present some details of the algorithm. The last section is devoted to the presentation and discussion of numerical results.

2 Overview of the numerical technique

2.1 Simulation scheme

The standard way to study SSB is to introduce first an explicit symmetry breaking term in the action. If we do that for the diquark in the SU(2) model we have to add a term[4] $\jmath(\psi\tau_2\psi + \bar{\psi}\tau_2\bar{\psi})$ and, after integrating the Grassmann field, the fermionic contribution for $N_f = 4$ quark flavours becomes proportional to the Pfaffian of a $4V \times 4V$ matrix[2]:

$$\mathcal{Z}_{ferm}(\jmath) = Pf(B + \jmath) = \pm\sqrt{\det(B + \jmath)} \tag{1}$$

where B is

$$B = \begin{pmatrix} 0 & \frac{1}{2}\Delta\tau_2 \\ -\frac{1}{2}\Delta^T\tau_2 & 0 \end{pmatrix}. \tag{2}$$

and Δ is the usual lattice Dirac operator (it contains the mass and μ dependence). Using the relation $\tau_2\Delta\tau_2 = \Delta^*$ we can easily prove that B is antihermitian and $\det(B + \jmath) \geq 0$ for any $\jmath$.

It can also be shown that the eigenvalues of B are doubly degenerate and B^2 is in a block diagonal form with two hermitian blocks on the diagonal having the same eigenvalues. It follows that to compute $\det(B + \jmath)$ for any $\jmath$ (i.e. to obtain all the eigenvalues of B) it is sufficient to diagonalize only one block of B^2 (reducing the problem to the diagonalization of a $2V \times 2V$ hermitian matrix) and then take the two pure imaginary square roots of the (real and negative) eigenvalues.

To avoid the sign ambiguity in (1) it is customary to consider a theory with $N_f = 8$ quark flavours where the fermionic partition function becomes $\mathcal{Z}_{ferm}(\jmath) = \det(B + \jmath)$. On the other side if we are able to work directly at zero diquark source things become much simpler. In the $\jmath = 0$ limit the sign ambiguity disappears and the Pfaffian is positive definite since $Pf(B) \equiv \det\Delta$ and the last quantity is real and positive for any value of μ. Then we can easily consider any value of N_f writing $\mathcal{Z}_{ferm}(\jmath = 0) = (\det(B))^{N_f/8}$. In the next section we show how it is possible to take advantage of this feature and extract the value of the order parameter for the diquark condensation for all N_f without any extrapolation to $\jmath = 0$.

We have studied the phase structure of the theory in the limit of infinite gauge coupling ($\beta = 0$). To simulate the $\beta = 0$ limit of the theory we have measured fermionic observables on gauge configurations generated randomly, i.e. with only the Haar measure of the gauge group as a weight. This choice implies a Gaussian distribution of the plaquette energy around zero which, according to the results of Morrison and Hands[4], has a net overlap with the importance sample of gauge configurations at the values of μ and m used in our calculations. The validity of this procedure has also been tested comparing different physical observables (number density and chiral condensate) with Hybrid Montecarlo results[5].

We have considered the theory in a 4^4 and 6^4 lattice diagonalizing 300 gauge configurations in the first lattice volume and 100 in the second one. As we pointed out in the introduction, to stay closer to the SU(3) case the simulations have been performed at non zero quark mass. We choosed $m = 0.025, 0.05, 0.20$ and values of the chemical potential ranging from $\mu = 0$ to $\mu = 1.0$.

All numerical simulations have been performed on a cluster of Pentium II and PentiumPro at the INFN Gran Sasso National Laboratory.

2.2 *Analysis of the probability distribution function*

The use of the p.d.f. to analyze the spontaneous symmetry breaking in spin system or Quantum Field Theories with bosonic degrees of freedom is a standard procedure. Less standard is its application to QFT with Grassmann fields where, for obvious reasons, the fermionic degrees of freedom have to be integrated analytically. Nevertheless this method has been developed to extract the chiral condensate in the chiral limit from simulations of QFT with fermions[6].

The same ideas can be used to study the vacuum structure of two colour QCD at non zero density and specifically to extract the diquark condensate at $j = 0$. We refer to the original paper for a full description of the p.d.f. technique and present a brief introduction focusing on the peculiarities of the diquark condensate case.

Let α be an index which characterizes all possible (degenerate) vacuum states and w_α the probability to get the vacuum state α when choosing randomly an equilibrium state. If c_α is the order parameter in the α state we can write

$$c_\alpha = \langle \frac{1}{V} \sum_x \psi \tau_2 \psi(x) + \bar{\psi} \tau_2 \bar{\psi}(x) \rangle_\alpha \tag{3}$$

where V is the lattice volume and the sum is over all lattice points. $P(c)$, the p.d.f. of the diquark order parameter c, will be given by

$$P(c) = \sum_{\alpha} w_\alpha \delta(c - c_\alpha)$$
$$= \lim_{V\to\infty} \frac{1}{\mathcal{Z}} \int [dU][d\bar\psi][d\psi] e^{-S_G(U)+\bar\psi\Delta\psi} \delta(\frac{1}{V}\sum_x \psi\tau_2\psi(x) + \bar\psi\tau_2\bar\psi(x) - c)$$

The main point is that while $P(c)$ is not directly accessible with a numerical simulation its Fourier transform

$$P(q) = \int dc e^{iqc} P(c) \tag{4}$$

can be easily computed. Inserting in (4) the definition of $P(c)$ and using an integral representation for the δ-function, after some algebra we arrive at the following formula

$$P_V(q) = \frac{1}{\mathcal{Z}(j=0)} \int [dU] e^{-S_G(U)} \frac{PfB(j=\frac{iq}{V})}{PfB(j=0)} [PfB(j=0)]^{\frac{N_f}{4}} \tag{5}$$

where $P_V(q)$ is the Fourier transformed p.d.f. of the diquark order parameter at zero external source and finite volume.

To take advantage of (5) we should be able to compute correctly the Pfaffians involved. This indeed turns out to be easy. First note that, as previously stated, $PfB(j=0)$ is a real and positive quantity so that the only ambiguity is related to the sign of $PfB(j=\frac{iq}{V}) = \pm\sqrt{\det B(j=iq/V)}$. If we remember that the $4V$ eigenvalues of the antihermitian matrix B are doubly degenerate and can be written as $\pm i\lambda_n$ $(n = 1, \cdots, V)$ with real and positive λ_n we arrive at the following expression

$$PfB(j=\frac{iq}{V}) = \pm \prod_{n=1}^{V} (\lambda_n^2 - \frac{q^2}{V^2}) \tag{6}$$

The sign ambiguity can now be solved noticing that (6) has to be positive at $j = 0$ and, increasing q, changes sign each time j is equal to one of the eigenvalues. The only possible exception would be an (accidental) extra degeneracy in the eigenvalues of B resulting in the Pfaffian not crossing zero but tangent to the horizontal axis. This situation never occurred in our simulations.

Once we have the $P_V(q)$ and hence, by simple Fourier transform, the p.d.f. of the order parameter we need to extract the correct value for the order parameter. To do that we first have to recall that diquark condensate is the order parameter of the $U_V(1)$ symmetry associated with baryon number conservation. Indeed $\langle\psi\tau_2\psi + \bar\psi\tau_2\bar\psi\rangle$ and $i\langle\psi\tau_2\psi - \bar\psi\tau_2\bar\psi\rangle$ are the components

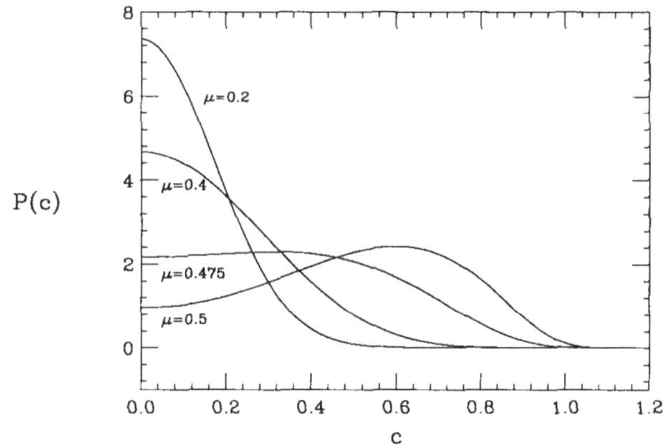

Figure 1. $P(c)$ for the 4^4 lattice at $\mu = 0.2, 0.4, 0.475, 0.5$ (from top to bottom).

of a vector (in a plane) which rotates by an angle 2θ when we do a global phase transformation of parameter θ on the fermionic fields. Therefore, if c_0 is the vacuum expectation value of the diquark condensate corresponding to the α-vacuum selected when switching-on a diquark source term, $P(c)$ can be computed as

$$P(c) = \frac{1}{2\pi} \int d\alpha \delta(c - c_0 \cos(2\alpha)) \tag{7}$$

which gives $P(c) = 1/(\pi(c_0^2 - c^2)^{1/2})$ for $-c_0 \leq c \leq c_0$ and $P(c) = 0$ otherwise[6]. In the symmetric phase $c_0 = 0$ and $P(c)$ reduces correctly to a δ-function in the origin.

The above results are valid in the thermodynamic limit while, at finite volume, the non analyticities of the p.d.f. are absent. Without entering in the details of a finite size scaling analysis we expect, for the finite volume p.d.f. $P_V(c)$, a function peaked in the origin in the symmetric phase and peaked at some non zero value in the broken phase. This is indeed the behaviour we can observe in fig. 1 where the the p.d.f. of the smallest volume is reported at different values of μ. It is clear that, increasing the chemical potential, the vacuum starts to be degenerate signalling a spontaneous breaking of the baryon number conservation.

We can also compare the $P_V(c)$ for two lattice volumes in the symmetric and broken phase. This is done in fig. 2 where we see clearly as, increasing the volume, the peak of the p.d.f. becomes sharper. To determine the value of the diquark condensate we used the position of the peak: a definition that clearly converges to the correct value in the thermodynamic limit.

From fig. (2) we see also that data of the larger volume are more noisy (indeed we also get negative values for the p.d.f. in the broken phase). To have an estimate of the errors we calculated several distribution functions for the 6^4 lattice using independent subsets of our data. We saw that the position of the peak was very stable, within 1 percent in the broken phase, and we used this quantity for the errorbars.

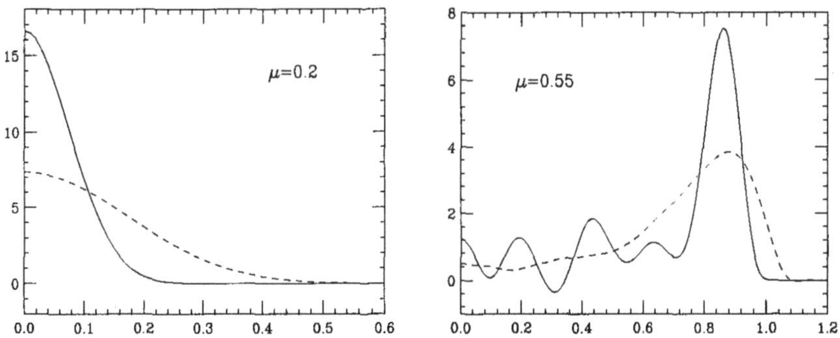

Figure 2. P.d.f. for 4^4 (dashed line) and 6^4 (continuous line) in the symmetric (left) and broken (right) phase.

3 Results

Here we present the results for the diquark condensate at $j = 0$ as a function of the chemical potential. Fig. 3 contains our data points for the 6^4 volume for several values of N_f at two different masses: $m = 0.025$ and $m = 0.2$.

First we can derive a peculiar phase diagram: two symmetric phases separated by a broken one and two possibly continuous transition points. This result is not new and has already been predicted by Mean Field calculations[7] as well as numerical calculations[2]. The high density symmetric phase has no physical relevance, it is simply the consequence of the saturation of all lattice sites with quarks. This saturation phenomena has nothing to do with continuum physics and is a pure lattice artifact. The physically interesting phase transition, i.e. the transition that has a continuum counterpart, is the first one.

To test our numerical results we have considered the analytical predictions of low energy effective Lagrangian. The $j = 0$ case is described by a simple formula where the only two parameters are the chiral condensate and one half the pion mass calculated at $j = 0, \mu = 0$[3]. We did independent simulations

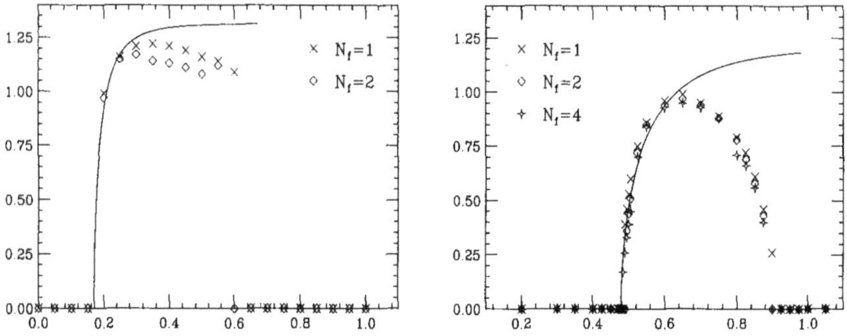

Figure 3. Diquark condensate as a function of μ for 6^4 lattice (points) and the continuum predictions (line) for $m = 0.025$ (left) and $m = 0.2$ (right).

Table 1. Parameters for the low energy effective Lagrangian predictions.

m	$\langle\psi\psi\rangle$	$m_\pi/2$
0.025	1.31(2).	0.1696(11)
0.2	1.23(2)	0.4841(7)

to calculate this quantities at the same values of m used in the previous analysis, obtaining the values in Table 1. Another prediction of the continuum calculation is that the value of the diquark condensate is independent of the flavour number.

The continuous line in fig. 3 is the analytical prediction for this model. We see a remarkable agreement with our strong coupling results up to values of the chemical potential where the saturation effects start to be relevant. We see also that the N_f dependence in the simulation points is small.

We checked explicitly the low energy effective Lagrangian prediction at finite j too. In that case we used standard techniques to compute chiral and diquark condensate as well as the number density[2] in a $N_f = 8$ simulation (fig. 4). Also in that case (where no phase transition is present and the predictions are still independent of N_f) the agreement is very good up to values of μ where the number density becomes a consistent fraction of its maximum lattice value (i.e. $\mu/m_\pi \simeq 0.6$).

It is surprising that a $\beta = 0$ calculation has a so good agreement with a continuum prediction. Since the analytical predictions are, at the values of μ presented, well inside the validity region of the low energy approximation ($\mu <<$ the mass of the first non goldstone excitation) we can conclude that this gives an indication of a very small dependence of lattice results on β.

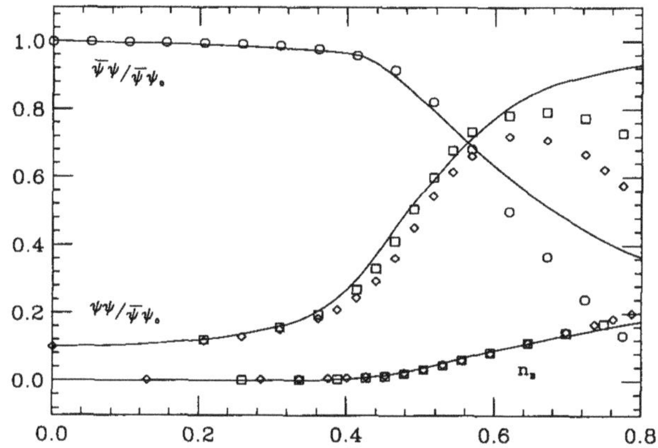

Figure 4. Low energy Lagrangian predictions (lines) and our $N_f = 8$ data (symbols) for chiral condensate, diquark condensate and number density at $m = 0.2$ and $j = 0.1m$ vs. μ/m_π. 6^4 data are squares and circles, 4^4 data are diamonds.

Acknowledgments

This work has been partially supported by CICYT (Proyecto AEN97-1680) and by a INFN-CICYT collaboration. The Consorzio Ricerca Gran Sasso has provided part of the computer resources needed for this work.

References

1. M. Alford, K. Rajagopal and F. Wilczek, *Phys. Lett.* B **422**, 247 (1998); *Nucl. Phys.* B **537**, 443 (1999); R. Rapp, T. Schaefer, E.V. Shuryak and M. Velkovsky, *Phys. Rev. Lett.* **81**, 53 (1998); T. Schafer and F. Wilczek, *Phys. Rev. Lett.* **82**, 3956 (1999).
2. R. Aloisio, V. Azcoiti, G. Di Carlo, A. Galante and A.F. Grillo, preprint DFTUZ 2000/03.
3. J.B. Kogut, M.A. Stephanov, D. Toublan, J.J.M. Verbaarschot and A. Zhitnitsky, hep-ph/0001171.
4. S. Morrison and S. Hands, in *Strong and Electroweak Matter 98*, Copenhagen, Dec. 1998, hep-lat/9902012.
5. R. Aloisio, V. Azcoiti, G. Di Carlo, A. Galante and A.F. Grillo, *Nucl. Phys.* B **564**, 489 (2000).
6. V. Azcoiti, V. Laliena and X.Q. Luo, *Phys. Lett.* B **354**, 111 (1995).
7. E. Dagotto, A. Moreo and U. Wolff, *Phys. Lett.* B **186**, 395 (1987).

SIMULATION OF SU(2) DYNAMICAL FERMIONS AT FINITE CHEMICAL POTENTIAL AND AT FINITE TEMPERATURE[a]

Y. LIU[1], O. MIYAMURA[1], A. NAKAMURA[1], T. TAKAISHI[2]
[1] *Hiroshima University, Higashi-Hiroshima 739-8521, Japan*
[2] *Hiroshima University of Economics, Hiroshima 731-01 Japan*

SU(2) lattice gauge theory with dynamical fermion at non-zero chemical potential and at finite temperature is studied. We focus on the influence of chemical potential for quark condensate and mass of pseudoscalar meson at finite temperature. Hybrid Monte Carlo simulations with $N_f = 8$ staggered fermions are carried out on $12 \times 12 \times 24 \times 4$ lattice. At $\beta = 1.1$ and m_q =0.05,0.07,0.1, we calculate the quark condensate and masses of pseudoscalar meson consisting of light and heavier quarks for chemical potential $\mu = 0.0,0.02,0.05,0.1,0.2$.

1 INTRODUCTION

The study of strong interactions at finite baryon density has a long history. Recent theoretical speculation is that the ground state of dense baryonic matter may be more exotic [1, 2]. Nambu-Jona- Lasinio type and instanton liquid models [2, 3] suggest that the QCD vacuum at sufficiently high baryon density could become a color superconductor [3].

Considerable progress has been made in simulating lattice quantum chromodynamics (LQCD) at nonzero temperature [4]. But, there has been little progress in understanding the theory at nonzero chemical potential even though a pioneering work was presented sixteen years ago[5]. The basic difficulty in simulation of LQCD at finite density is that the effective action becomes complex due to the introduction of chemical potential. Standard algorithms such as hybrid molecular dynamics (HMD)[6] or hybrid Monte Carlo (HMC)[7] can not be applied in this situation. In order to avoid the complex determinant, early simulations used quenched approximation, but they do not correspond to the real world [8]. In this situation, a possible way is to examine response of physical quantities to chemical potential at $\mu = 0$. Quark-number susceptibilities have been studied by S.Gottlieb *et al*[9] and give us the signals for the chiral- symmetry-restoration phase transition. Study of response of hadron masses to chemical potential is now under investigation by $QCD - TARO$ collaboration [10].

Another way is to study LQCD for color SU(2) system where action remains real. Hamiltonian approach at zero temperature and finite density has

---

[a] Talk presented by Y.Liu
E-mail: liu@hirohe.hepl.hiroshima-u.ac.jp

been studied by E.B.Gregory *et al*[11], and gives us some results for vacuum energy, chiral condensate, baryon-number density and its susceptibility. Numerical simulation at non zero chemical potential has been performed by S.Hands *et al*[12], by using several algorithms; HMD algorithm ; HMC algorithm ; Two-Step Multi-Boson (TSMB) algorithm[13] on rather small lattices (mostly on 6^4 lattice) at zero temperature. They studied relation of chiral condensate, pion mass and chemical potential at $\mu > \frac{m_\pi}{2}$ as well as $\mu < \frac{m_\pi}{2}$.

In this paper, we focus on the effects of chemical potential to masses and condensate at finite temperature (μ=0.02, 0.05, 0.1 and 0.2) on larger lattices $12^3\times4$ and $12^2\times24\times4$. One of interests is precise study of masses of meson with nonzero chemical potential since QCD Sum Rule suggests linear dependence on density. [14] Firstly, we shall observe chiral condensate with nonzero chemical potential near critical temperature. Secondly, we shall examine meson masses consisting of light-light quarks and light-heavy quarks. The organization of the paper is as follows: In Section 2 we state our procedure. Simulation is described in section 3. We show some preliminary results of the present simulation in Section 4. Summary and conclusions are stated in Section 5.

2 STAGGERED FERMION ACTION AT NONZERO CHEMICAL POTENTIAL

At non zero chemical potential, staggered fermion matrix M is given by

$$\begin{aligned} M_{x,y}(U,\mu) = 2m\delta_{xy} &+ \sum_{\nu=1}^{3} \eta_\nu(x)[U_{x,\nu}\delta_{y,x+\hat\nu} - U^\dagger_{y,\nu}\delta_{y,x-\hat\nu}] \\ &+ \eta_4(x)[e^{\mu}U_{x,4}\delta_{y,x+\hat4} - e^{-\mu}U^\dagger_{y,4}\delta_{y,x-\hat4}], \end{aligned} \tag{1}$$

where U is SU(2) link variables and μ is chemical potential. η_ν is staggered phase and m is quark mass parameter.

The matrix M satisfies the following relation

$$M_{x,y}(U,\mu) = (-1)^{sig(x)-sig(y)} M^\dagger_{x,y}(U,-\mu). \tag{2}$$

The path integral for the partition function is given by

$$Z = \int[DU]det(M)^{N_f/4}e^{-S_g}, \tag{3}$$

where S_g is plaquette action written as ,

$$S_g = \beta\sum_P[1-\tfrac{1}{4}Tr(U_P+U_P^\dagger)], \tag{4}$$

where $\beta = \frac{4}{g^2}$.

In cases of simulations by HMC algorithm, we need positive definiteness of hamiltonian. Since det(M) is real for SU(2) system, the positive definiteness condition is achieved by taking det$(M)^2$ which includes $n_f = 8$ dynamical fermions. Note that the standard even-odd partitioning of the fermion matrix which is commonly used to reduce the number of flavors can not be applied for simulations at finite chemical potential. Thus $n_f = 8$ is the minimum number of flavors for $SU(2)$ HMC simulations with dynamical fermions. The present study is performed with $n_f = 8$.

3 SIMULATIONS

Firstly, we have done simulation on 6^4 lattice and confirmed that our results are consistent with the previous work by S.Hands *et al*[12]. After that, numerical simulations on $12^3 \times 4$ and $12^2 \times 24 \times 4$ lattices have been carried out for parameters summarized in Table 1.

Table 1: List of the data sample

N_F	m_q	μ	lattice size	β
	0.05	0.0 , 0.02	$6 \times 6 \times 6 \times 6$	0.1~2.5 (for test)
8	0.07	0.05 , 0.1	$12 \times 12 \times 12 \times 4$	1.0~2.0(to determine β_c)
	0.1	0.2	$12 \times 12 \times 24 \times 4$	1.1 (for simulations)

Number of configuration for every parameter is also given in Table 2.

Table 2: Statistics of the data

	$\mu = 0.0$	$\mu = 0.02$	$\mu = 0.005$	$\mu = 0.1$	$\mu = 0.2$
$m_q = 0.05$	120	180	160	80	30
$m_q = 0.07$	140	170	160	160	30
$m_q = 0.10$	180	170	160	140	70

Most of the simulations are carried out on small chemical potential and at fixed $\beta = 1.1$ which is little below β_c. Time step of HMC algorithm is dt=0.01. After 1000 trajectories of thermalization, data are taken at every 100 trajectories. Since our simulation is restricted in a region of small chemical potential, the algorithm works without difficulty. But, necessary CPU time increases with the chemical potential. A typical trend is shown in table 3. At $\mu = 0.2$, average CPU time is 3.5 times longer than that at $\mu = 0.0$.

Table 3: CPU time of getting one configuration at $\beta = 1.1$, $m_q = 0.05$

	0.0	0.02	0.05	0.1	0.2
time(second)	2223.3	2305.7	2536.0	3122.5	7568.0

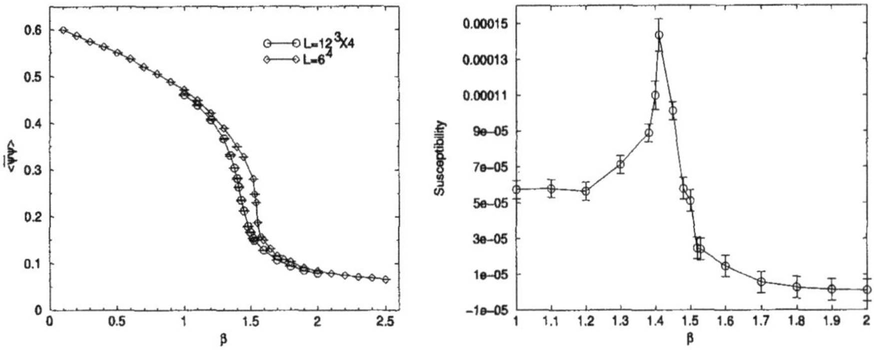

(a) (b)

Figure 1: (a) $< \bar{\psi}\psi >$ vs. β for $L = 6^4$(diamonds) and $L = 12^3 \times 4$(circles), (b) Susceptibility of $< \bar{\psi}\psi >$ vs. β for $L = 12^3 \times 4$.

4 RESULTS

Firstly we determine phase transition point β_c at $\mu = 0$ on $12^3 \times 4$ lattice.

Chiral condensate $< \bar{\psi}\psi >$ versus β is shown in figure 1(a). Chiral restoration occurs at around $\beta \approx 1.4$. In order to determine the critical point more precisely, susceptibility of the condensate is analyzed as shown in figure 1(b). By this analysis, the critical point for $m = 0.07$ is determined as

$$\beta_c = 1.41 \pm 0.03 \tag{5}$$

This value is slightly smaller than that of finite size crossover transition point $\beta_c = 1.54$ on 6^4 lattice.

Nextly, we study quark condensate and pseudoscalar meson mass at small but finite μ. We stay in confinement phase ($\beta = 1.1$). In figure 2(a), we plot the $< \bar{\psi}\psi >$ as a function of μ ($0.0 \sim 0.2$) at several m (0.05,0.07,0.1). As shown in the figure, the condensate decreases with the chemical potential. But it varies very weakly around $\mu = 0$.

Pseudoscalar meson mass is extracted in a standard way. Typical examples of pseudoscalar correlator is shown in figure 2(b). In the present region of study, single pole fit gives reasonable fitting and results for pion mass squared are shown in figure 3(a). In order to examine the influence of chemical potential on hadron mass, we examine pseudoscalar meson consisting of light and heavier quarks. figure 3(b) shows their masses at small μ. As shown in the figure, they are very stable and no appreciable dependence on chemical potential is seen in this region.

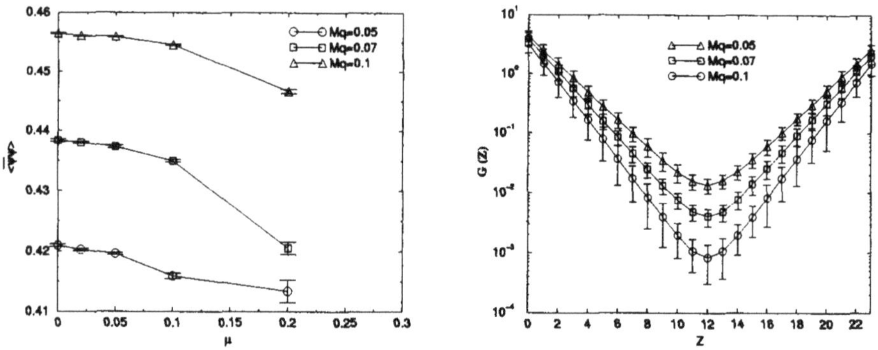

(a) (b)

Figure 2: (a) $<\bar{\psi}\psi>$ vs. μ for $m_q = 0.05, 0.07, 0.1$; $\beta = 1.1$. (b) correlation function of $<\bar{q}q>$ vs. β for $\beta = 1.1$. $\mu = 0.1$, $m_q = 0.05, 0.07, 0.1$.

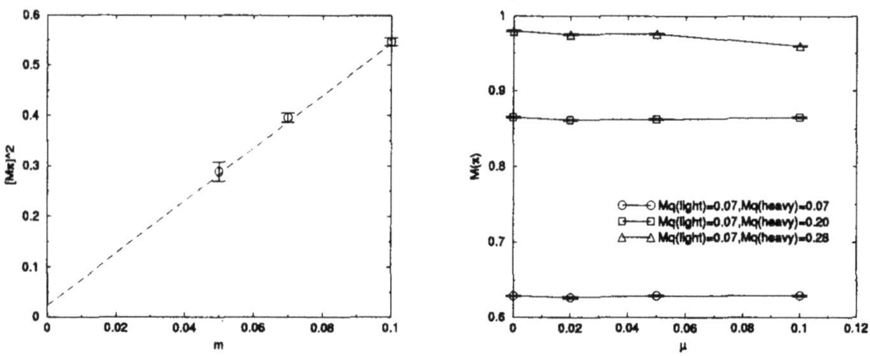

(a) (b)

Figure 3: (a) ${M_\pi}^2$ vs. m_q for $\beta = 1.1$, $\mu = 0$; (b) M_π vs. μ for $\beta = 1.1$,

5 SUMMARY AND CONCLUSIONS

In this paper, we study 8 flavors SU(2) QCD at finite temperature and small μ on $12^3 \times 4$ and $12^2 \times 24 \times 4$ lattices. The chiral transition at $\mu = 0$ is determined to be $\beta_c = 1.41$. Chiral condensate decreases with μ but variation is very weak around $\mu = 0$. Mass of pseudoscalar meson consisting of light and heavier quarks is stable against chemical potential.

References

1. M. Alford and F. Wilczek, *Phys. Lett.* B **422**, 247 (1999) ;
   M. Alford and F. Wilczek, *Nucl. Phys.* B **537**, 443 (1999).
2. R. Rapp, T. Schäfer, E. V. Shuryak, M. Velkovsky, *Phys. Rev. Lett.* **81**, 53 (1998).
3. D. Bailin, and A. Love , *Phys. Rept.* **107**, 325 (1984);
   M. Alford, K.Rajagopal, F. Wilczek, *Nucl. Phys.* B **537**, 215 (1999).
4. *see for a review* K. Kanaya, *Nucl. Phys.* B **47**, 47 (1996).
5. A. Nakamura, *Phys. Lett.* B **149**, 391 (1984).
6. S. Duane and J. B. Kogut, *Nucl. Phys.* B **275**, 398 (1986).
7. S. Duane, A. D. Kennedy, B. J. Pendleton, D. Roweth, *Phys. Lett.* B **195**, 216 (1987).
8. C.T.H. Davies, E. G. Klepfish, *Phys. Lett.* B **256**, 68 (1991).
9. S. Gottlieb, W. Liu, R. L. Renken, R. L. Sugar, D. Toussaint *Phys. Rev.* D **38**, 2888 (1988).
10. QCD-TARO Collaboration, *Nucl. Phys.* B **83-84**, 408 (2000).
11. E. B. Gregory, S-H. Guo, H. Kröger, X-Q. Luo, *Phys. Rev.* D **62**, 054508 (2000).
12. S. Hands, J. B. Kogut, M.-P. Lombardo, S. E. Morrison, hep-lat/9902034; S. Hands, S. E. Morrison, hep-lat/9905021; S. Hands, I. Montvay, S. E. Morrison, M. Pevers, L. Scorzato, J. Skullerud, hep-lat/0006018.
13. I. Montvay, *Nucl. Phys.* B **466**, 259 (1996).
14. T.Hatsuda, S-H.Lee, *Phys. Rev.* C **46**, 34 (1992).

QCD AT FINITE DENSITY

XIANG-QIAN LUO, ERIC B. GREGORY, SHUO-HONG GUO
Department of Physics, Zhongshan University Guangzhou 510275, China

HELMUT KRÖGER
Département de Physique, Université Laval, Québec, Québec G1K 7P4, Canada

At sufficiently high temperature and density, quantum chromodynamics (QCD) predicts phase transition from the hadronic phase to the quark-gluon plasma phase. Lattice QCD is the most useful tool to investigate this critical phenomenon, which status is briefly reviewed. The usual problem in the Lagrangian formulation at finite density is either an incorrect continuum limit or its complex action and a premature onset of the transition as the chemical potential is raised. We show how the difficulties are overcome in our Hamiltonian approach.

1 Introduction

1.1 Motivation

According to the standard model of cosmology, 12-15 billion years ago, the early universe underwent a series of drastic changes. Several microseconds after the big bang it was in a hot and dense quark-gluon plasma (QGP) state, where quarks and gluons were deconfined. Several minutes later, it changed to a phase, with the quarks and gluons confined inside the hadrons. Today it remains in the low temperature and low density hadronic phase. The ultimate goal of machines such as the Relativistic Heavy Ion Collider (RHIC) at BNL and the Large Hadron Collider (LHC) at CERN is to create the high temperature QGP phase and replay the birth and evolution of the universe. In February 2000, physicists at CERN declared that they had experimentally found a new state of matter [1], which had never seen before. Determining whether it is a QGP state requires further investigation from the future RHIC or LHC experiments. QGP may also exist at low temperature, in the core of very dense stars such as neutron stars.

Quantum chromodynamics (QCD) is the fundamental theory of quarks and gluons. Figure 1 shows a recently proposed[2] two-flavor QCD phase diagram on the temperature T and chemical potential μ plane. It assembles the information accumulated from lattice QCD, random matrix model, and other approximation methods. There is confinement and spontaneous chiral-symmetry breaking in the hadronic phase, which is separated from other phases by a chiral phase transition. The transition changes from second to first order at a tricritical point. At high densities and low temperatures, a superconducting phase "2SC" was recently found[2] in which up and down quarks with two out of three colors form Cooper-pairs characterized by a diquark condensate $<\psi\psi>$. The transition between this 2SC phase and the QGP phase is likely first order. The chiral phase transition on the horizontal axis between the hadronic and 2SC phases is first order.

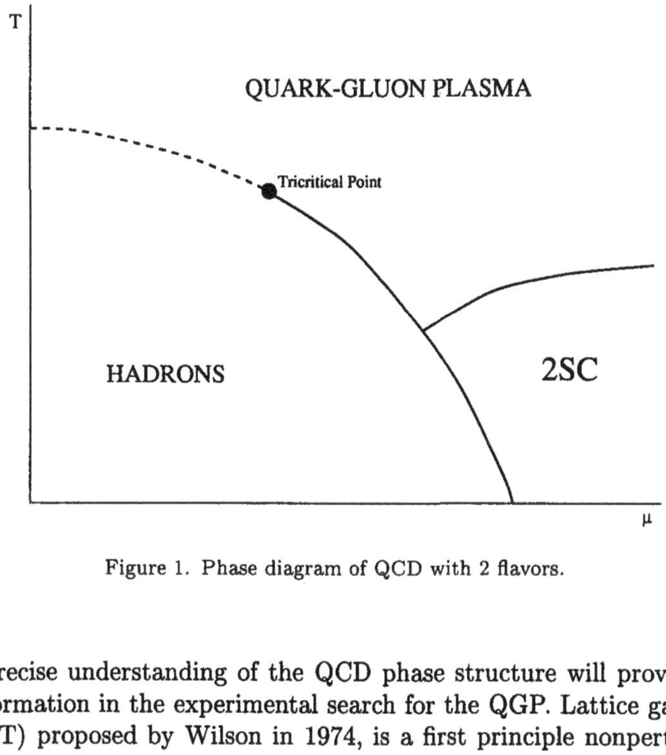

Figure 1. Phase diagram of QCD with 2 flavors.

A precise understanding of the QCD phase structure will provide valuable information in the experimental search for the QGP. Lattice gauge theory (LGT) proposed by Wilson in 1974, is a first principle nonperturbative technique for the investigation of phase transitions. There are no free parameters in LGT when the continuum limit is taken, in contrast with other nonperturbative techniques. Although the standard lattice Lagrangian Monte Carlo method works very well for QCD at finite temperature, it unfortunately breaks down at non-zero chemical potential (due to the so-called complex action problem). This is briefly reviewed in Sect. 1.2. In Sect. 2, we present our recently developed Hamiltonian approach [3] to lattice QCD at finite density, by introducing the chemical potential in a natural way, which avoids those usual problems.

1.2 Present status

In the continuum, the grand canonical partition function of QCD at finite temperature T and chemical potential μ is given by

$$Z = \mathrm{Tr}\ \mathrm{e}^{-\beta(\mathrm{H}-\mu\mathrm{N})}, \quad \beta = (\mathrm{k_B T})^{-1}, \tag{1}$$

where k_B is the Boltzmann constant, H is the Hamiltonian, and N is particle number operator

$$N = \int d^3x\ \psi^\dagger(x)\psi(x). \tag{2}$$

The energy density of the system with free quarks is given by [4]

$$\epsilon = \frac{1}{V}\frac{1}{Z}\mathrm{Tr\ H\ e}^{-\beta(\mathrm{H}-\mu\mathrm{N})} = -\frac{1}{\mathrm{V}}\frac{\partial \ln \mathrm{Z}}{\partial\beta}|_{\mu\beta}. \tag{3}$$

Going over to $T \to 0$ and chiral limit $m \to 0$, one gets the energy density (where the contribution of $\mu = 0$ is subtracted)

$$\epsilon_{sub} = \frac{\mu^4}{4\pi^2}. \tag{4}$$

Unfortunately, in Lagrangian LGT, a naive discretization of the chemical potential term does not lead to the correct continuum relation Eq.(4). Let us take the naive fermions as an example. The action reads

$$S_f = a^4\sum_x m\bar{\psi}(x)\psi(x) + \frac{a^3}{2}\sum_{x,k}\bar{\psi}(x)\gamma_k\psi(x+\hat{k}) + a^4\mu\sum_x \psi^\dagger(x)\psi(x), \tag{5}$$

where $\gamma_{-k} = -\gamma_k$. In the chiral limit $m \to 0$ and the continuum limit $a \to 0$, the subtracted energy density from this action is $\epsilon_{sub} \propto (\mu/a)^2$, i.e. becoming quadratically divergent [5], and therefore it is inconsistent with the continuum result of Eq.(4). This problem is not due to the species doubling of naive fermions, because the case of Kogut-Susskind fermions or Wilson fermions is similar.

Hasenfratz and Karsch [5] proposed to introducing the chemical potential exponentially

$$S_f = a^4\sum_x m\bar{\psi}(x)\psi(x) + \frac{a^3}{2}\sum_x\sum_{j=1}^{3}\left[\bar{\psi}(x)\gamma_j\psi(x+\hat{j}) - \bar{\psi}(x+\hat{j})\gamma_j\psi(x)\right]$$
$$+ \frac{a^3}{2}\sum_x\left[e^{\mu a}\bar{\psi}(x)\gamma_4\psi(x+\hat{4}) - e^{-\mu a}\bar{\psi}(x+\hat{4})\gamma_4\psi(x)\right]. \tag{6}$$

The chemical potential can be introduced analogously for KS as well as for Wilson fermions. Such treatment of the chemical potential is numerically feasible in the quenched approximation (where the fermionic determinant $\det \Delta$ is constraint to be 1, and quark loops are suppressed). However, there is evidence [6] that the quenched approximation produces an unphysical onset of the critical chemical potential at the value $\mu_C = M_\pi(m \neq 0)/2$, being in conflict with other theoretical predictions $\mu_C \approx M_N^{(0)}/3$ ($M_N^{(0)}$ is the nucleon mass at $\mu = 0$ and $M_\pi(m \neq 0)$ is the pion mass at finite bare quark mass m. A finite bare quark mass must be introduced in most of the numerical simulations).

For full QCD, the fermionic degrees of freedom have to be integrated out. In the measure occurs the fermionic determinant $\det \Delta$. For finite chemical potential $\det \Delta$ from Eq. (6) becomes complex, i.e, the effective fermionic action becomes complex, which renders numerical simulations extremely difficult. Much effort has been made to solve this notorious complex action problem:

(1) The Glasgow group has suggested to treat $\det \Delta$ as observable[7]. This method requires a very large number of configurations, in particular for $\mu \approx \mu_C$. Even on a very small lattice $V = 4^4$, the computational costs exceed the current computer capacity [8]. Even in this case, the unphysical onset of μ_C still exists. Therefore, it is unclear whether the onset is an intrinsic problem of the proposal Eq. (6).

(2) In the imaginary chemical potential method [9] $\det \Delta$ becomes real, which works well for numerical simulations at high temperature and low density. But it might not work at low temperature and high density.

(3) It has been proposed to utilize a special symmetry [10]. This is the only successful method in Lagrangian formulation, but it works only for the SU(2) gauge group.

(4) In [11], the probability distribution function method [12] is applied to 2-color QCD (free of complex action problem), and very interesting results for the diquark condensate $< \psi\psi >$ are obtained.

2 Hamiltonian Approach

2.1 *Free fermions at zero chemical potential*

The lattice Hamiltonian describing noninteracting Wilson fermions in $d+1$ dimensions at $\mu = 0$ is

$$H = \sum_x m\bar{\psi}(x)\psi(x) + \sum_{x,k} \frac{1}{2a}\bar{\psi}(x)\left(\gamma_k\psi(x+\hat{k}) + r\left(\psi(x) - \psi(x+\hat{k})\right)\right). \quad (7)$$

The up and down components of ψ are coupled via the γ_k matrices, and one can use a unitary transformation [13]

$$H' = \exp(-iS)\ H\ \exp(iS) \tag{8}$$

to decouple them. The operator S can be computed explicitly [13] in momentum space:

$$S = -\sum_p \frac{\theta_p}{A_p} \sum_{j=1}^{d} \psi_p^\dagger \gamma_j \psi_p \frac{\sin p_j a}{a}, \quad A_p = \left(\sum_{j=1}^{d} \left(\frac{\sin p_j a}{a} \right)^2 \right)^{1/2} . \tag{9}$$

The physical vacuum state of H reads

$$|\Omega\rangle = \exp(iS)|0\rangle, \tag{10}$$

where $|0>$ is the bare vacuum state defined as $\xi|0\rangle = \eta|0\rangle = 0$. The parameter θ_p in Eq. (9) is so determined that the vacuum energy $E_\Omega = \langle\Omega|H|\Omega\rangle = \langle 0|H'|0\rangle$ is minimized. After the unitary transformation, the fermionic field ψ can be simply expressed by up and down 2-spinors ξ and $\eta^\dagger$ as $\psi = \begin{pmatrix} \xi \\ \eta^\dagger \end{pmatrix}$ and H' is diagonal

$$H' = \sum_p A'_p \bar{\psi}_p \psi_p = \sum_p \left(\left(m + \frac{2r}{a} \sum_{j=1}^{d} \sin^2 (p_j a/2) \right)^2 + A_p^2 \right)^{1/2} \bar{\psi}_p \psi_p . \tag{11}$$

For Wilson fermions, in the continuum limit $a \to 0$, for any finite momentum p, we have $E_\Omega = -2N_c N_f \sum_p \sqrt{m^2 + p^2}$, giving the correct dispersion relation. Here N_c and N_f, respectively, are the number of colors and number of flavors.

2.2 Free fermions at nonzero chemical potential

Now we naturally introduce the chemical potential

$$H_\mu = H - \mu N, \tag{12}$$

where H is given by Eq.(7) and N is given by Eq.(2). Let us define the state $|n_p, \bar{n}_p\rangle$ by

$$\xi_p|0_p, \bar{n}_p\rangle = 0, \quad \xi_p^\dagger|0_p, \bar{n}_p\rangle = |1_p, \bar{n}_p\rangle, \quad \xi_p|1_p, \bar{n}_p\rangle = |0_p, \bar{n}_p\rangle, \quad \xi_p^\dagger|1_p, \bar{n}_p\rangle = 0,$$

$$\eta_p|n_p, 0_p\rangle = 0, \quad \eta_p^\dagger|n_p, 0_p\rangle = |n_p, 1_p\rangle, \quad \eta_p|n_p, 1_p\rangle = |n_p, 0_p\rangle, \quad \eta_p^\dagger|n_p, 1_p\rangle = 0.$$

The numbers n_p and $\bar{n}_p$ take the values 0 or 1 due to the Pauli principle. By definition, the up and down components of the fermion field are decoupled.

Obviously, this is not an eigenstate of H_μ due to the non-diagonal form of H. However, they are eigenstates of H'_μ, which are related to H_μ by a unitary transformation

$$H'_\mu = \exp(-iS)\ H_\mu\ \exp(iS) = H' - \mu N. \tag{13}$$

For the vacuum eigenstate of H_μ we make an ansatz of the following form

$$|\Omega\rangle = \exp(iS) \sum_p f_{n_p,\bar{n}_p} |n_p, \bar{n}_p\rangle. \tag{14}$$

S is given by Eq.(9), and H' is given by Eq. (11). The vacuum energy is $E_\Omega = 2N_cN_f \sum_p C_{n_p,\bar{n}_p} \left[\left(A'_p - \mu\right) n_p + \left(A'_p + \mu\right) \bar{n}_p - A'_p - \mu \right]$. Here we have introduced the notation $C_{n_p,\bar{n}_p} = f^2_{n_p,\bar{n}_p}$, which have not yet been specified. For this purpose we use the stability condition of the vacuum. Because $\mu > 0$, the vacuum energy increases with n_p. This means the vacuum is unstable unless $\bar{n}_p = 0$. This simplifies the vacuum energy to $E_\Omega = 2N_cN_f \sum_p \left[C_{1_p} \left(A'_p - \mu\right) - A'_p - \mu \right]$, where we use the abbreviation $C_{n_p} = C_{n_p,0}$ and the normalization condition. The dependence of C_{1_p} on the value of μ can be seen by inspection of the derivative $\partial E_\Omega / \partial C_{1_p} = 2N_cN_f \left(A'_p - \mu\right)$. For $\mu > A'_p$, the right-hand side is negative. Maximizing C_{1_p} means minimizing the vacuum energy. Therefore, $C_{1_p} = 1$. On the other hand, for $\mu < A'_p$, the right-hand side is positive and for any C_{1_p} the vacuum is unstable. Therefore, $C_{1_p} = 0$. We can summarize these properties by writing $C_{1_p} = \Theta\left(\mu - A'_p\right)$. Thus the vacuum energy becomes $E_\Omega = 2N_cN_f \sum_p \left(C_{1_p} A'_p - A'_p\right)$. The subtracted energy density reads

$$\epsilon_{sub} = \frac{E_\Omega - E_\Omega|_{\mu=0}}{N_cN_fN_s} = \frac{\mu^4}{4\pi^2}. \tag{15}$$

Here N_s is the number of spatial lattice sites. Thus we have proven that our Hamiltonian approach to free quarks at finite chemical potential leads to the correct continuum result for the vacuum energy density, Eq. (4). From this relation, we can easily see that the free quark number density is proportional to μ^3. For naive fermions, in the continuum limit $a = 0$, there will be an extra factor of 2^d.

2.3 Strong coupling QCD at nonzero chemical potential

As is well known, lattice QCD at $\mu = 0$ confines quarks and spontaneously breaks chiral symmetry. For a sufficiently large chemical potential, this picture may change. Here we set out to investigate finite density QCD in the strong

Γ_A	1	γ_j	γ_4	γ_5	$i\gamma_4\gamma_5$	$i\gamma_4\gamma_j$	$i\epsilon_{jj_1j_2}\gamma_{j_1}\gamma_{j_2}$	$i\epsilon_{jj_1j_2}\gamma_4\gamma_{j_1}\gamma_{j_2}$
L_A	1	$2\delta_{k,j}-1$	-1	-1	1	$1-2\delta_{k,j}$	$2\delta_{k,j}-1$	$1-2\delta_{k,j}$

Table 1. Γ matrices and coefficients.

coupling regime $1/g^2 << 1$. Following reference[13], H' in Eq. (13) is now replaced by

$$H' = \left[m\left[1-(2\theta_0)^2 d\right]+\frac{(2\theta_0)\, d}{a}\right]\sum_x \bar{\psi}(x)\psi(x) + \frac{g^2 C_N d\,(2\theta_0)^2}{4a}\sum_x \psi^\dagger(x)\psi(x)$$
$$- \frac{g^2 C_N\,(2\theta_0)^2}{32aN_c}\sum_{x,k} L_A\psi^\dagger_{f_1}(x)\Gamma_A\psi_{f_2}(x)\psi^\dagger_{f_2}(x+k)\Gamma_A\psi_{f_1}(x+k). \qquad (16)$$

The additional four-fermion term is induced by gauge interactions with fermions. There $d=3$ denotes the spatial dimension, f_1, f_2 are flavor indices (summation over repeated indices is understood), $\theta_0 = 1/(4ma+g^2C_N)$, and $C_N = (N_c^2-1)/(2N_c)$. The matrices Γ_A and their coefficients L_A are given in Tab.[1].

The unitary transformed Hamiltonian Eq. (16) can also be re-expressed in terms of the following pseudoscalar and vector operators [13]

$$\Pi = \frac{1}{2\sqrt{-\bar{v}}}\psi^\dagger\,(1-\gamma_4)\,\gamma_5\psi,$$
$$V_j = \frac{1}{2\sqrt{-\bar{v}}}\psi^\dagger\,(1-\gamma_4)\,\gamma_j\psi, \qquad (17)$$

as [3]

$$H' = E_\Omega^{(0)} + G_1\sum_p\left(\Pi^\dagger(p)\Pi(p) + \sum_j V_j^\dagger(p)V_j(p)\right)$$
$$+ G_2\sum_p\left(\Pi^\dagger(p)\Pi^\dagger(-p)+h.c.\right)\sum_j \cos p_j a$$
$$+ G_2\sum_{p,j}\left(V_j^\dagger(p)V_j^\dagger(-p)+h.c.\right)\left(\sum_{j'}\cos p_{j'}a - 2\cos p_j a\right), \qquad (18)$$

where

$$E_\Omega^{(0)} = N_f N_s\left[m\left[1-(2\theta_0)^2 d\right] + \frac{(2\theta_0)\, d}{a}\right]\bar{v}$$

$$+ N_f N_s \frac{g^2 C_N d (2\theta_0)^2}{4a} v^\dagger - N_f N_s \frac{g^2 C_N (2\theta_0)^2 d}{16 a N_c} \left(v_2^\dagger - \bar{v}_2 \right),$$

$$G_1 = 2\left[m \left[1 - (2\theta_0)^2 d \right] + \frac{(2\theta_0) d}{a} \right] + \frac{g^2 C_N d (2\theta_0)^2}{4 a N_c} \bar{v},$$

$$G_2 = -\frac{g^2 C_N (2\theta_0)^2}{8 a N_c} \bar{v},$$

$$v^\dagger = \frac{2N_c}{N_s} \sum_p C_{n_p, \bar{n}_p} (n_p - \bar{n}_p + 1),$$

$$\bar{v} = \frac{2N_c}{N_s} \sum_p C_{n_p, \bar{n}_p} (n_p + \bar{n}_p - 1),$$

$$v_2^\dagger = \frac{(2N_c)^2}{N_s} \sum_p C_{n_p, \bar{n}_p} (n_p - \bar{n}_p + 1)^2,$$

$$\bar{v}_2 = \frac{(2N_c)^2}{N_s} \sum_p C_{n_p, \bar{n}_p} (n_p + \bar{n}_p - 1)^2.$$

Eq. (18) can be diagonalized by a Bogoliubov transformation [13]

$$P(p) = \cosh u(p) a(p) + \sinh u(p) a^\dagger(-p),$$
$$V_j(p) = \cosh v_j(p) b(p) + \sinh v_j(p) b^\dagger(-p),$$

$$\tanh 2u(p) = \frac{-2G_2}{G_1} \sum_j \cos p_j a,$$

$$\tanh 2v_j(p) = \frac{-2G_2}{G_1} \left(\sum_{j'} \cos p_{j'} a - 2 \cos p_j a \right).$$

The Bogoliubov transformed Hamiltonian eventually becomes

$$H'' = E_\Omega^{(0)} + \frac{N_f^2}{2} G_1 \sum_p \left[\sqrt{1 - \tanh^2 2u(p)} - 1 \right]$$
$$+ \frac{N_f^2}{2} G_1 \sum_{p,j} \left[\sqrt{1 - \tanh^2 2v_j(p)} - 1 \right]$$
$$+ G_1 \sum_p \sqrt{1 - \tanh^2 2u(p)} a^\dagger(p) a(p)$$
$$+ G_1 \sum_{p,j} \sqrt{1 - \tanh^2 2v_j(p)} b_j^\dagger(p) b_j(p). \tag{19}$$

Using the notation, normalization condition and arguments for the coefficient as in Sect. 2.2, we obtain $\bar{n}_p = 0$ and $C_{1_p} = \Theta\left(\mu - m_{dyn}^{(0)}\right)$. Here $m_{dyn}^{(0)} = d/(ag^2 C_N)$ is the dynamical quark mass at $\mu = 0$. Therefore, in the chiral limit $m = 0$, the vacuum energy is

$$\begin{aligned} \frac{E_\Omega}{2N_c N_f N_s} &= \left(m_{dyn}^{(0)} - \mu\right)\Theta\left(\mu - m_{dyn}^{(0)}\right) - m_{dyn}^{(0)} - \mu \\ &+ \frac{N_f}{2N_s} G_1 \sum_p \left[\sqrt{1 - \tanh^2 2u(p)} - 1\right] \\ &+ \frac{N_f}{2N_s} G_1 \sum_{p,j} \left[\sqrt{1 - \tanh^2 2v_j(p)} - 1\right]. \end{aligned} \tag{20}$$

According to the Feynman-Hellmann theorem, the chiral condensate is related to the ground state energy by

$$\langle\bar{\psi}\psi\rangle = \frac{1}{N_f N_s} \lim_{m\to 0} \frac{\partial E_\Omega\,(m \neq 0)}{\partial m} = \langle\bar{\psi}\psi\rangle^{(0)}\left[1 - \Theta\left(\mu - m_{dyn}^{(0)}\right)\right], \tag{21}$$

where $\langle\bar{\psi}\psi\rangle^{(0)}$ is the chiral condensate at $\mu = 0$

$$\langle\bar{\psi}\psi\rangle^{(0)} = -2N_C\left(1 - \frac{4d}{g^4 C_N^2}\right)\left(1 - \frac{N_f}{N_c} I_1 - \frac{N_f}{N_c} I_2\right) \tag{22}$$

and for $d = 3$, $I_1 = 0.078354 \pm 2 \times 10^{-6}$, $I_2 = 0.235075 \pm 4 \times 10^{-6}$. According to Eq. (21), for $\mu < m_{dyn}^{(0)}$, $\langle\bar{\psi}\psi\rangle = \langle\bar{\psi}\psi\rangle^{(0)} \neq 0$, i.e., chiral symmetry is spontaneously broken. For $\mu > m_{dyn}^{(0)}$, $\langle\bar{\psi}\psi\rangle = 0$, i.e., chiral symmetry is restored. Therefore, there is a first order chiral phase transition and the critical value of μ is given by

$$\mu_C = m_{dyn}^{(0)} = \frac{d}{g^2 C_N a}, \tag{23}$$

The critical chemical potential μ_C is equal to the dynamical quark mass at $\mu = 0$, which agrees with the result from an entirely different method [15] in Hamiltonian formulation. (The authors argued this was a second order phase transition, in contrast we clearly observe a first order transition). Our result is consistent with other theoretical predictions $\mu_C \approx M_N^{(0)}/3$, because (see below) at $\mu = 0$ holds $M_N^{(0)} \approx 3m_{dyn}^{(0)}$.

We can compute now the quark number density in the chiral limit $m = 0$, which yields

$$n_q = \frac{-1}{2N_c N_f N_s}\frac{\partial E_\Omega}{\partial \mu} - 1 = \frac{\langle\Omega|\sum_x \psi^\dagger(x)\psi(x)|\Omega\rangle}{2N_c N_f N_s} - 1 = \Theta\left(\mu - \mu_C\right), \tag{24}$$

This is consistent with the $\beta = 0$ simulation results described in [16], and however, is different from the large μ behavior in the continuum (*i.e.* the Stefann-Boltzmann law $n_q \propto \mu^3$ for free quarks). It remains to be seen whether higher order $1/g^2$ calculations will improve this behavior.

The quark number susceptibility, standing for the response of the quark number density to infinitesimal changes in μ, is

$$\chi_q = \frac{\partial n_q}{\partial \mu} = \delta(\mu - \mu_C). \tag{25}$$

Finally, let us look at some implications on the thermal mass spectrum of the pseudoscalar meson, vector meson and nucleon. The thermal mass is defined by $M_h^\star = \langle h|H - \mu N|h\rangle - E_\Omega$. For the pseudoscalar meson, in the chiral limit $m = 0$,

$$M_\pi^\star = G_1\sqrt{1 - \tanh^2 2u(p=0)} = \begin{cases} 0 & \text{for} \quad \mu < \mu_C, \\ 4m_{dyn}^{(0)} & \text{for} \quad \mu > \mu_C. \end{cases} \tag{26}$$

Therefore, in the broken phase, the pseudoscalar is a Goldstone boson ($M_\pi^\star \propto \sqrt{m} \to 0$), and in the symmetric phase, it is no longer a Goldstone boson. For the vector meson,

$$M_V^\star = G_1\sqrt{1 - \tanh^2 2v_j(p=0)} = \begin{cases} M_V^{(0)} & \text{for} \quad \mu < \mu_C, \\ 4m_{dyn}^{(0)} & \text{for} \quad \mu > \mu_C. \end{cases} \tag{27}$$

where $M_V^{(0)} = 4\sqrt{d-1}/(ag^2 C_N)$ is the vector mass at $\mu = 0$. Therefore, $\partial M/\partial\mu \propto \delta(\mu - \mu_C)$ for the pseudoscalar and vector mesons. To see the critical behavior at zero temperature, one should be very close to μ_C. This behavior is consistent with that of the quark number density. To see whether the meson thermal masses depend on μ, higher order $1/g^2$ corrections must be included.

For the nucleon, we obtain the expected behavior

$$M_N^\star = M_N^{(0)} - 3\mu \tag{28}$$

for $\mu < \mu_C$, where $M_N^{(0)} \approx 3m_{dyn}^{(0)}$. This leads to $M_N^\star = 0$ at $\mu = \mu_C$.

3 Outlook

In this paper, we have developed a Hamiltonian approach to lattice QCD at finite density. The chemical potential is introduced in a very natural way as in the continuum. It avoids the persisting problem in the Lagrangian approach of an incorrect continuum limit, or complex action or a premature onset of the transition to nonzero quark density as μ is raised. The main result in the free case is given by Eq. (15), and those in the strong coupling regime are given by Eqs. (20)-(28). We have seen that the approach works well in the free case and also in the strong coupling regime. We predict that at strong coupling, the chiral transition is of first order, and the critical chemical potential $\mu_C \approx M_N^{(0)}/3$.

We have not yet specified the nature of the chiral-symmetric phase for $\mu > \mu_C$. Is it a QGP phase or a color-superconduction phase [2] (see Fig. 1)? Up to now, there has been no first principle investigation of such a phase. Better understanding of it will affect our knowledge of the formation of the neutron star.

At this point, we should mention that the results of Hamiltonian formulation and Lagrangian formulation [17] might be quite different. It is interesting to see whether they describe the same physics in the continuum. For pure gauge theory, within a Hamiltonian approach, we can extend to the intermediate coupling and obtain meaningful results for the glueballs [18]. For fermions, the calculation is far from trivial. Recently we proposed a Monte Carlo technique in the Hamiltonian formulation [19] for the purpose to do nonperturbative numerical simulations, by combining the virtues of the Monte Carlo algorithm with importance sampling and the Hamiltonian approach. We hope to apply it to QCD and with the aim to obtain useful information for RHIC and LHC physics and neutron star phenomenology.

Acknowledgments

We are grateful to the attendees of the workshop for useful discussions. X.Q.L. is supported by the National Science Fund for Distinguished Young Scholars (19825117), National Natural Science Foundation (19677205), the Ministry of Education, the Doctoral Program of Higher Education and Guangdong Provincial Natural Science Foundation (990212) of China. X.Q.L. and E.B.G. are supported by the Guangdong National Communication Ltd. H.K. has been supported by NSERC Canada.

References

1. http://press.web.cern.ch/.
2. K. Rajagopal, Nucl. Phys. **A661**, 150 (1999), and references there in.
3. E.B. Gregory, S. Guo, H. Kröger, X.Q. Luo, Phys. Rev. **D62**, 054508 (2000).
4. J.R. Li, *Introduction to Quark Matter Theory*, (Hunan Education Press, 1989); J. Kapusta, *Finite Temperature Field Theory*, (Cambridge Uni-

versity Press, Cambridge, England, 1989); B. Freedman and L. Mclerran, *Phys. Rev.* **D16**, 1130, (1977); **16** 1147 (1977); **16** 1169 (1977).
5. P. Hasenfratz and F. Karsch, Phys. Lett. **B125**, 308 (1983).
6. J. Kogut, M.P. Lombardo, and D. Sinclair, Phys. Rev. **D51**, 1282 (1995); I. Barbour, S. Morrison, E. Klepfish, J. Kogut and M. Lombardo, Phys. Rev. **D56**, 7063 (1997); M.A. Stephanov, Phys. Rev. Lett. **76**, 4472 (1996).
7. I. Barbour, C. Davies, and Z. Sabeur, Phys. Lett. **B215**, 567 (1988).
8. R. Aloisio, V. Azcoiti, G. DiCarlo, A. Galante, and A. Grillo, Phys. Lett. **B453**, 275 (1999).
9. M. Alford, A. Kapustin and F. Wilczek, Phys. Rev. **D59**, 054502 (1999); M.P. Lombardo, Nucl. Phys. **B(Proc. Suppl.)83**. 375 (2000).
10. S. Hands, J. Kogut, M.P. Lombardo and S. Morrison, Nucl. Phys. B **558**, 327 (1999).
11. R. Aloisio, A. Galante, V. Azcoiti, G. DiCarlo, and A. Grillo, these proceedings, hep-lat/0007018.
12. V. Azcoiti, V. Laliena, and X.Q. Luo, Phys. Lett. **B354**, 111 (1995).
13. X.Q. Luo and Q. Chen, Phys. Rev. **D46** (1992) 814; Q. Chen, and X. Luo, *ibid.* **42**, 1293 (1990).
14. J. Bjorken and S. Drell, *Relativistic Quantum Mechanics* (McGraw-Hill, New York); P. Roman, *Advanced Quantum Theory* (Addison-Wesley, New York).
15. A. Le Yaouanc, L. Oliver, O. Pene, J. Raynal, M. Jarfi and O. Lazrak, Phys. Rev. **D37**, 3691 (1988).
16. R. Aloisio, V. Azcoiti, G. DiCarlo, A. Galante and A. Grillo, Nucl. Phys. **B564**, 489 (2000).
17. N. Bilic, K. Demeterfi, B. Petersson, Nucl. Phys. **B377**, 651 (1992).
18. X.Q. Luo and Q. Chen, Mod. Phys. Lett. **A11**, 2435 (1996); X.Q. Luo, Q. Chen, S. Guo, X. Fang and J. Liu, Nucl. Phys. **B(Proc. Suppl.)53**, 243 (1997).
19. H. Jirari, H. Kröger, X.Q. Luo and K. Moriarty, Phys. Lett. **A258**, 6 (1999). X.Q. Luo, H. Jirari, H. Kröger, K. Moriarty, these proceedings.

HIGH TEMPERATURE QCD AND DIMENSIONAL REDUCTION

BENGT PETERSSON

University of Bielefeld, Faculty for Physics, P.O.Box 10 01 31, D-33501 Bielefeld, Germany

E-mail: bengt@physik.uni-bielefeld.de

1 Abstract

In this talk I will first give a short discussion of some lattice results for QCD at finite temperature. I will then describe in some detail the technique of dimensional reduction, which in principle is a powerful technique to obtain results on the long distance properties of the quark-gluon plasma. Finally I will describe some new results, which test the technique in a simpler model, namely three dimensional gauge theory.

2 Introduction

The physical challenges for lattice computations of high temperature QCD are to obtain information, which can be of use for the planning and analysis of recent experiments, as well as to give theoretical insight into the non perturbative properties of QCD. Some of the more important goals are the determination of the value of the transition temperature in physical units, the equation of state, the order or even the existence of a phase transition to a high temperature quark-gluon plasma, and the long distance properties of the plasma phase.

A simple picture of the finite temperature properties of QCD, as first discussed by Cabibbo and Parisi [1], predicts a low temperature confined phase, and a high temperature phase which is essentially an ideal quark-gluon gas. The transition temperature is expected to be of the order of the Hagedorn temperature, i.e. around 140 MeV. Up to now only lattice calculations can give more quantitative information. Reviews of the lattice results can be found in the proceedings of the yearly lattice conferences, the most recent published review is reference [2]. Here I will only emphasize some of the results on the quantities mentioned above.

In the pure gluon theory it has been shown that the thermodynamics can be essentially completely solved, by continuum extrapolation of the lattice results. In this way, one can estimate the ratio between the critical temperature and the square root of the string tension to be 0.630(5).[3] A similar extrapolation using a different lattice action gives 0.650(5).[4] The difference seems to come from the measurement of the string tension, which is difficult to estimate precisely, and where systematic errors come from the uncertainties in the non asymtotic terms in the spatial distance. Still, the estimates agree within 3 %. Using a value $\sqrt{\sigma} = 425$ MeV, one obtains a rather high critical temperature of around 270 MeV.

Unfortunately, as far as the order of the transition and the critical temperature goes, the pure gluon theory is not a good approximation to full QCD with dynamical quarks. Direct lattice calculations with systematic continuum extrapolation in this case are far too time consuming for present computers, and would, with the methods currently available, need computing power in the range of 10 to 100 Teraflops. From the present results one expects the critical temperature e.g. in ratio to the string tension to be considerably lower. In physical units one estimates the critical temperature to be in the range of 150 - 190 MeV, where the latest results are near 170 MeV.[2,5] It is difficult to estimate the systematic errors. The order of the transition may be very sensitive to the strange quark. A crossover as a shadow of a nearby second or first order transition is favoured by the data, and also by general universality arguments.

In the pure gluon theory, the pressure and energy density as a function of temperature can also be extrapolated to the continuum limit, and also here results from different calculations agree within about 3%.[3,4] The main message from these calculations is that the free gluon gas limit is approached rather slowly, and that even at temperatures of two times the critical temperature, which corresponds to energy densities at least 16 times the energy density at the transition, there are 15 - 20% corrections to the ideal gas limit. Qualitatively, one observes a similar behaviour in full QCD.[6] In this case a continuum extrapolation has not been attempted, and one is restricted to results on rather coarse lattices. One should remark that the experiments planned at Brookhaven and even LHC will operate in a range of temperatures, where the deviations from the ideal gas behaviour should be substantial. A better understanding of the non perturbative effects is certainly needed. Resummation techniques of the perturbative series, essentially introducing a gluon mass, do agree with the lattice results for sufficiently high temperature (from about $2T_c$). [7,8] The systematic errors in this procedure are also not very well known, and futher investigation of the long distance properties of the plasma is certainly needed. Dimensional reduction is a technique, which is well suited for studying these long range properties. In the following I will give a short review of the general technique, and report on some recent results, which test the approach in detail.[9]

3 Dimensional Reduction

Dimensional reduction is in principle a powerful technique that was introduced in the early eighties as a mean to treat high temperature field theories. In the Euclidean formulation such theories are defined in a volume with one compact dimension of extent $1/T$. As the temperature T becomes large, one may expect that the non-static modes in the temperature direction can be neglected, and that one is left with a theory in one dimension less [10,11]. This naive reduction is, however, only exact in the classical limit. In general, one has to integrate over the nonstatic modes to obtain the effective action [12,13,14]. As was shown in Refs [12,14] the effective action for the long distance phenomena will, however, only contain a limited number of local terms at sufficiently high temperature, because the integral over the nonstatic modes does not contain infrared divergencies. Furthermore, the coefficients can be determined from a perturbative expansion of the integral over the non-static modes. This effective action is then expected to describe correctly the infrared behaviour of the full theory. It has been successfully applied to the electroweak phase transition.[15] In QCD the action is of the form

$$S = \frac{1}{g^2}\int_0^{1/T} d\tau \int d^dx \left[F_{\mu\nu}^2(\tau,\vec{x}) + \bar{\psi} D\psi\right] \tag{1}$$

where possible ghost and gauge fixing terms have been suppressed. It is convenient to analyse the theory in the static Landau gauge, defined by

$$A_0(\tau,\vec{x}) = A_0(\vec{x}) \tag{2}$$

$$\int_0^{1/T} d\tau\, \partial_i A_i = 0. \tag{3}$$

The bosonic and fermionic fields are periodic and antiperiodic respectively in the temperature direction, and the corresponding Matsubara frequencies are

$$\begin{aligned} \omega_n &= 2\pi n T \quad n = 0, \pm 1, \pm 2, \ldots && \text{Bosons} \\ \omega_n &= (2n+1)\pi T && \text{Fermions} \end{aligned} \tag{4}$$

Note that all modes become infinitely heavy in the high temperature limit, apart from the static modes in the bosonic case. If one neglects the non static modes, or at least integrate them out perturbatively, a much simpler theory describes the long distance properties of the plasma. At the tree level, one is left with the effective action in one less dimension

$$S \to \frac{1}{g^2 T} \int d^d x \left[F_{ij}^{St\,2}(\vec{x}) + \left(D_i A_0^{St}(\vec{x}) \right)^2 \right] . \tag{5}$$

In the quantum theory, one should integrate over the non static modes. This gives in principle a very complicated non local effective action. If one observes that the integration over these modes is infrared finite, one may, however, develop the action in a series of local terms, which describe the leading behaviour of correlation functions at large T and small momenta.

If we introduce a field ϕ instead of A_0

$$A_0^{St}(\vec{x}) = \sqrt{T}\,\phi(\vec{x}) \tag{6}$$

we can see essentially from power counting and symmetry arguments that the dominating terms in perturbation theory at high temperature and long distances are the first orders in a polynomial in ϕ:

$3+1 \to 3$	$2+1 \to 2$
$g^2 T^2\ \phi^2$	$g_3^2 T\ \phi^2$
$g^4 T\ \phi^4$	$g_3^4\ \phi^4$
$g^6\ \phi^6$	$g_3^6/T\ \phi^6$

The numerical value of the coefficients can be calculated in perturbation theory.

Although this is a very nice result, which means essentially that e.g. full QCD in 3+1 dimensions is replaced by a purely bosonic theory in three dimensions, the range of validity of the approximation is not yet known, although several results indicate that it may be valid down to a temperature of about two times the critical temperature. [16,17,18,19,20]

In a recent work we have investigated how dimensional reduction works, both for correlations between Polyakov loops, which are related to chromoelectric screening and for spatial Wilson loops, which are related to the chromomagnetic sector. We have performed this investigation on a simpler model, three dimensional $SU(3)$ gauge theory. Thereby we can obtain a high statistical accuracy in the comparison. The reduced model is a two dimensional adjoint Higgs model. This model is interesting in its own right, unfortunately it has not been analytically solved.

Three dimensional gauge theories have several similarities with the full four dimensional case, as e.g. confinement and asymptotic freedom. They are, however, superrenormalizable, and have infrared singularities, which are stronger than in four dimensions. Even the leading perturbative definition of the Debye screening mass is not well defined, because of a infrared logarithmic singularity.

I will not go into the technical details of the reduction, but instead I refer to our publication [9]. We have simulated the three dimensional theory in a wide range of temperature, for lattices with four steps in the temperature direction, using the Wilson action.[21] We also calculated in the one loop approximation, the effective action in the lattice regularization. Again the action can be written in terms of lattice variables. In the scaling limit we get the continuum form

$$\mathcal{L}_{eff} = \frac{1}{4}\sum_{c=1}^{8} F_{ij}^c F_{ij}^c + \mathrm{tr}\,[D_i\phi]^2 + \frac{g_2^2}{32\pi}\left(\frac{g_2}{T}\right)^2 \mathrm{tr}\ \phi^4 + \mathcal{L}_{CT}, \tag{7}$$

$$D_i\phi = \partial_i\phi + ig_2[A_i,\phi],$$

$$F_{ij} = \partial_i A_j - \partial_j A_i + ig_2[A_i, A_j],$$

$$\mathcal{L}_{CT} = -\frac{3g_2^2}{2\pi}\left[-\log(aT) + 5/2\log 2 - 1\right] \mathrm{tr}\ \phi^2. \tag{8}$$

This is a 2D, SU(3) gauge invariant Lagrangian for an adjoint scalar ϕ, but it is far from being the most general one. The gauge coupling $g_2 = \sqrt{g_3^2 T}$, with its canonical dimension one in energy, sets the scale. The non kinetic quadratic term is the counterterm $\mathcal{L}_{CT}$, suited to a lattice UV regularization with spacing a. As is explained in our paper [9], it is very important to include properly the regularized divergence in the ϕ^2 term.

On the lattice, the threedimensional model is defined on a $L_0 \times L_s^2$ lattice and the two dimensional model on a L_s^2 lattice. In the reduced model, the spatial vector fields A_i are replaced by the group elements U_i, as usual, and the space part of the kinetic term is replaced by the Wilson action. The field ϕ is kept as a scalar field defined on the sites of the lattice, and the appropriate covariant definition of the covariant derivative is used. Thereby the explicit Z_3 symmetry of the original three dimensional theory is lost. It can be restored formally, e.g. by substituting ϕ by a group element. This substitution is, however, not uniquely determined by perturbation theory. One proposal has been made recently by Pisarski.[22] We have investigated some similar Z_3 symmetric models. The hope would be to get a description,

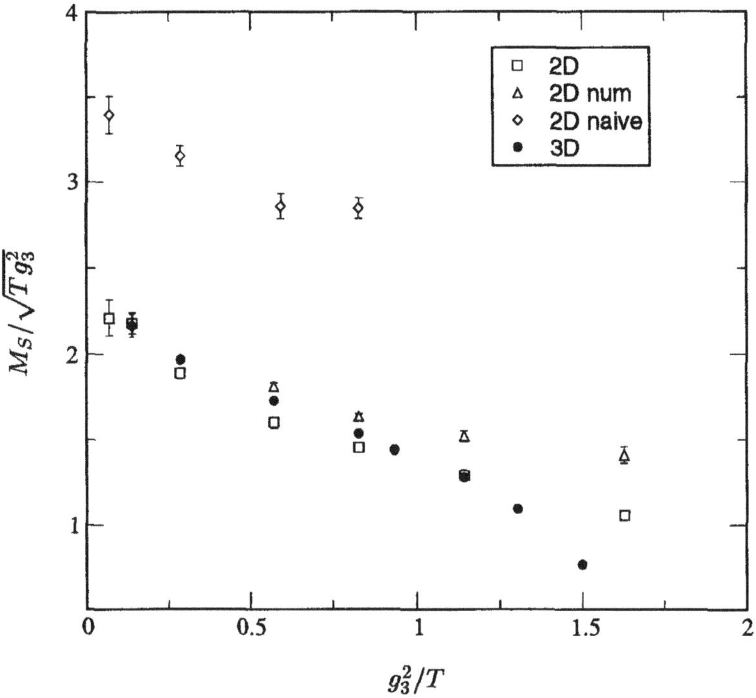

Figure 1. Physical screening masses M_S in units of $g_3\sqrt{T}$ versus g_3^2/T, in $(2+1)D$ (black points) and $2D$ (squares). Also shown are the masses obtained with the numerical value of the two point coupling (triangles) instead of its scaling form, and with the tree level reduced action(diamonds).

which is valid all the way down to the phase transition.

To the physical variables g_3 and T correspond the lattice parameters $L_0 = 1/aT$ and $\beta_3 = 6/ag_3^2$. It follows that as $a \to 0$, scaling (constant physics) corresponds to

$$L_0 \to \infty, \quad \beta_3 \to \infty, \qquad \tau \equiv \frac{\beta_3}{6L_0} = \frac{T}{g_3^2} = constant. \tag{9}$$

The dimensionless quantity τ thus measures temperature in units of the scale g_3^2, high temperature means large τ values. From numerical simulations we obtain that $T_c/g_3^2 \approx 0.61$ [21]. Also, on the lattice, L_s must be kept much larger than the largest spatial correlation length in lattice units.

The two quantities, which we have studied with precision are the correlation between Polykaov loops and the spatial string tension. In Fig. 1 is shown the screening mass, as defined from the inverse correlation length of the Polyakov loops. The details of the fit are discussed in the article. It is important to mention that the best fit is obtained by a simple pole, although in resummed perturbation theory one might expect a cut, because two gluons have to be exchanged, as the Polyakov loops are colour singlets. As can be seen from the figure the agreement between the reduced and the full model is very good down to about $1.5T_c$. Going to the critical temperature, the mass in the full model goes to zero, while in the reduced model, which does not have the corresponding phase transition, it stays finite. The data for the full model comes from lattice calculations with $L_0 = 4$. We have, however, investigated scale breaking effects in the reduced model, and found them to be small.

Another important message is that the terms coming from the quantum effects in the reduction are essential for the agreement between the reduced and the full model. The naively reduced model gives masses, which are more than 50% larger.

The spatial Wilson loop in three dimensions is not a static operator, and thus not predicted to assume the same value at high temperature and in the reduced two dimensional model. However, there is a finite string tension in both models, and one may expect that the non static corrections to this observable are small. We have also compared the string tension in the full model to the string tension in two dimensional $SU(3)$ theory, which can be easily calculated analytically. In fact one expects

$$\frac{\sigma_2^0}{g_3^2 T} = \frac{2}{3} + \frac{7}{36}\frac{g_3^2}{T}(aT)^2 + \mathcal{O}\left[(\frac{g_3^2}{T})^2 (aT)^4\right], \tag{10}$$

showing that σ_2^0 scales as $2g_3^2T/3$, up to scaling violations at finite T of order $(aT)^2 = 1/L_0^2$, where σ_2^0 is the string tension in two dimensional $SU(3)$. In Fig. 2 is shown the comparison between the spatial string tension in the full and the reduced model, as well as the analytical result for two dimensional SU(3). One should note that this is done only for a fixed value of $L_0 = 4$. In this case we have not estimated scale breaking effects.

4 Conclusions

Although the thermodynamics of pure gluon theory is essentially solved numerically, the same is not true for full QCD. Furthermore, the long distance properties are even more difficult to extract. Dimensional reduction gives potentially a simpler and systematic alternative to direct QCD calculations at

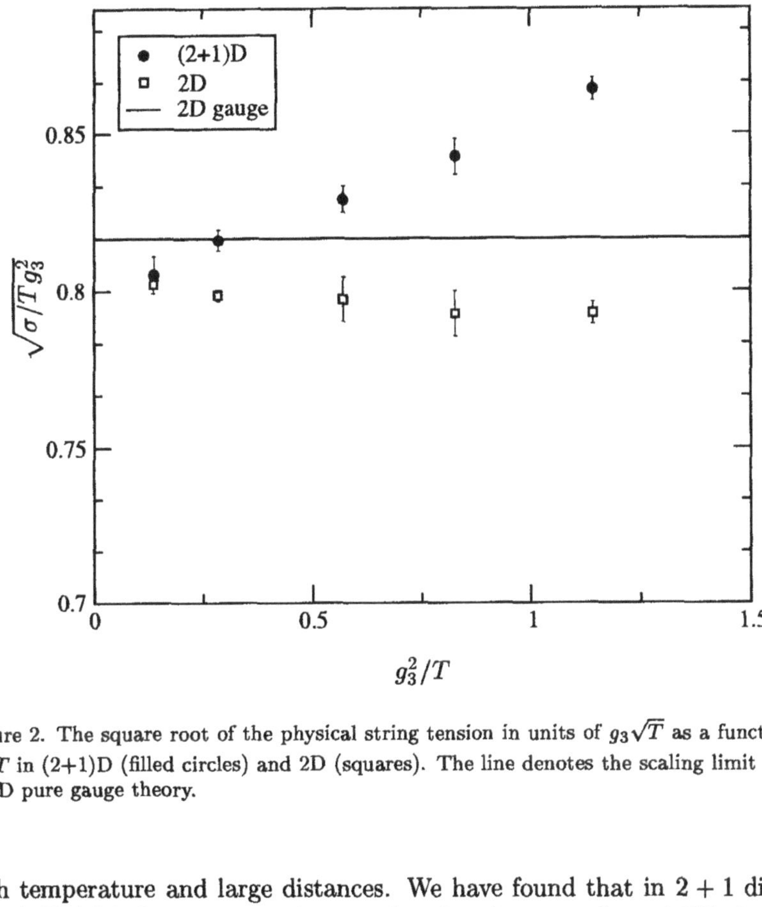

Figure 2. The square root of the physical string tension in units of $g_3\sqrt{T}$ as a function of g_3^2/T in (2+1)D (filled circles) and 2D (squares). The line denotes the scaling limit $\sqrt{2/3}$ in 2D pure gauge theory.

high temperature and large distances. We have found that in $2+1$ dimensional $SU(3)$ it is a very good approximation down to about $1.5T_c$, but the dynamics of the phase transition is not captured by the present approach. It is, however, important to notice that we can describe both chromoelectric and chromomagnetic quantities consistently, without extra parameters, which has not been shown in the $3+1$ dimensional case. Systematic extensions of the method would be very important, in particular since the potential gain in computer time is enormous.

Acknowledgments

I want to thank my collaborators P. Bialas, C. Legeland, A. Morel, K. Petrov and T. Reisz for a very stimulating collaboration, which led to the results described here. I also thank Xiang-Qian Luo and Eric Gregory for a very interesting conference.

References

1. N. Cabibbo and G. Parisi, *Phys. Lett.* B **59**, 67 (1975).
2. F. Karsch Nucl.Phys.Proc.Suppl. **83**, 14 (2000).
3. G. Boyd et al., *Phys. Rev. Lett.* **75**, 4196 (1995) and *Nucl. Phys.* B **469**, 429 (1996) B. Beinlich et al., Eur.Phys.J.C **6**, 133 (1999)
4. M. Okamoto et al., *Phys. Rev.* D **60**, 94510 (1999)
5. A. Ali Khan et al., hep-lat/0008011
6. F. Karsch, E. Laermann and A. Peikert, *Phys. Lett.* B **478**, 447 (2000).
7. J.-P. Blaizot, E. Iancu and A. Rebhan, *Phys. Rev. Lett.* **83**, 2906 (1999); *Phys. Lett.* B **470**, 181 (1999)
8. J.O. Andersen, E. Braaten and M. Strickland, *Phys. Rev. Lett.* **83**, 2139 (1999);*Phys. Rev.* D **61**, 014017,074016 (2000)
9. P. Bialas, A. Morel, B. Petersson, K. Petrov and T. Reisz, *Nucl. Phys.* B **581**, 477 (2000)
10. P. Ginsparg, *Nucl. Phys.* B **170**, 388 (1980)
11. T. Appelquist and R. Pisarski, *Phys. Rev.* D **23**, 2305 (1981)
12. S. Nadkarni, *Phys. Rev.* D **27**, 917 (1983); *Phys. Rev.* D **38**, 3287 (1988)
13. N. P. Landsman, *Nucl. Phys.* B **322**, 498 (1989)
14. T. Reisz, *Z. Phys.* C **53**, 169 (1992)
15. K. Kajantie, M. Laine, K. Rummukainen, M. Shaposhnikov, *Phys. Lett.* B **77**, 2887 (1996), *Nucl. Phys.* B **493**, 413 (1997)
    M. Gurtler, E.-M. Ilgenfritz, A. Schiller, *Phys. Rev.* D **56**, 3888 (1997)
    F. Karsch, T. Neuhaus, A. Patkos, J. Rank, Nucl.Phys.Proc.Suppl. **53**, 623 (1997)
16. P. Lacock, D. E. Miller and T. Reisz, *Nucl. Phys.* B **369**, 501 (1992)
17. L. Kärkkäinen, P. Lacock, D. E. Miller, B. Petersson and T. Reisz, *Phys. Lett.* B **282**, 121 (1992)
18. F. Karsch, M. Oevers and P. Petreczky, *Phys. Lett.* B **442**, 291 (1998)
19. S. Datta, S. Gupta, *Phys. Lett.* B **534**, 392 (1998), *Phys. Lett.* B **471**, 382 (2000)
20. A. Hart and O. Philipsen *Nucl. Phys.* B **572**, 243 (2000)
21. C.Legeland, PhD. Thesis (University of Bielefeld, Germany, September 1998).
22. R. D. Pisarski, hep-ph/0006205

5 QCD Vacuum and Topological Issues

Descending one of the Seven Star Crags in Zhaoqing.

QCD WITH A θ-VACUUM TERM: A COMPLEX SYSTEM WITH A SIMPLE COMPLEX ACTION

V. AZCOITI* AND V. LALIENA
Departamento de Física Teórica, Facultad de Ciencias, Universidad de Zaragoza, 50009 Zaragoza, Spain
E-mail: azcoiti@azcoiti.unizar.es, laliena@posta.unizar.es

A. GALANTE
Dipartimento di Fisica dell'Università di L'Aquila, 67100 L'Aquila, Italy
E-mail: galante@lngs.infn.it

We reanalyze in the first part of this paper the old question of P and CT realization in QCD. The second part is devoted to establish general results on the phase structure of this model in the presence of a θ-vacuum term.

1 Introduction

Path integral formulation of Quantum Field Theories associates, in the more standard cases, a four-dimensional classical statistical mechanics system to a given quantum system. The euclidean partition function shows a well-defined Boltzmann weight, the correlation functions are the euclidean Green functions and the free energy density is equivalent to the vacuum energy density in the hamiltonian approach.

Even if the previous statement is true for almost all physically relevant systems, there are some notorious and important exceptions. The first relevant exception is lattice formulation of Quantum Chromodynamics at finite baryon density. Since the determinant of the Dirac operator at finite chemical potential is a complex number, the euclidean action of this model is complex. This is the most standard example of a complex system with a complex action. The meaning of "complex action" in this case is twofold: first, as previously discussed, the action is a complex number and second the complex part of the action involves fermionic degrees of freedom which are Grassmann variables. It is well known that these two features have enormously delayed all attempts to analyze the behaviour of high density matter from first principles.

The second relevant exception to the equivalence between a quantum system and a classical statistical mechanics system is QCD with a topological term in the action. Due to the fact that the local density of topological

*TALK PRESENTED BY V. AZCOITI

charge picks-up a factor of i under Wick rotation, the euclidean action of this model is also complex. This is an example of a complex system with a simple complex action. In fact even if the system is complex as opposite to simple, the complex part of the action is at least much simpler that the one which appears in QCD at finite baryon density since it involves only gauge degrees of freedom. It is therefore natural to think that if we have some hope to understand QCD at finite baryon density from first principles, we should be able to understand previously a simpler case of a complex system with a complex action: QCD with a topological term in the action. This was our last but not the least motivation to analyze this system.

This paper contains two differentiate parts. In the first one we will give a new look to the old Vafa-Witten theorem on the impossibility to break spontaneously parity and CT in vector-like theories as QCD[1] and will show how an essential ingredient in the Vafa-Witten demonstration, the assumption that the free energy density exists in the presence of any external symmetry breaking field, requires the previous assumption that the symmetry is realized in the vacuum[2]. In other words, the thesis of the theorem is in some sense assumed as an hypothesis.

The second part of the paper is devoted to the analysis of the phase structure of QCD with a θ-vacuum term in the action. We will demonstrate a theorem which stays that if the θ-vacuum term is relevant, either QCD has a phase transition at $\theta = \pi$ which breaks spontaneously parity or the free energy density has some non-analyticity at some $\theta < \pi$.

2 New look to the Vafa-Witten theorem.

Let us start this section by recalling the main ingredients and steps of Vafa-Witten theorem[1] which stays that parity and CT cannot be spontaneously broken in vector-like parity conserving theories as QCD.

The theorem is based on the following two main points:

(i) The crucial observation that any arbitrary hermitian local order parameter X for parity constructed from Bose fields have to be proportional to an odd power of the four indices antisymmetric tensor $\epsilon^{\mu\nu\rho\eta}$,

(ii) The assumption that the free energy density is well defined in the presence of a symmetry breaking source λX.

The first of these two ingredients implies that any arbitrary order parameter X should contain and odd number of time derivatives plus time components of the gauge field and should pick-up therefore a factor of i under Wick rotation. This observation plus assumption (ii) makes the demonstration of the thesis actually very simple. In fact let $\mathcal{Z}$ be the euclidean partition function of QCD (or any parity conserving vector-like theory) in the presence of a local symmetry breaking source X,

$$\mathcal{Z} = \int dA^a_\mu d\bar{\psi} d\psi \exp\left(-\int d^4x \left(\mathcal{L}(x) + i\lambda X(x)\right)\right) \tag{1}$$

where $\mathcal{L}(x)$ is the standard QCD Lagrangian and $X(x)$ is a real number since we have exhibited explicitly the factor of i which arises from Wick rotation. As can be seen in (1) the symmetry breaking term is a pure phase factor in the integrand of the partition function.

Since the determinant of the Dirac operator is positive definite in a vector-like theory as QCD, the presence of a pure phase factor in the integrand of the partition function can only decrease the value of $\mathcal{Z}$. Therefore the free energy density $f(\lambda)$ defined as

$$\mathcal{Z} = e^{-Vf(\lambda)} \tag{2}$$

where V is the space-time volume, will increase with λ. The vacuum energy density, which is given by the free energy density, will also increase with λ and therefore the symmetric vacuum will be stable under small perturbations.

Before going on in a deeper analysis let us say that it is actually surprising the fundamental role played in the mechanism previously discussed by the factor of i picked-up by the order parameter under Wick rotation. In fact the standard way to analyze spontaneous symmetry breaking is to add a symmetry breaking source to the symmetric action, compute the mean value of the order parameter, take the infinite volume limit and then the limit of vanishing symmetry breaking source. If at the end of this procedure we get a non-vanishing value for the order parameter the symmetry is spontaneously broken. But it is also well known that the symmetric action contains enough information on the realization of the symmetry in the vacuum. Indeed the analysis of the probability distribution function (p.d.f.) of the order parameter in the symmetric model has been extensively and successfully employed in the investigations on spontaneous symmetry breaking in spin systems like the Ising model[3], in spin-glass models in order to analyze the complicated structure of equilibrium states not connected by symmetry transformations[4], and more recently this formalism has also been extended to quantum systems with fermionic degrees of freedom[5].

But what is even more surprising is the contradictory result we obtain if we apply Vafa-Witten argumentation to the Ising model. The Ising model is not a quantum field theory but it verifies the main requirement in Vafa-Witten theorem since the integration measure in this model is positive definite. If we add an imaginary external magnetic field to the hamiltonian of the Ising model and apply Vafa-Witten argumentation, we should conclude that the Z_2 symmetry of this model is not spontaneously broken. But this is obviously wrong since it is well known that in the low temperature phase the Ising model shows spontaneous magnetization. The solution to this paradox lies in the fact that the free energy density in the low temperature phase and for an imaginary magnetic field is not defined (it is singular on the imaginary axis of the complex magnetic field plane). The Lee-Yang zeroes of the partition function[6] live on the imaginary axis and approach the origin with velocity V, the lattice volume, forbidding to get a well defined thermodynamical limit.

We will show in the following that this is not a pathology of the Ising model but a general feature of any model with a discrete Z_2 symmetry. In other words, we will show that to assume the euclidean free energy density $f(\lambda)$ is well defined in the presence of any external symmetry breaking source requires the previous assumption that the symmetry is realized in the vacuum.

To start the proof let us come back to equation (1) which defines the euclidean path-integral formula for the partition function. Using the p.d.f. of the order parameter X we can write it as

$$\mathcal{Z}(\lambda) = \mathcal{Z}(0) \int d\tilde{X} P(\tilde{X}, V) e^{-i\lambda V \tilde{X}} \tag{3}$$

where V in (3) is the space-time volume, $P(\tilde{X}, V)$ is the p.d.f. of X at a given volume

$$P(\tilde{X}, V) = \frac{\int dA_\mu^a d\bar{\psi} d\psi e^{-\int d^4x \mathcal{L}(x)} \delta\left(\bar{X}(A_\mu^a) - \tilde{X}\right)}{\int dA_\mu^a d\bar{\psi} d\bar{\psi} e^{-\int d^4x \mathcal{L}(x)}} \tag{4}$$

and

$$\bar{X}(A_\mu^a) = \frac{1}{V} \int d^4x X(x)$$

Notice that, since the integration measure in (4) is positive or at least semi-positive definite, $P(\tilde{X}, V)$ is a true well normalized p.d.f.

Let us assume that parity is spontaneously broken. In the simplest case in which there is no an extra vacuum degeneracy due to spontaneous breakdown of some other symmetry, we will have two vacuum states as corresponds to a discrete Z_2 symmetry. Since X is an intensive operator, the p.d.f. of X will be, in the thermodynamical limit, the sum of two δ distributions:

$$\lim_{V\to\infty} P(\tilde{X},V) = \frac{1}{2}\delta(\tilde{X}-a) + \frac{1}{2}\delta(\tilde{X}+a) \tag{5}$$

At any finite volume, $P(\tilde{X},V)$ will be some symmetric function ($P(\tilde{X},V) = P(-\tilde{X},V)$) developing a two peak structure at $\tilde{X} = \pm a$ and approaching (5) in the infinite volume limit.

Due to the symmetry of $P(\tilde{X},V)$ we can write the partition function as

$$\mathcal{Z}(\lambda) = 2\mathcal{Z}(0) Re \int_0^\infty P(\tilde{X},V) e^{-i\lambda V\tilde{X}} d\tilde{X} \tag{6}$$

and if we pick up a factor of $e^{-i\lambda V a}$

$$\mathcal{Z}(\lambda) = 2\mathcal{Z}(0) Re \left(e^{-i\lambda V a} \int_0^\infty P(\tilde{X},V) e^{-i\lambda V(\tilde{X}-a)} d\tilde{X} \right) \tag{7}$$

which after simple algebra reads as follows:

$$\begin{aligned} \mathcal{Z}(\lambda)/(2\mathcal{Z}(0)) = \quad & \cos(\lambda V a) \int_0^\infty P(\tilde{X},V) \cos\left(\lambda V(\tilde{X}-a)\right) d\tilde{X} \\ & - \sin(\lambda V a) \int_0^\infty P(\tilde{X},V) \sin\left(\lambda V(\tilde{X}-a)\right) d\tilde{X} \end{aligned} \tag{8}$$

The relevant zeroes of the partition function in λ can be obtained as the solutions of the following equation:

$$\cot(\lambda V a) = \frac{\int_0^\infty P(\tilde{X},V) \sin\left(\lambda V(\tilde{X}-a)\right) d\tilde{X}}{\int_0^\infty P(\tilde{X},V) \cos\left(\lambda V(\tilde{X}-a)\right) d\tilde{X}} \tag{9}$$

Let us assume for a while that the denominator in (9) is constant at large V. Since the absolute value of the numerator is bounded by 1, the partition function will have an infinite number of zeroes approaching the origin ($\lambda = 0$) with velocity V. In such a situation the free energy density does not converge in the infinite volume limit.

But this is essentially what happens in the actual case. In fact if we consider the integral in (9)

$$f(\lambda V, V) = \int_0^\infty P(\tilde{X}, V) \cos\left(\lambda V(\tilde{X} - a)\right) d\tilde{X} \tag{10}$$

as a function of λV and V it is easy to check that the derivative of $f(\lambda V, V)$ respect to λV vanishes in the large volume limit due to the fact that $P(\tilde{X}, V)$ develops a $\delta(\tilde{X} - a)$ in the infinite volume limit. At fixed large volumes V, $f(\lambda V, V)$ as function of λV is an almost constant non-vanishing function (it takes the value of $1/2$ at $\lambda V = 0$). The previous result on the zeroes of the partition function in λ remains therefore unchanged; it generalizes the Lee-Yang theorem on the zeroes of the grand canonical partition function of the Ising model in the complex fugacity plane[6] to any statistical model with a discrete Z_2 symmetry.

To illustrate this result with an example, let us take for $P(\tilde{X}, V)$ a double gaussian distribution

$$P(\tilde{X}, V) = \frac{1}{2}\left(\frac{V}{\pi}\right)^{1/2}\left(e^{-V(\tilde{X}-a)^2} + e^{-V(\tilde{X}+a)^2}\right) \tag{11}$$

which gives for the partition function

$$\mathcal{Z}(\lambda) = Z(0)\cos(\lambda V a) e^{-\frac{1}{4}\lambda^2 V} \tag{12}$$

and for the mean value of the order parameter

$$< iX >= \frac{1}{2}\lambda + \tan(\lambda a V) a \tag{13}$$

The zeroes structure of the partition function is evident in (12) and consequently the mean value of the order parameter (13) is not defined in the thermodynamical limit. Notice also that if $a = 0$ (symmetric vacuum), the free energy density is well defined at any λ and then Vafa-Witten's argument applies.

2.1 Conclusions

We have shown in the previous section that an essential ingredient in the Vafa-Witten theorem on the impossibility to break spontaneously parity in a parity-conserving vector-like theory as QCD, the existence of the free energy density $f(\lambda)$ in the presence of any symmetry breaking external source λX, does not work if the symmetry is spontaneously broken. The requirement that the free energy density is well defined implies the previous assumption that the symmetry is realized in the vacuum. This does not necessarily implies that Vafa-Witten conjecture is wrong but at least a theorem on it is still lacking.

From a speculative point of view let us assume for a while that strong interaction shows a weak spontaneous parity breaking. In such a case we would expect a non-vanishing vacuum expectation value for the density of topological charge at $\theta = 0$ and, as we have shown, the theory would be ill-defined at $\theta \neq 0$. Actually this issue seems not very likely from a phenomenological point of view, but this is not the only scenario in which the theory is ill-defined at $\theta \neq 0$. In fact even if the symmetry is realized in the vacuum, the partition function could have zeroes in θ approaching the $\theta = 0$ point with velocity less than the space-time volume V and also in this case the theory would be ill-defined at $\theta \neq 0$.

A simple model which shows this feature is the lattice free-fermion theory, the partition function of which shows zeroes on the imaginary axis of the complex fermion-mass plane approaching the origin with velocity less than the lattice volume V. The chiral symmetry is realized in the free fermion model and notwithstanding that, the zeroes structure of the partition function forbids to define the theory at imaginary values of the fermion mass.

This speculative mechanism would provide us with a simple explanation for the θ – *vacuum* or strong CP problem: θ is zero because otherwise the theory is ill-defined. As stated before this is a pure speculation at present but we think it is worthwhile to investigate such an issue.

3 QCD at finite θ.

The partition function of QCD with a topological term in the action reads as follows:

$$\mathcal{Z} = \int [dA_\mu^a]\,[d\psi]\,[d\bar{\psi}]\, \exp\left\{ - \int d^4x\,[\mathcal{L}(x) - \frac{i\theta}{16\pi^2} X(x)] \right\} \qquad (14)$$

where $\mathcal{L}(x)$ is the standard QCD Lagrangian and

$$X(x) = \epsilon_{\mu\nu\rho\sigma}\,\mathrm{Tr}\,F_{\mu\nu}F_{\rho\sigma} \qquad (15)$$

is, up to a normalization constant, $1/16\pi^2$, the euclidean local density of topological charge. The normalization constant has been chosen in such a way that the topological charge

$$\frac{1}{16\pi^2} \int d^4x \, X(x) \tag{16}$$

is an integer.

Assuming that the theory has a non-trivial θ dependence, we will show now the following theorem: *"QCD with a topological term in the action, either breaks parity spontaneously at $\theta = \pi$, or has a phase transition at some critical θ_c less than π".*

To start with the proof, let us write the partition function in a finite space-time volume, V, as a sum over all topological sectors, labeled by the integer n that gives the topological charge of the partition functions of each sector, weighted by the proper topological phase:

$$\mathcal{Z}_V(\theta) \; = \; \sum_n g_V(n) \, e^{i\theta n} \tag{17}$$

where

$$g_V(n) \; = \; \int_n [dA^a_\mu] \, [d\psi] \, [d\bar{\psi}] \; \exp[- \int d^4x \, \mathcal{L}(x)] \tag{18}$$

is the standard partition function computed over the gauge sector with topological charge equal to n. The function $g_V(n)$ is, up to a normalization factor, $\sum_n g_V(n)$, the probability, $p_V(n)$, of the topological sector n at $\theta = 0$:

$$p_V(n) \; = \; \frac{g_V(n)}{\sum_n g_V(n)} \, . \tag{19}$$

If we define the mean topological charge density

$$x_n \; = \; \frac{n}{V} \tag{20}$$

we can write the previous partition function as

$$\mathcal{Z}_V(\theta) \; = \; \sum_{x_n} h_V(x_n) \, e^{i\theta V x_n} \, , \tag{21}$$

where $h_V(x_n) = g_V(n)$ and the step Δx_n in the topological charge density is $1/V$.

Let the new $h_V(x)$ be a continuous interpolation of $h_V(x_n)$ and let us define a new function of θ in the following way:

$$\mathcal{Z}_{c,V}(\theta) = \int dx\, h_V(x)\, e^{i\theta V x}\,. \tag{22}$$

Summing up the "pseudo-partition" functions $\mathcal{Z}_{c,V}(\theta + 2\pi m)$ for all integers m we get

$$\sum_m \mathcal{Z}_{c,V}(\theta + 2\pi m) = \int dx\, h_V(x)\, e^{i\theta V x} \sum_m e^{i2\pi m V x}\,. \tag{23}$$

Using now the following representation of the periodic delta function:

$$\sum_m e^{i2\pi m V x} = \frac{1}{V} \sum_m \delta(x - \frac{m}{V})\,, \tag{24}$$

we get the following identity:

$$\mathcal{Z}_V(\theta) = V \sum_m \mathcal{Z}_{c,V}(\theta + 2\pi m)\,, \tag{25}$$

which relates our QCD partition function $\mathcal{Z}_V(\theta)$ to the "pseudo-partition" functions $\mathcal{Z}_{c,V}(\theta + 2\pi m)$. The usefulness of this identity will become evident in a while. Before, let us come back to expression (21). Under a parity transformation, the density of topological charge, x_n, changes sign, whereas $h_V(x_n)$ is an even function of x_n since at $\theta = 0$ the QCD action is parity invariant. Taking also into account that the full topological charge $V x_n$ is an integer, we conclude that at $\theta = \pi$ each term entering the sum of Eq. 21 is parity invariant. Therefore, at $\theta = \pi$ the theory recovers the symmetry under parity transformations, and if the symmetry is also realized in the vacuum we must obtain a vanishing value for the expectation value of the topological charge density: $\langle x \rangle = 0$.

Using Eq. (25) we can write for the mean value of the density of topological charge:

$$\langle x \rangle = \sum_m \langle x \rangle_{c,m} \left(\frac{\mathcal{Z}_{c,V}(\theta + 2\pi m)}{\sum_n \mathcal{Z}_{c,V}(\theta + 2\pi n)} \right)\,, \tag{26}$$

with

$$\langle x \rangle_{c,m} = \frac{\int dx\, x\, h_V(x)\, e^{i(\theta + 2\pi m) V x}}{\int dx\, h_V(x)\, e^{i(\theta + 2\pi m) V x}}\,, \tag{27}$$

Since x in (27) is a continuous variable (i.e. there is no symmetry forcing the numerator to be zero), we expect a non-vanishing value for $\langle x \rangle_{c,m}$ at $\theta = \pi$. However, it is simple to see that parity symmetry is realized at $\theta = \pi$ in a finite volume. In fact, the contribution of the $m = 0$ sector compensates the contribution of $m = -1$ in (26), the contributions $m = 1$ cancels that of $m = 2$, and so on. Therefore, at $\theta = \pi$ different sectors compensate to give $\langle x \rangle = 0$.

Equation (25) is valid for any value of θ. At $\theta = 0$ it is simple to verify that in the infinite volume limit the sector $m = 0$ gives all the contribution to the partition function. This is a simple feature which follows from the fact that at $\theta = 0$ and in the infinite volume limit the exact solution for the free energy density is given by the saddle point solution, which is also the solution for the $m = 0$ sector. If this sector dominates for any value of θ, we will get a first order phase transition at $\theta = \pi$, with a non-vanishing value of the topological charge density (remember the discussion following Eq. (27)), and the theory will undergo spontaneous parity breaking. Otherwise, there will be some critical value, θ_c, at which other sectors start to give a contribution to the partition function, and in such a case we will get a phase transition at this θ_c. On general grounds, we can say nothing about the number and order of phase transitions expected in this case.

To see how this mechanism works in practical cases and to get intuition of what we can expect in physical systems, let us analyze in the following two simple examples: the Ising model within an imaginary external magnetic field and a gaussian model.

3.1 The one-dimensional Ising model in an imaginary external magnetic field

In the previous demonstration we have not made use of any specific property of QCD, except the quantization of the topological charge. Therefore, the result applies to any model with a quantized topological charge which appears as an imaginary contribution to the euclidean action. A simple example of that is the one-dimensional Ising model in an imaginary external magnetic field. The hamiltonian of this model can be written as

$$H_N = -J \sum_{i=1}^{N} S_i S_{i+1} - \mathrm{i}\frac{\theta k_B T}{2} \sum_{i=1}^{N} S_i \,, \tag{28}$$

where J is the coupling constant between nearest neighbours, k_B is the Boltzmann constant, T the physical temperature, N the number of spins, and we assume periodic boundary conditions.

The partition function is given by

$$\mathcal{Z}_N = \sum_{\{S\}} e^{F\sum_i S_i S_{i+1} + i\theta\frac{1}{2}\sum_i S_i} \tag{29}$$

where $F = J/k_B T$ and the sum is over all spin configurations.

For an even number of spins, the quantity $1/2 \sum_i S_i$ which appears in the imaginary part of the hamiltonian is an integer taking values between $-N/2$ and $N/2$, and therefore it can be seen as a quantized "topological" charge. Furthermore, the theory has a Z_2 symmetry at $\theta = 0$ and $\theta = \pi$ which, in the sense of this work, is the analogous of parity in QCD.

The transfer matrix technique allows to compute exactly the partition function defined in Eq. (29). The final result is

$$\mathcal{Z}_N = \lambda_+^N + \lambda_-^N , \tag{30}$$

where $\lambda_\pm$ are the two eigenvalues of the transfer matrix, which are given by the following equation:

$$\lambda_\pm(\theta) = e^F \cos\frac{\theta}{2} \pm \left(-e^{2F}\sin^2\frac{\theta}{2} + e^{-2F}\right)^{1/2} . \tag{31}$$

We see that the external field θ is an angle taking values between $-\pi$ and π, as expected.

Let us discuss first the simpler infinite temperature case. At $T = \infty$, $\lambda_- = 0$ and $\lambda_+ = 2\cos(\theta/2)$. The partition function has the simple form:

$$\mathcal{Z}_N = 2^N \left[\cos(\theta/2)\right]^N , \tag{32}$$

from which it follows that the free energy density is

$$f = \frac{1}{N}\log \mathcal{Z}_N = \log 2 + \log\cos\frac{\theta}{2} . \tag{33}$$

For the mean magnetization we get

$$\langle m \rangle = \left\langle \frac{1}{N}\sum_i S_i \right\rangle = i\tan\frac{\theta}{2} . \tag{34}$$

This last expression shows that a first order phase transition takes place at $\theta = \pm\pi$, with divergent spontaneous magnetization. The origin of this phase transition is clarified by noticing that the sector $m = 0$ in Eqs. (25) and (26) dominates the thermodynamical limit for any θ.

The finite temperature case is somehow different. The eigenvalues of the transfer matrix, given by Eq. (31), are real and positive if

$$\sin^2 \frac{\theta}{2} \leq \mathrm{e}^{-4F} . \tag{35}$$

In this case the free energy density has a well defined thermodynamical limit, and the mean magnetization is given by

$$\langle m \rangle = \mathrm{i} \frac{\sin(\theta/2)}{[\mathrm{e}^{-4F} - \sin^2(\theta/2)]^{1/2}} , \tag{36}$$

which shows a first order phase transition with a divergent mean magnetization at

$$\theta_c^{\pm} = \pm 2 \arcsin \mathrm{e}^{-2F} . \tag{37}$$

For $\theta_c^+ < \theta < \pi$ or $-\pi < \theta < \theta_c^-$, the two eigenvalues of the transfer matrix are complex conjugate numbers. The partition function $\mathcal{Z}_N$ is real but not positive definite, and oscillates in sign with N, making it impossible to define a thermodynamical limit. For instance, the mean magnetization does not converge as $N \to \infty$. Thus, the theory is ill-defined in these intervals of θ. For $\theta_c^- < \theta < \theta_c^+$, the sum in the right-hand side of Eqs. (25) and (26) is dominated by the term with $m = 0$, and the mean magnetization (36) is the analytic continuation of the saddle point solution obtained with real magnetic field.

3.2 Gaussian distribution.

The second illustrative example we want to discuss here are models with a density of topological charge at $\theta = 0$ distributed according to a gaussian distribution. Thus, let us assume that the function $h_V(x_n)$ which enters Eq. (21) has the form

$$h_V(x_n) = \mathrm{e}^{-Vax_n^2} , \tag{38}$$

where a is a parameter related to the width of the distribution. This form of $h_V(x_n)$ is a natural assumption from a physical point of view as a first approximation to the actual distribution of nearly any model. In fact, outside second order phase transitions, the probability distribution function of intensive operators as the density of topological charge is expected to be gaussian in the vicinity of its maximum. Of course, deviations from the gaussian behaviour far from the maximum can induce important changes in the θ-dependence of the theory in the large θ regime (the previous example illustrates very well this point). However, the gaussian distribution provides us with a simple model that can be analytically solved and gives useful insights on the general problem that we are addressing.

The partition function of the model is then

$$\mathcal{Z}_V(\theta) = \sum_{x_n} \mathrm{e}^{-Vax_n^2} \mathrm{e}^{\mathrm{i}\theta V x_n} . \tag{39}$$

The pseudo-partition function $\mathcal{Z}_{c,V}(\theta + 2\pi m)$ entering Eq. (25) can be analytically computed, and reads:

$$\mathcal{Z}_{c,V}(\theta + 2\pi m) = \int dx\, \mathrm{e}^{-Vax^2} \mathrm{e}^{\mathrm{i}(\theta+2\pi m)Vx} = \left(\frac{\pi}{aV}\right)^{1/2} \mathrm{e}^{-\frac{1}{4a}(\theta+2\pi m)^2 V} . \tag{40}$$

Using Eq. (25) we can write

$$\mathcal{Z}_V(\theta) = \left(\frac{\pi V}{a}\right)^{1/2} \sum_m \mathrm{e}^{-\frac{1}{4a}(\theta+2\pi m)^2 V} , \tag{41}$$

and, if $|\theta| < \pi$, it is simple to verify that the free energy density $f(\theta)$ is given by

$$f(\theta) = \lim_{V\to\infty} \frac{1}{V} \log \mathcal{Z}_V(\theta) = -\frac{1}{4a}\theta^2 . \tag{42}$$

The $m = 0$ sector dominates for every θ between $-\pi$ and π. The vacuum expectation value of the density of topological charge, $\langle x \rangle$, is

$$\langle x \rangle = \mathrm{i}\frac{\theta}{2a} , \tag{43}$$

and the model breaks spontaneously parity at $\theta = \pi$.

This simple model which, as we have seen, illustrates very well the theorem demonstrated in the first part of this section, has also further relevance. In fact, using a duality relation of gauge theories with a large number of colours and string theory on a certain space-time manifold, Witten has recently studied the θ dependence of pure gauge theories in four dimensions[7] and has found Eqs. (42) and (43). Combining both results, we conclude that the probability distribution function of the topological charge density in pure $SU(N)$ gauge theory is gaussian in the large N limit. It is interesting to note that also another analytically solvable model (although not physically relevant) has the same θ dependence: the quantum rotor in one dimension[8].

3.3 Conclusions.

We have demonstrated in the previous sections a theorem which states that if the θ dependence of QCD is non-trivial, either the theory has a first order phase transition at $\theta = \pi$ which breaks parity spontaneously, or the model shows a phase transition at some θ_c less than π. In the last case, we are not able to establish the order of the phase transition. Furthermore, the proof makes no use of any specific property of QCD, apart from the quantization of the topological charge. This implies that the result applies to any vector-like model with a quantized charge that appears as an imaginary contribution to the euclidean action.

To illustrate this result in practical cases and to get intuition on what we can expect in physical models, we have analyzed two simple examples: the one-dimensional Ising model in an imaginary external magnetic field, and a model in which the probability distribution function of the density of topological charge at $\theta = 0$ is assumed to be gaussian.

In the first case, we have found a first order phase transition with divergent spontaneous magnetization, at a critical imaginary magnetic field that depends on the temperature. At infinite temperature the transition is located at $\theta_c = \pi$ and the system breaks spontaneously its Z_2 symmetry at this θ value. At finite temperature the transition appears at $\theta_c < \pi$, and the theory is ill-defined for larger values of θ (modulo 2π).

The model with gaussian distribution can be analyzed simply, and we have found a first order phase transition at $\theta_c = \pi$, which breaks spontaneously parity. The free energy density and the density of topological charge, which are quadratic and linear functions of θ respectively, agree exactly with the θ-dependence of $SU(N)$ gauge theories in the large N limit found recently by Witten. This implies that the topological charge density of this theory has a gaussian distribution at $\theta = 0$.

References

1. C. Vafa, E. Witten, *Phys. Rev. Lett.* **53**, 535 (1984).
2. V. Azcoiti, A. Galante, *Phys. Rev. Lett.* **83**, 1518 (1999).
3. K. Binder, *Z. Phys.* C**43**, 119 (1981).
4. M. Mezard, G. Parisi and M.A. Virasoro in *Spin Glass Theory and Beyond*, World Scientific Lecture Notes in Physics Vol. **9** (1987).
5. V. Azcoiti, V. Laliena, X.Q. Luo, *Phys. Lett.* B **354**, 111 (1995).
6. C.N. Yang and T.D. Lee, *Phys. Rev.* **87**, 410 (1952).
7. E. Witten, *Phys. Rev. Lett.* **81**, 2862 (1998).
8. W. Bietenholz, R. Brower, S. Chandrasekharan, U.-J. Wiese, *Phys. Lett.* B **407**, 283 (1997).

GLUONS IN THE LATTICE SU(2) CLASSICAL GAUGE FIELD

YING CHEN, JI-MIN WU, BING HE
Institute of High Energy Physics, Chinese Academy of Sciences,
P.O.Box 918-4, Beijing 100039, P. R. China
E-mail: cheny@hptc5.ihep.ac.cn

The SU(2) gluonic correlation functions, glueball effective masses in the $J^P = 0^+$, 2^+ and 0^- channels were calculated from the lattice classical gauge configurations which were obtained by smoothing the thermal gauge configurations through the improved cooling method. The instanton-induced attractive force in the 0^+ channel and the repulsive force in the 0^- channel are confirmed in the Monte Carlo simulation. There is evidence that the instanton vacuum contribution to the 0^+ glueball mass is significant.

1 Introduction

QCD predicts there exist glueballs. The structure and formation mechanism of glueballs are so extricated that it must be explored nonperturbatively. Extensive Monte Carlo simulations have been performed and a few statements seem to be firmly established in the glueball sector[1]: i) the scalar glueball is the lightest with the mass in the 1.6-1.8 GeV range; ii) the tensor glueball and the pseudoscalar one are significantly heavier than the scalar one with the mass ratio $m_{2^{++}}/m_{0^{++}} \simeq 1.4$ and $m_{0^{-+}}/m_{0^{++}} \simeq 1.5 - 1.8$, respectively[2]; iii) The scalar is much smaller than other glueballs($r_{0^{++}} \simeq 0.2fm$, $r_{2^{++}} \simeq 0.8fm$[3]). From these statements one can find that glueballs are much heavier than typical quark-model hadrons and the size difference betweensi the scalar and the tensor glueball is much larger than that of the similar mesons(a similar measurement for the π and ρ mesons gives $0.32fm$ and $0.45fm$). This indicates that the spin-dependent forces between gluons are stronger than that between quarks.

The information about the gluonic interactions can be obtained by studying the correlation functions of gluonic operators with the relevant quantum numbers. Schäfer and Shuryak[5] calculated the correlation functions and the Bethe Salpeter amplitudes in an instanton-based model of the QCD vacuum. Their results show that instantons lead to a strong attractive force in the $J^{PC} = 0^{++}$ channel, thus the scalar glueball is much smaller than other glueballs. In the 0^{-+} channel the corresponding force is repulsive, and in the 2^{++} case it is absent.

Recently the cooling method, a smoothing algorithm for the gauge configurations, has been developed to extract classical contents(such as instantons) of QCD vacuum on the lattice. This motivates us to study the relationship of instantons and glueballs by the lattice simulation. The goal of this work is to study the instanton effects in the glueball masses and gluonic correlation functions in the lattice SU(2) gauge theory. The tadpole improved Symanzik's action[7] is used on an anisotropic lattice to perform the Monte Carlo simulation as accurate as possible on coarse lattices. The smearing and fuzzy scheme[8] are applied to the glueball operators to increase the signal-to-noise ratio. These schemes have been verified to be successful in Morningstar and Peardon's calculation[9] and in our work. We employ the improved cooling to extract the instanton configurations. The gluonic correlators are measured during the cooling procedure and the glueball masses are extracted by the exponential fall-off of these correlators with respect to the Euclidean time. To investigate the spin dependence of the instanton effects, the behaviors of the gluonic vacuum correlation functions are also explored in this work.

The simulation details, the results and the discussion are given in section II. Section III is the conclusion and summary.

2 Lattice Calculation

If the lattice action is properly improved, the costs of the Monte Carlo simulation will be greatly reduced and the numerical study can be performed on much coarser lattice system. In this work, the gluonic correlation functions and the masses of the lowest-lying glueballs in the SU(2) gauge theory are calculated from uncooled and cooled configurations. The simulations are performed on an $8^3 \times 24$ anisotropic lattice by using tadpole-improved Symanzik's action[7,10]. The lattice spacing in the temporal direction is smaller than that in the spatial direction with the anisotropy ratio $\xi = a_s/a_t = 3$, so that the physically useful signals can be obtained before they would be undermined by the statistical fluctuations. The form of the lattice gauge action we use is expressed as

$$\begin{aligned} S_1 = \beta \sum_{x,s>s'} \left[\frac{5}{3}\frac{P_{ss'}}{\xi u_s^4} - \frac{1}{12}\frac{R_{ss'}}{\xi u_s^6} - \frac{1}{12}\frac{R_{s's}}{\xi u_s^6} \right] \qquad (1) \\ + \beta \sum_{x,s} \left[\frac{4}{3}\frac{P_{st}\xi}{u_s^2 u_t^2} - \frac{1}{12}\frac{R_{st}\xi}{u_s^4 u_t^2} \right], \end{aligned}$$

where $\beta = \frac{4}{g^2}$, g is the gauge coupling. $P_{\mu\nu}$ is the plaquette operator and $R_{\mu\nu}$ 2×1 rectangular operator, u_s and u_t are tadpole improvement parameters. u_s is defined by $u_s =< (1/2)TrW_{sp} >^{1/4}$, while u_t is set to be 1 because of $u_t = 1 - O(\xi^{-2})$ (ξ is much bigger than one).

In the continuum theory, the gluonic operators with quantum numbers 0^+, 0^-,2^+ are defined by the field strength squared, the topological charge density, and the energy density, respectively,

$$O_S = (gG^a_{\mu\nu})^2, O_P = \frac{1}{2}\epsilon_{\mu\nu\rho\sigma} g^2 G_{\mu\nu} G_{\rho\sigma}, \quad (2)$$
$$O_T = \frac{1}{4}(gG^a_{\mu\nu})^2 - g^2 G^a_{0\alpha} G^a_{0\alpha}.$$

The correlation functions $C(x)$ for the Euclidean separation x are defined as

$$C_\Gamma(x) = \langle 0|O_\Gamma(x) O_\Gamma(0)|0\rangle. \quad (3)$$

The glueball spectra are obtained by the spatial integral of two points functions of operators with definite J^{PC} quantum numbers, such as

$$C_\Gamma(t) = \int d\vec{x} \langle 0|O_\Gamma(x) O_\Gamma(0)|0\rangle \quad (4)$$
$$= \sum_{n_\Gamma} \frac{1}{2m_{n_\Gamma}} |\langle 0|O_\Gamma|n_\Gamma, \vec{p}=0\rangle|^2 e^{-m_{n_\Gamma}|t|}$$
$$\propto |\langle 0|O_\Gamma|0_\Gamma, \vec{p}=0\rangle|^2 e^{-m_{0_\Gamma}|t|} (t \to \infty).$$

The lattice simulation of masses of lowest-lying hadrons is well established. The lattice version of O_Γ are defined as [11]

$$O_{0^+}(t) = \sum_{\vec{x}} [P_{12}(\vec{x},t) + P_{23}(\vec{x},t) + P_{31}(\vec{x},t)], \quad (5)$$

$$O_{2^+}(t) = \sum_{\vec{x}} [P_{12}(\vec{x},t) - P_{13}(\vec{x},t)]. \quad (6)$$

The corresponding glueball masses are extracted from the exponential fallof of the lattice correlation functions.

$$m_\Gamma(t) = \frac{1}{a_t} ln \left(\frac{C(t)}{C(t-a_t)} \right) \quad (7)$$

We first calculate the lowest-lying glueball masses on the uncooled configurations at $\beta = 1.0, 1.1, 1.2$ through the standard procedure, where the fuzzy and smearing method [8,9] are used to increase the signal-to-noise ratio. The configurations are thermalized through 5000 Monte Carlo sweeps(each sweep is composed of four heatbath sweeps and one micro-canonical iteration) and the measurements are performed on 10000 configurations. We analyze the statistical error by dividing the 10000 configuration into 20 bins. The physical scale is set by calculating the heavy quark potential. With the value of the string tension $\sqrt{\sigma} = 440MeV$, a_s takes the value $0.277fm$, $0.233fm$, and $0.183fm$[12] respectively for $\beta = 1.0, 1.1, 1.2$. The glueball masses at $\beta = 1.0$ are extracted to be

$$\begin{aligned} m_{0^+} &= 1634 \pm 60MeV \\ m_{2^+} &= 2305 \pm 72MeV \\ m_{0^-} &= 2574 \pm 224MeV \end{aligned}$$

Our results are in good agreement with that of other work[12].

To explore the role of instantons in the gluonic correlation functions and glueball masses, the cooling method is used to get the instanton configurations. The link-updating of the cooling iteration is defined as below

$$U_\mu \to U'_\mu = c\Sigma^+_\mu,$$

where c is the normalization factor so that U'_μ is an element of SU(2) group. Σ_μ is defined by the local action

$$s(U_\mu) = 1 - \frac{1}{2} Tr U_\mu \sum_{\nu \neq \mu} (staples) \equiv 1 - \frac{1}{2} Tr U_\mu \Sigma_\mu,$$

This transformation justifies that the local action will be minimized. We take the measurement on a sample of 500 configurations generated and cooled by the lattice action S_1. The gluonic correlators are measured every five cooling sweeps and the spectra are extracted by the same approach as that was used in the uncooled case.In order to avoid the auto-correlation, every two adjacent configuration are separated by 20 Monte Carlo sweeps. Each configuration is cooled by 100 steps. The cooling procedure is monitored by calculating topological charges at each cooling step. The lattice charge density operator($Q_{cont.}(x) = \frac{1}{16\pi^2} Tr F_{\mu\nu}(x) \tilde{F}_{\mu\nu}(x)$ in continuum theory) used in this work is the improved version,

$$\begin{aligned} Q_L(x) &= -\frac{1}{512\pi^2} \left(\frac{5}{3} Q_P(x) - \frac{1}{6} Q_S(x) \right) \\ &= a^4 (Q_{cont.}(x) + O(a^4)), \end{aligned} \tag{8}$$

where

$$Q_P(x) = \sum_{\mu\nu\rho\sigma=\pm 1}^{\pm 4} \tilde{\epsilon}_{\mu\nu\rho\sigma} Tr \left(P_{\mu\nu}(x) P_{\rho\sigma}(x) \right) \tag{9}$$

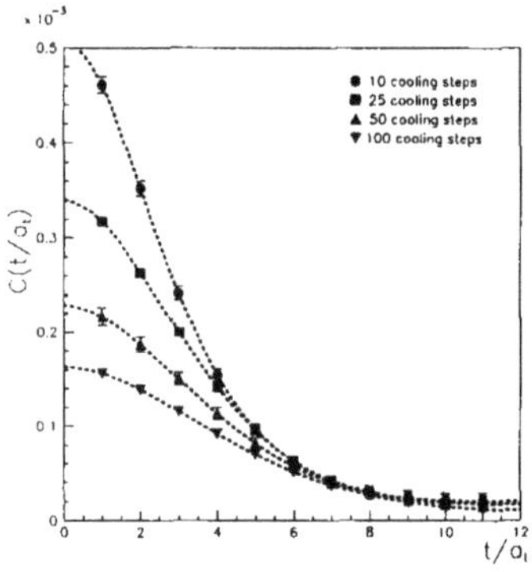

Figure 1: The correlation function C(t) in 0^+ channel at various cooling steps are illustrated. The dots, the rectangulars, the triangles and the upside-down triangles correspond to the cooling steps 10, 25, 50 and 100, respectively.

$$Q_S(x) = \sum_{\mu\nu\rho\sigma=\pm 1}^{\pm 4} \tilde{\epsilon}_{\mu\nu\rho\sigma} Tr\left[\left(S_{\mu\nu}^{1\times 2}(x) + S_{\mu\nu}^{2\times 1}(x)\right)\left(S_{\rho\sigma}^{1\times 2}(x) + S_{\rho\sigma}^{2\times 1}(x)\right)\right].$$

Here $\tilde{\epsilon}_{\mu\nu\rho\sigma}$ is the standard Levi-Civita tensor for positive directions while for negative ones the relation $\tilde{\epsilon}_{\mu\nu\rho\sigma} = -\tilde{\epsilon}_{-\mu\nu\rho\sigma}$ holds. $P_{\mu\nu}$ is the plaquette operator in the $\mu - \nu$ plane, while $S_{\mu\nu}^{i\times j}$ is $i \times j$ rectangular operator. We find that after several cooling sweeps the charges will reach meta-stable plateaus with approximate integer values.

Fig. 1, 2, 3 show the correlation functions of the three channels. In each channel, the correlation function is plotted at 10, 25, 50 and 100 cooling steps. The glueball "masses" are extracted in the same way as in the uncooled case and are plotted in Fig. 4, 5, 6.

Shuryak and Schäfer has calculated the gluonic correlation functions in the single-instanton background field. Taking the free gluon propagator into consideration, the correlators are expressed as[5],

$$C_S(x) = \frac{384g^4}{\pi^4 x^8} + n\rho^4 C_{inst}(x), \tag{10}$$

$$C_P(x) = -\frac{384g^4}{\pi^4 x^8} + n\rho^4 C_{inst}(x), \tag{11}$$

$$C_T(x) = \frac{24g^4}{\pi^4 x^8}, \tag{12}$$

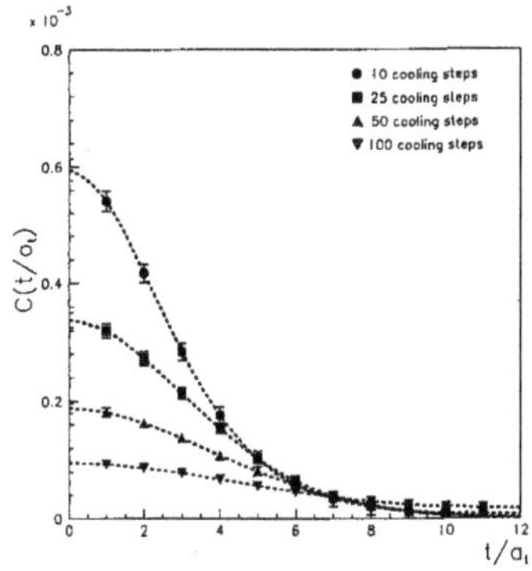

Figure 2: The correlation functions C(t) in 2^+ channel at various cooling steps are illustrated. The dots, the rectangulars, the triangles and the upside-down triangles correspond to the cooling steps 10, 25, 50 and 100, respectively.

where g is the running coupling constant and $C_{inst}(x)$ is the instanton contribution. $C_{inst}(x)$ is definitely positive and its detailed expression is abbreviated here. The signs in Eqn.(10) and Eqn.(11) are impressive: the two contributions reinforce in the scalar channel(attractive) but cancel to some extent in the pseudoscalar channel(repulsive). There is no effect in the tensor channel.

Our results shown in the figures are qualitatively in agreement with these comments. At small t, the amplitudes of the correlation functions decrease rapidly with the increase of the cooling steps, so do the extracted masses. In the intermediate and large t range, the correlation functions are less sensitive to the cooling. The "masses" take out short plateaus in the 0^+ channel and the 2^+ channel in the intermediate t. These are the results of that the ultraviolet components of the gauge fields are removed by the cooling but their topological structure is kept more stable. Small instantons are smoothed out by the cooling iterations(we monitor the cooling procedure and find that only the instantons with sizes greater than $2a_s$ survive), so the correlation functions and the "masses" at small t may be underestimated. The fact that in the 2^+ channel the correlation function tends to be flat(Fig. 2) and the mass-plateau decreases rapidly (Fig. 5) imply that there is no instanton effect in this channel. In contrast to this, the fall-off the correlation functions with t in the 0^+ and 0^- channels are obvious even after 100 cooling sweeps. The height of the mass plateau in the 0^+ channel is less sensitive to the cooling steps in the intermediate t range(Fig. 4). This means the contribution of instantons to the glueball mass in the 0^+ channel is significant. As illustrated in Fig. 3, the

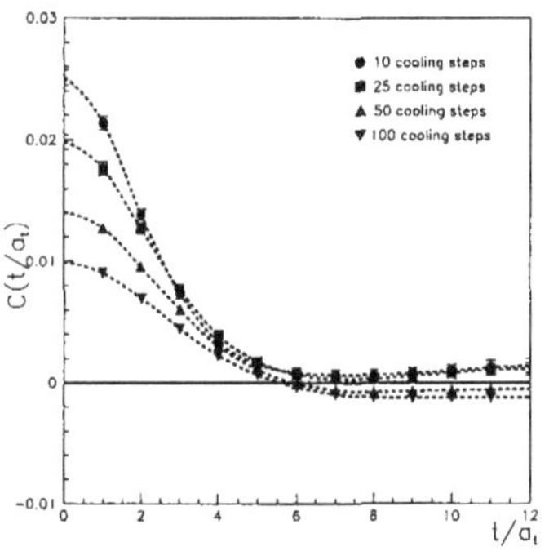

Figure 3: The correlation functions C(t) in 0^- channel at various cooling steps are illustrated. The dots, the rectangulars, the triangles and the upside-down triangles correspond to the cooling steps 10, 25, 50 and 100, respectively.

correlation function in the 0^- channel even changes sign at $t = 5a_t - 8a_t$ after 50 cooling steps. This is an evidence of that the gluonic interaction induced by instantons is repulsive in the pseudoscalar channel. The mass information can not be clearly extracted in this channel (Fig. 6).

3 Conclusion

We simulated the SU(2) glueball masses in the cooled and cooled gauge configurations. The results in the uncooled case are in agreement with the existing work. In the cooled case there is evidence of the instanton-induced attractive and repulsive interactions in the 0^+ channel and the 0^- channel, respectively, which are predicted by Shuryak and Schäfer based on the instanton liquid models. The contribution of instantons to the 0^+ glueball mass is significant.

Acknowledgement

The authors thank Dr. Jianbo Zhang and Dr. Daren Ji for their kindly providing us part of the Fortran code. Some of the calculations were performed on the Dawning-2000I parallel computer of the National Research Center for Intelligent Computing System. Y. Chen thanks Prof. Mingfa Zhu, Prof. Xiangzhen Qiao, Dr. Shilen Yao, and Prof. Jincai Shi for their kind help.

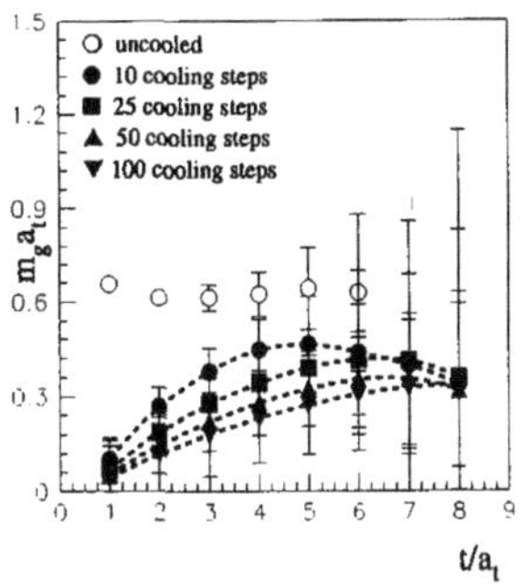

Figure 4: The 0^+ channel 'glueball masses' are extracted from the exponential decay of the correlators in cooled case. The masses vs. Euclidean time at various cooling steps are plotted.

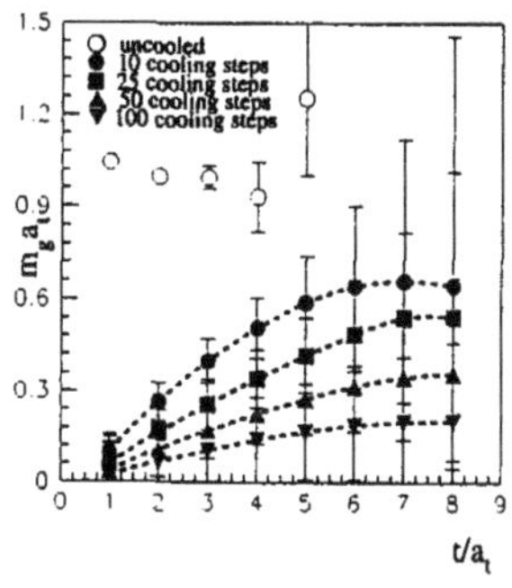

Figure 5: The 2^+ channel 'glueball masses' are extracted from the exponential decay of the correlators in cooled case. The masses vs. Euclidean time at various cooling steps are plotted.

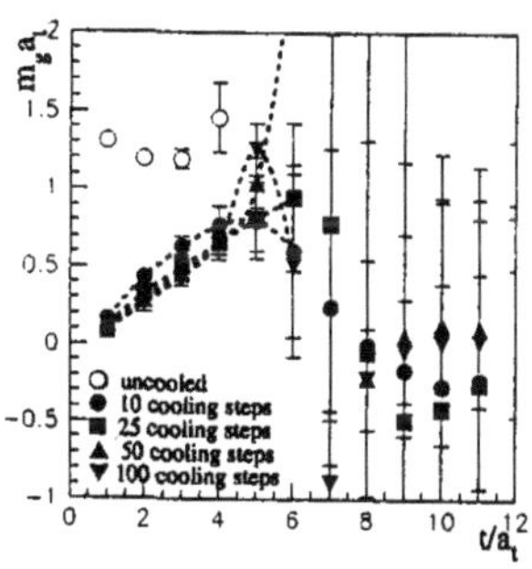

Figure 6: The 0^- channel 'glueball masses' are extracted from the exponential decay of the correlators in cooled case. The masses vs. Euclidean time at various cooling steps are plotted. No plateaus appear. The mass information can not be obtained in this channel.

References

1. D. Weingarten, Nucl. Phys. B(Proc. Suppl.)34(1994) 29.
2. G. S. Bali, K. Schilling, A. Hulsebos, A. C. Irving, C. Michael, and P. W. Stephenson, Phys. Lett. B243(1993) 413.
3. F. de Forcrand, and K. -F. Liu, Phys. Rev. Lett. 69(1992) 245.
4. V. A. Novikov, M. A. Shifman, A. I. Vainshtein, and V. I. Zakharov, Nucl. Phys. B165(1979)67; Nucl. Phys. B174(1980)378; Nucl. Phys. B191(1981)301.
5. T. Schäfer and E. V. Shuryak, Phys. Rev. Lett. 75(1995) 1707.
6. M. -C. Chu, J. M. Grandy, S. Huang, and J. W. Negele, Phys. Rev.Lett.70(1993) 225.
7. G. P. Lepage, P. B. Mackenzie, Phys. Rev. D48(1993)2250.
8. C. Michael and M. Teper, Phys. Lett. B 199(1987)95; M. Albanese et al., Phys. Lett. B192 (1987) 163.
9. C. J. Morningstar and M. Peardon, Nucl. Phys. B(Proc. Suppl.)47(1996)258.
10. M. Lüscher and P. Weisz, Commun. Math. Phys. 97(1983)59.
11. B.Berg and A. Billoirs, Nucl. Phys. B221(1992)109.
12. Jian Bo Zhang, Min Jin and Da Ren Ji, Chin. Phys. Lett.15(1998)865.

DO INSTANTONS OF THE CP(N-1) MODEL MELT?

M. MAUL

Department of Theoretical Physics, Lund University Sölvegatan 14A,
S - 223 62 Lund, Sweden
E-mail: maul@thep.lu.se

D. DIAKONOV

Nordic Institute for Theoretical Physics (NORDITA),
Blegdamsvej 17, DK-2100 Copenhagen, Denmark
E-mail: diakonov@nordita.dk

In the two-dimensional CP^{N-1} model one can parametrize exact many-instanton solutions via N 'constituents' (called 'zindons'). This parameterization allows, in principle, a complete 'melting' of individual instantons. The model is therefore well suited to study whether dynamics prefers a dilute or a strongly overlapping ensemble of instantons. We study the statistical mechanics of instantons both analytically and numerically. We find that at $N = 2$ the instanton system collapses into zero-size instantons. At $N = 3, 4$ we find that well-isolated instantons are dynamically preferred though 15-25% of instantons have a considerable overlap with others.

1 Introduction

Instantons, the specific fluctuations of the gluon field, carrying topological charge, play an important role in explaining many features of QCD, like the spontaneous breaking of chiral symmetry [1,2]. Furthermore, the instanton vacuum calculations are capable of providing the non-perturbative input to a variety of observables in Deep Inelastic Scattering (DIS) like polarized and unpolarized parton distributions [3], two-hadron distribution amplitudes skewed parton distributions and [4], higher-twist matrix elements [5] and other observables.

The role of instantons in the confinement phenomenon is still not clear. In general, a rigorous proof of a linear confining potential between static probe quarks in a 4-dimensional pure Yang–Mills theory from first principles is still missing [6], while the extraction of that potential from the current lattice data is subject to large systematic uncertainties [7].

It has been noticed some time ago [8] that an infinitely rising linear potential may be achieved if the instanton size distribution falls off as $\nu(\rho) \sim 1/\rho^3$ at large ρ. Such a regime would mean that large instantons overlap, and that the widely used sum ansatz of single instanton solutions is not too meaningful. If instantons are of any relevance for confinement, it cannot be seen in the dilute-gas approximation. Unfortunately, the true multi-instanton solution is not available in QCD: the long-known ADHM multi-instanton solution [9] is not an explicit one.

This motivates to investigate the overlap of instantons in a theory which is more simple than QCD. Such a theory is the two-dimensional CP^{N-1} model. The model contains asymptotic freedom, confinement and instantons whose explicit form is known for any topological number and any number of colors N [10,11]. The model is solvable at large N [12,13] and, most important, the true multi-instanton measure of the theory is known analytically [14,15].

This paper reports on some of the results of our study of the statistical mechanics of instantons in the CP^{N-1} model, by combining analytical and numerical methods [16]. Our conclusion is that, though the bulk of instantons appears to be well isolated, some 15-25% of them have a significant overlap.

2 The CP^{N-1} model

The CP^{N-1} model is defined in two dimensions which can be represented by the complex plain. The dynamical variables are the N complex fields u_A, $A = 1, \ldots, N$, which are normalized to unity:

$$u_A = \frac{v_A}{|v|}; \quad |v|^2 = \sum_{A=1}^{N} |v_A|^2 . \tag{1}$$

We shall call the index A 'color' in analogy to QCD. From the fields u_A a vector potential A_μ can be constructed:

$$A_\mu = \frac{i}{2}\left(u_A \partial_\mu u_A^* - u_A^* \partial_\mu u_A\right), \quad (\mu = 1, 2) . \tag{2}$$

The theory is defined by the partition function:

$$\mathcal{Z} = \int \mathcal{D}u_A(x)\mathcal{D}u_A^*(x)\mathcal{D}A_\mu(x)\delta(|u|^2 - 1)\exp\left(-\frac{1}{g^2}\int d^2x|\nabla_\mu u_A|^2\right), \tag{3}$$

with the covariant derivative being

$$\nabla_\mu = \partial_\mu - iA_\mu . \tag{4}$$

The fact that the fields u_A are normalized to unity makes the theory non-linear. The theory possesses the Abelian gauge invariance. From the vector potential A_μ a topological charge density $q_T(x)$ can be defined as:

$$q_T(x) = \frac{1}{4\pi}\epsilon_{\mu\nu}F_{\mu\nu}, \quad F_{\mu\nu} = \partial_\mu A_\nu - \partial_\nu A_\mu . \tag{5}$$

Here $\epsilon_{\mu\nu}$ is the antisymmetric tensor, i.e. $\epsilon_{12} = 1, \epsilon_{21} = -1$, and 0 for the other two index combinations. The multi–instanton (multi–anti-instanton) solution of the theory is known exactly [10,11] and can be expressed in terms of the unnormalized fields v_A up to an inessential constant as a product of simple monomials:

$$\textbf{instantons}: \quad v_A = \prod_{i=1}^{N_+}(z - a_{Ai}); \quad z = x + iy ,$$
$$\textbf{anti}-\textbf{instantons}: \quad v_A = \prod_{j=1}^{N_-}(z^* - b^*_{Aj}); \quad z^* = x - iy . \tag{6}$$

N_+ is the number of instantons and N_- the number of anti-instantons. A single instanton solution is therefore given by a single monomial and characterized by N 2-dimensional points a_A, which are called 'instanton zindons'. In the same way the 2-dimensional coordinates b_A are called the positions of 'anti-instanton zindons'. The word zindon is Persian or Tadjik and means 'prison' or 'castle'. There are, thus, N types of 'colors' of instanton zindons (denoted by a_A) and N types of anti-instanton zindons (denoted by b_A). It is essential, that the true multi-instanton solution is a product and not a sum of single-instanton solutions. As in QCD, single instanton solutions show up as well defined peaks in the topological charge density:

$$v_A = (z - a_A) \longrightarrow q_T(x) = \frac{1}{\pi}\frac{\rho^2}{((x - x_0)^2 + \rho^2)^2} , \tag{7}$$

where x_0 is the instanton center coinciding with the center of mass of N zindons of different 'colors' and ρ, the instanton size, is given by the spatial dispersion of zindons:

$$x_0 = \frac{1}{N}\sum_A a_A; \quad \rho^2 = \sum_A \frac{1}{N}|x_0 - a_A|^2 . \tag{8}$$

The corresponding single anti-instanton topological charge density has the same form, but with a negative sign, so it forms a local minimum in the topological charge density. For the combination of multi-instantons and multi-anti-instantons one conventionally uses the product ansatz [17]:

$$v_A = \prod_{i=1}^{N_+}(z - a_{iA}) \prod_{j=1}^{N_-}(z^* - b^*_{jA}) . \tag{9}$$

Naturally, it is not an exact solution (it becomes such only in the limit of large separations between instanton and anti-instanton zindons), therefore the action computed on this ansatz is not a sum of the individual actions. The corresponding interaction of instantons and anti-instantons formulated in terms of zindons has been found in Ref. [16], see the factor w_{ab} below. Combining it with the known multi-instanton (w_a) and multi-anti-instanton (w_b) weights [14,15] describing the interaction of 'same-kind' zindons, one writes the partition function in the form of statistical mechanics of interacting particles (N kinds of instanton zindons and N kinds of anti-instanton zindons):

$$Z = \sum_{N_+ + N_-} \frac{e^{i\theta N_+}}{(N_+!)^N} \frac{e^{-i\theta N_-}}{(N_-!)^N} \int \mathcal{D}a\mathcal{D}b\Lambda^{2N(N_+ + N_-)} w_a w_b w_{ab} \,. \qquad (10)$$

Λ is the only dimensional constant of the theory and can be set to 1. In the parameterization of ref. [16] w_a is given by:

$$w_a = \exp\left[\sum_{i<i'}^{N_+}\sum_{A} \ln((a_{Ai} - a_{Ai'})^2\Lambda^2) - \frac{N}{2}\sum_{i,i'}^{N} \ln\left[\sum_{A<B}^{N}(a_{Ai} - a_{Bi'})^2\Lambda^2\right]\right]$$
$$\times \exp\left[N\frac{N_+(N_+ - 1)}{2}\ln\frac{N(N-1)}{2}\right] \,. \qquad (11)$$

The corresponding weight for the anti-zindon interaction w_b is defined similarly. The instanton–anti-instanton interaction is described by the factor [16]:

$$w_{ab} = \exp\left\{2\beta\sum_{i=1}^{N_+}\sum_{i'=1}^{N_-}\sum_{A,B}^{N} \mathcal{P}_{AB}\ln\left[(a_{Ai} - b_{Bj})^2\Lambda^2\right]\right\},$$
$$\mathcal{P}_{AB} = \begin{cases} \frac{N-1}{N};\ A = B \\ -\frac{1}{N};\ A \neq B \end{cases} \,. \qquad (12)$$

$\beta = 2\pi/(g^2N)$ is the coupling between instantons and anti-instantons. The partition function describes two systems of zindons, namely instanton and anti-instanton ones, experiencing logarithmic interactions, whose strength is $N-1$ times stronger for same-color zindons than for different-color zindons. One has attraction for zindons/anti-zindons of different color and repulsion for zindons/anti-zindons of the same color. At $N = 2$ corresponding to the $CP^1 = O(3)$ model one can think of the ensemble as of that of e^+, e^-, μ^+, μ^- particles [17]. The interaction of opposite-kind zindons are suppressed by an additional factor $\beta = 2\pi/(g^2N_c)$. Since it is a classical system and not a quantum-mechanical one where stable atoms do exist, such an ensemble, i.e. the CP^1 model is unstable, as we will see in the next section.

3 Instanton size distribution

The multi-instanton ansatz (9) allows for a complete 'melting' of instantons. Indeed, if zindons a_A and b_A are evenly distributed in space individual instantons loose any meaning. In principle, another scenario could take place: a clustering of N-plets of zindons of N different 'colors' into 'color-neutral' objects. If such clusters are well isolated from other color-neutral clusters, they form well-separated or dilute instantons. We recall that a single instanton consists of N zindons a_A with different colors A. Which scenario takes place in reality is a matter of the dynamics of the ensemble given by the partition function (10).

The partition function (10) describing the instanton–anti-instanton ensemble in the zindon parameterization has been simulated with a Metropolis algorithm in Ref. [16]. One of the main objectives has been to find the size distribution of instantons.

The basic question is how to identify instantons and anti-instantons. This is also a serious problem for lattice QCD (see e.g. [18]). In the case of the CP^{N-1} model we can compare two ways of extracting the instanton content. The first one, which we call 'geometrical', is inspired by the zindon parameterization. Given a configuration of zindons a_{Ai} on the plain being at the thermodynamical equilibrium according to the partition function Z one can group them into instantons using the following procedure:

- Take the group of N zindons of N different colors, which has the smallest dispersion $\rho^2 = \frac{1}{N}\sum_A |a_A - x_0|^2$ out of the ensemble and call this group an instanton of size ρ located at the position $x_0 = \frac{1}{N}\sum_A a_A$.

- From the rest of the ensemble take out the next group of N zindons with the smallest dispersion ρ, and so on until the whole ensemble has been grouped into instantons (and anti-instantons).

This 'geometrical' identification of instantons assumes that the overlap does not affect much the peak structure of the topological charge density.

The second way of looking at the instanton size distribution is through the topological charge density $q_T(x)$. To that end we compute q_T on a grid from an equilibrium distribution of zindons obtained from running the Metropolis algorithm. The grid size limits the resolution of small size fluctuations, but not of large size ones. The method used to identify instantons is the following:

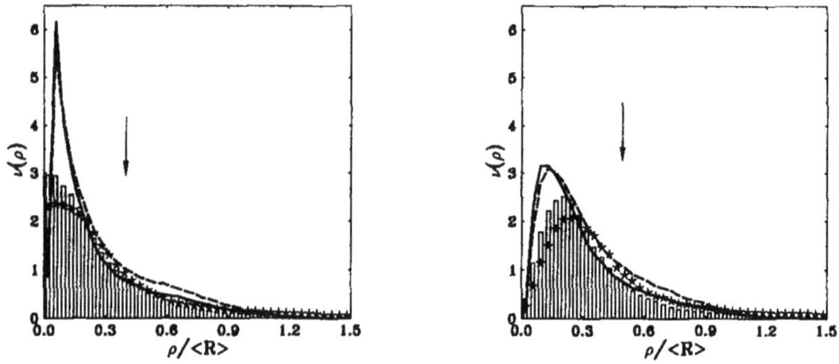

Figure 1: Instanton size distributions are displayed for $N = 3$ (left), and $N = 4$ (right) for $N_+ = N_- = 8$. The histograms show the 'geometric' size distributions, and the solid lines the size distributions seen by the 'lattice' method, using a 100×100 grid. The instanton–anti-instanton coupling constant for the solid lines and the histograms is $\beta = 0$. The stars and the dashed lines show the 'geometric' and 'lattice' size distributions for the case of $\beta = 0.5$. Instanton sizes are plotted in units of the average separation $\langle R \rangle$. The arrows show the maxima of the 'geometric' size distributions obtained in the case of purely random space distribution of zindons.

- Find a local maximum x_0, i.e. a grid point where the topological charge is larger than the one of the eight surrounding neighbors on the 9-plet centered at this grid point. In the spirit of the single instanton solution take then the first approximation for the instanton size to be $\rho_0^2 = 1/(\pi q_T(x_0))$.

- Interpolate the topological charge density on the 9 plet quadratically and calculate the two curvatures λ_1 and λ_2. If they are both negative then the local maximum is confirmed and we obtain at the same time two further estimates for the density via $\rho_i^2 = \sqrt{4/(\pi|\lambda_i|)}$, $i = 1, 2$.

- If all three values ρ_0, ρ_1, ρ_2 are smaller than $L/2$ where L is the size of the box where we place the ensemble, then the local maximum is accepted to be an instanton and the size is given by the geometric mean of all three estimates, i.e. by $\rho = (\rho_0\rho_1\rho_2)^{1/3}$.

Fig. 1, which has been taken from [16], shows the size distribution of instantons for $N = 3$ and $N = 4$. In the case $N = 2$ the zindons tend to condense into color neutral pairs and the ensemble collapses, so from that point of view the theory does not exist at all for $N = 2$. If one disregards the interaction of instantons with anti-instantons then the attraction of different-color zindons leads to a decrease of the average size of the instantons. This can be seen by comparing the maximum of the size distribution for $\beta = 0$ with the maximum for an interaction-less purely random distribution of zindons, which is represented by the arrows.

Switching in instanton–anti-instanton interactions, i.e. moving from $\beta = 0$ [histogram/solid line] to $\beta = 0.5$ [stars/dashed line] one observes that instantons are on the average shifted to larger sizes, however the effect is rather weak. The effect of the size shrinkage owing to the multi-(anti)instanton weight $w_{a(b)}$ is much more pronounced.

Remarkably, one observes a large discrepancy between the instantons identified by the 'lattice' method versus the 'geometric' method at small instanton sizes. The discrepancy is prominent at $N = 3$ but becomes considerably smaller at $N = 4$, so that a tendency is visible that it may die out as N is increased. This phenomenon of unphysical small size fluctuations is similar to the well known 'dislocation' phenomenon observed in lattice studies [19,20].

For instantons with a size larger than half of the average separation, i.e., $\rho/\langle R\rangle > 0.5$ one can say that the overlap becomes essential. This is the case for 15 - 25% of instantons, displayed in the figure. So one can say that the dilute gas ansatz is justified for many purposes, but that there is a considerable amount of instantons where the overlap cannot be neglected.

4 Summary and conclusions

We have formulated the statistical mechanics of instantons and anti-instantons in the $d = 2$ CP^{N-1} model in terms of their 'constituents' which we call 'zindons'. We have derived the interactions of same-kind and opposite-kind zindons for arbitrary N.

Though the zindon parameterization of instantons and of their interactions allow for complete 'melting' of instantons and is quite opposite in spirit to dilute gas Ansätze, we observe that zindons, nevertheless, tend to form 'color-neutral' clusters which can be identified with well-isolated instantons. This effect is due to a combination of two different factors both supporting clustering. One factor is the interactions: same-color zindons are strongly repulsive while different-color zindons are attractive. The second factor is pure geometry: even with a purely random distribution of zindons in space the probability to combine N zindons into a neutral cluster smaller than the average separation is quite sizeable. Both these factors are expected to be even stronger in four dimensions appropriate for the Yang-Mills instantons.

Despite an apparent tendency for clustering of zindons into well-isolated instantons, there always exist a portion of instantons which are strongly overlapping with the others. Depending on what one calls a 'strong' overlap we estimate the portion of such instantons to be about 15-25%. If a similar effect takes place in the Yang–Mills case, it means that there are long-range color correlations, which might be relevant to confinement. At the same time if the bulk of instantons are well separated (as we have found for the $CP^{(N-1)}$ model) it would explain the success of instantons in describing physics related to chiral symmetry breaking.

Acknowledgments

One of us (M.M.) thanks the Knut and Alice Wallenberg Foundation for financial support.

References

1. D. Diakonov and V. Petrov, *Nucl. Phys.* B **272**, 457 (1986); D. Diakonov, hep-ph/9602375.
2. T. Schafer and E. V. Shuryak, *Rev. Mod. Phys.* **70**, 323 (1998) [hep-ph/9610451].
3. D. Diakonov, V. Petrov, P. Pobylitsa, M. Polyakov and C. Weiss, *Nucl. Phys.* B **480**, 341 (1996) [hep-ph/9606314]; *Phys. Rev.* D **56**, 4069 (1997) [hep-ph/9703420].
4. M. V. Polyakov and C. Weiss, *Phys. Rev.* D **59**, 091502 (1999) [hep-ph/9806390]; *ibid. Phys. Rev.* D **60**, 114017 (1999) [hep-ph/9902451].
5. B. Dressler, M. Maul and C. Weiss, *Twist-4 contribution to unpolarized structure functions* F_L *and* F_2 *from instantons*, hep-ph/9906444.
6. R. W. Haymaker, *Phys. Rept.* **315**, 153 (1999) [hep-lat/9809094].
7. D. Diakonov and V. Petrov, *Phys. Scripta* **61**, 536 (2000) [hep-lat/9810037].
8. D. Diakonov and V. Petrov, in: *Nonperturbative Approaches to QCD*, Proc. Internat. workshop at ECT*, Trento, July 1995, ed. D. Diakonov, PNPI (1995) p. 239.
9. M. F. Atiyah, N. J. Hitchin, V. G. Drinfeld and Y. I. Manin, *Phys. Lett.* A **65**, 185 (1978).
10. V.L.Golo and A.M.Perelomov, *Phys. Lett.* **B79**, 112 (1978).
11. A. D'Adda, M. Luscher and P. Di Vecchia, *Nucl. Phys.* B **146**, 63 (1978).
12. A. D'Adda, P. Di Vecchia and M. Luscher, *Nucl. Phys.* B **152**, 125 (1979).
13. E. Witten, *Nucl. Phys.* B **149**, 285 (1979).

14. V. A. Fateev, I. V. Frolov and A. S. Schwarz, *Nucl. Phys.* B **154**, 1 (1979); *Sov. J. Nucl. Phys.* **30**, 590 (1979).
15. B. Berg and M. Luscher, *Commun. Math. Phys.* **69**, 57 (1979).
16. D. Diakonov and M. Maul, *Nucl. Phys.* B **571**, 91 (2000) [hep-th/9909078].
17. A. P. Bukhvostov and L. N. Lipatov, *Nucl. Phys.* B **180**, 116 (1981).
18. J. W. Negele, *Nucl. Phys. Proc. Suppl.* **73**, 92 (1999) [hep-lat/9810053].
19. B. Berg and M. Luscher, *Nucl. Phys.* B **190**, 412 (1981).
20. M. Luscher, *Nucl. Phys.* B **200**, 61 (1982).

A STUDY OF CENTER VORTICES IN SU(2) AND SU(3) GAUGE THEORIES [a]

MICHELE PEPE
Inst. für Theoretische Physik, ETH Hönggerberg, CH-8093 Zürich, Switzerland
e-mail: pepe@itp.phys.ethz.ch

PHILIPPE DE FORCRAND
Inst. für Theoretische Physik, ETH Hönggerberg, CH-8093 Zürich, Switzerland
and
CERN, Theory Division, CH-1211 Genève 23, Switzerland
e-mail: forcrand@itp.phys.ethz.ch

We show how center vortices and Abelian monopoles both appear as local gauge ambiguities in the Laplacian Center gauge. Numerical results, for $SU(2)$ and $SU(3)$, support the view that the string tension obtained in the center-projected theory matches the full string tension when the continuum limit is taken.

1 Introduction

There are several approaches addressing the problem of understanding the mechanism of color confinement in non-Abelian gauge theories. The most popular share the idea that only a subset of the degrees of freedom are relevant for confinement. In the Abelian projection approach[1,2], one takes into account the maximal Abelian subgroup $U(1)^{N-1}$ of the gauge group $SU(N)$ and monopoles are the effective degrees of freedom under study. In the Center projection approach[3,4], it is the center group Z_N which is considered and center vortices are the effective degrees of freedom under study.

These two schemes are commonly believed to give alternative descriptions of confinement. Many analytical and numerical studies have been and are being performed using these two approaches. They give clear evidence that both Abelian and center degrees of freedom play a relevant role.

The reduction of the gauge symmetry and the selection of the effective degrees of freedom is usually carried out by a partial fixing of the initial $SU(N)$ gauge freedom. We show that, in the Laplacian Center gauge, monopoles and center vortices are closely related in a unified description and that the latter are the effective degrees of freedom relevant for confinement. Indeed center vortices are related to a more reduced gauge symmetry than monopoles. A fundamental issue of this study is the use of the Laplacian Center gauge, which

[a] presented by M. Pepe.

is free from the problem of the lattice Gribov copies affecting the widely used Maximal Center gauge.

2 The center degrees of freedom

Let us consider an $SU(2)$ gauge field defined in a plane and expressed in terms of the radial and angular components (A_r, A_φ). Suppose that the radial component is vanishing, $A_r = 0$, and the angular one is given by $A_\varphi = \frac{1}{2r}\sigma_3$. This gauge field configuration describes a magnetic flux tube crossing the plane at the origin. If we consider a Wilson loop encircling the origin, we see that it has a non–trivial value $e^{i\pi\sigma_3} = -1$ with respect to the center group Z_2 of $SU(2)$. Conversely a Wilson loop non encircling the origin has a trivial value $+1$ with respect to Z_2. This is an example of what a center vortex is.

Consider now an $SU(2)$ gauge theory on the lattice and decompose the link variable $U_\mu(x)$ as the product of two parts

$$U_\mu(x) = Z_\mu(x)\, U'_\mu(x) \tag{1}$$

where $Z_\mu(x)$ lives in the center group Z_2 and $U'_\mu(x)$ is the coset part belonging to $SU(2)/Z_2$. For instance, this splitting can be carried out defining $U'_\mu(x)$ as having positive trace. If $W(C)$ is a Wilson loop along the closed path C, making use of the decomposition (1), we can write

$$W(C) = \sigma(C)\, W'(C) = \left[\prod_{p\in\Sigma} \sigma(p)\right] W'(C) \tag{2}$$

$W'(C)$ and $\sigma(C) \equiv \prod_{p\in\Sigma} \sigma(p)$ are respectively the Wilson loops evaluated with the coset links $U'_\mu(x)$ and with the center links $Z_\mu(x)$. $\prod_{p\in\Sigma}$ is the product over all the plaquettes p belonging to a surface Σ supported by C; the value of $\sigma(C)$ does not depend on the choice of Σ and, for simplicity, we can choose it as the planar surface bounded by C. If we fix a gauge where U'_μ is smooth, then $W(C)$ has a non–trivial value with respect to Z_2 if $\sigma(C)$ does. Moreover, considering (2) for a single plaquette, $\sigma(p) = -1$ is a "signal" for a center vortex. In 4–dimensional space–time, center vortices form closed surfaces in the dual lattice.

It is possible to give a handwaving and qualitative argument showing how the described center vortices can give rise to an area law behavior for the Wilson loops. According to the previous argument, we are interested in the plaquettes p having $\sigma(p) = -1$. Suppose that t is the probability for a plaquette to have $\sigma(p) = -1$, consequently $(1-t)$ is the probability to have $\sigma(p) = +1$. Consider a time–slice of a lattice $SU(2)$ gauge configuration, then center vortices form 1–dimensional closed strings in the dual lattice. The following figure displays an example of such a time–slice

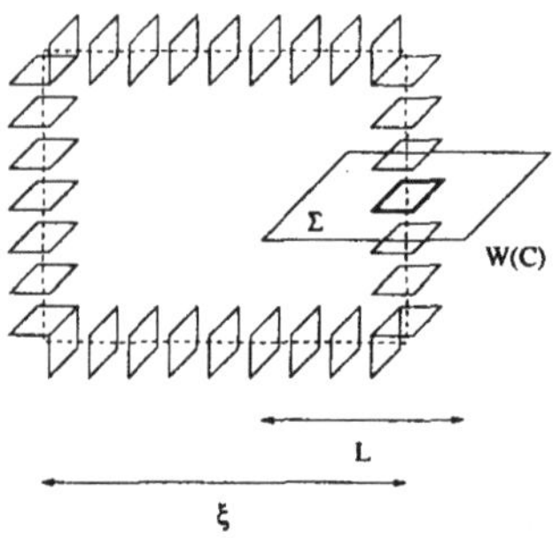

The drawn plaquettes have $\sigma(p) = -1$ and the broken line is the closed string in the dual lattice representing the center vortex. Consider a Wilson loop $W(C)$ which, in the figure, is represented by the continuous line; the plaquette in bold line belongs to the planar surface Σ spanned by C. Let L be the linear extension of C and ξ the average linear size of the center vortex strings. If $\xi \gg L$, the value $\sigma(p)$ of each plaquette p in Σ is independent of the others; then, with respect to the center degrees of freedom, one can write

$$< W(C) > \sim < \sigma(p) >^{A} = \mathrm{e}^{A \log(1-2t)} \tag{3}$$

where A is the area of Σ. With this qualitative argument we have obtained on one hand that the center degrees of freedom can give rise to an area law behavior for the Wilson loop and, on the other hand, that the string tension is about twice the probability t for a plaquette to have $\sigma(p) = -1$. We have assumed that $\xi \gg L$: in order for this inequality to be satisfied for arbitrarily large Wilson loops, ξ must be divergent, that is center vortex strings must percolate through the lattice. Conversely, suppose that the center vortices do not percolate, so that ξ has a finite value, then for L sufficiently large, one must have $\xi \ll L$. The next figure displays such a case

The values $\sigma(p)$ of the plaquettes in Σ well inside the boundary C are strongly correlated by pairs; moreover these pairs do not give a net contribution to $\sigma(C)$. So only the plaquettes belonging to a ring of thickness ξ around the boundary of Σ can give a non–trivial contribution. According to this reasoning and, with respect to the center degrees of freedom, we can write

$$< W(C) > \sim < \sigma(p) >^{\xi P} = \mathrm{e}^{\xi P \log(1-2t)} \tag{4}$$

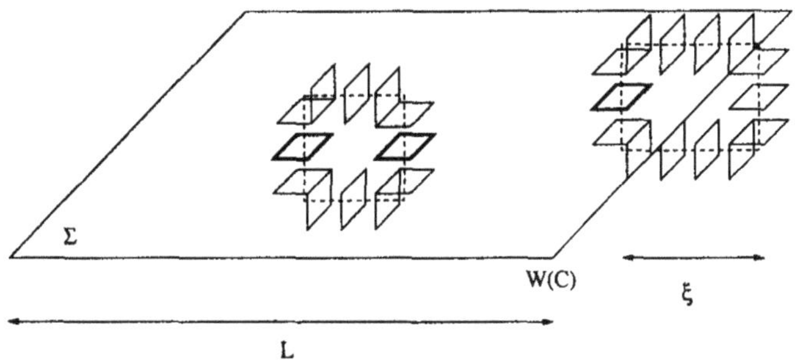

where P is the perimeter of Σ. Thus, we have obtained a perimeter law behavior for the Wilson loop. The conclusion from this handwaving argument is that the center degrees of freedom can give rise to an area law for the Wilson loop and that the area/perimeter behavior can be recast in terms of percolation/non-percolation of center vortices.

3 Laplacian gauge fixing

Many lattice simulations – starting with the initial results by Greensite and collaborators[5] – have been performed in order to investigate the role of the center degrees of freedom in the non perturbative features of the non-Abelian gauge theories. In these studies the selection of the center degrees of freedom is carried out by the numerical partial gauge fixing of an ensemble of configurations. The most widely used gauge is the Maximal Center gauge. In this gauge, it has been shown[6,7] that the removal of the center vortices leads to the loss of the area law for the Wilson loop, to the restoration of chiral symmetry and to the disappearance of non–trivial topological features. The numerical implementation of the Maximal Center gauge fixing is performed by a local iterative procedure and, as there are many local extrema, lattice Gribov copies are present. This is a serious problem[8,9] that can lead to a complete loss of meaningful information in the Z_2 projected model. Thus, it is important to study the role of the center degrees of freedom in the confinement mechanism considering a smooth gauge not affected by this problem of the lattice Gribov copies. In the Laplacian gauge the reduction of the gauge degrees of freedom is carried out in an unambiguous way. This gauge was proposed by Vink and Wiese[10]: they suggested to use the eigenvectors of the covariant Laplacian operator to fix the gauge. Since we are interested in reducing unambiguously the symmetry of the gauge group from $SU(N)$ to its center Z_N, it is useful to consider the Laplacian operator in the adjoint representation. In fact, as the adjoint representation is invariant under gauge transformations in Z_N, the adjoint Laplacian procedure fixes unambiguously the gauge up to the center symmetry.

Consider the 4–dimensional lattice $SU(N)$ gauge theory. The adjoint covariant Laplacian operator $\Delta^{ab}_{yx}(\dot{U})$ is given by

$$-\Delta^{ab}_{yx}(\dot{U}) = \sum_{\mu} \left(2\,\delta_{y,x}\,\delta^{ab} - \dot{U}^{ab}_{\mu}(x-\hat{\mu})\,\delta_{y,x-\hat{\mu}} - \dot{U}^{ba}_{\mu}(x)\,\delta_{y,x+\hat{\mu}} \right) \tag{5}$$

where a,$b = 1,\ldots,(N^2-1)$ are color indices and x,y are space–time lattice coordinates. The dotted $\dot{U}_\mu(x)$ are the link variables in the adjoint representation and are related to the links $U_\mu(x)$ in the fundamental by

$$\dot{U}^{ab}_{\mu}(x) = \frac{1}{2}\mathrm{Tr}\left(\lambda_a U_\mu(x) \lambda_b U^{\dagger}_{\mu}(x)\right) \tag{6}$$

λ_i, $i = 1,\ldots,(N^2-1)$ being the generators of $SU(N)$ with the normalization $\mathrm{Tr}(\lambda_a\lambda_b) = 2\delta_{ab}$. If V is the volume of the lattice, $\Delta(\dot{U})$ is a $[(N^2-1)V] \times [(N^2-1)V]$ real symmetric matrix which depends on the gauge field. The eigenvalues μ_j of Δ are real and the eigenvector equation is

$$\Delta^{ab}_{yx}(\dot{U})\phi^{(j)}_b(x) = \mu_j\,\phi^{(j)}_a(y) \tag{7}$$

where $\phi^{(j)}$, $j = 1,\ldots,[(N^2-1)V]$ are the (real) eigenvectors. So we can associate (N^2-1)–dimensional real vectors $\phi^{(j)}(x)$ to every lattice site x. Consider now a gauge transformation on the links in the fundamental representation $U'_\mu(x) = \Omega(x)U_\mu(x)\Omega^{\dagger}(x+\hat{\mu})$; the eigenvector equation (7) becomes

$$\dot{\Omega}^{\dagger\,ai}(y)\Delta^{ik}_{yx}(\dot{U}')\dot{\Omega}^{kb}(x)\phi^{(j)}_b(x) = \mu_j\,\phi^{(j)}_a(y) \tag{8}$$

where $\dot{\Omega}$ is the gauge transformation in the adjoint representation. This relation shows that the eigenvalues are gauge invariant and the eigenvectors transform according to $\dot{\Omega}^{ab}(x)\phi^{(j)}_b(x) = \phi^{(j)}_a(x)'$. This transformation law can be rewritten as follows

$$\Omega(x)\Phi^{(j)}(x)\Omega^{\dagger}(x) = \Phi^{(j)}(x)' \tag{9}$$

where we have defined the $su(N)$ matrices – i.e. in the $SU(N)$ algebra – $\Phi^{(j)}(x) = \sum_{a=1}^{N^2-1}\phi^{(j)}_a(x)\lambda_a$ and $\Phi^{(j)}(x)' = \sum_{a=1}^{N^2-1}\phi^{(j)}_a(x)'\lambda_a$. Gauge transformations rotate the vectors $\phi^{(j)}(x)$ in color space and we can fix the gauge by requiring a conventional arbitrary orientation for the $\phi^{(j)}(x)$. To perform the reduction of the gauge symmetry from $SU(N)$ to Z_N we only need to fix the orientation of two eigenvectors of the Laplacian operator. As we are interested in the non–perturbative features and in fixing a smooth gauge, we consider the two lowest-lying eigenmodes $\phi^{(1)}$ and $\phi^{(2)}$.

The gauge fixing procedure can be split into two steps. In the first, one rotates $\Phi^{(1)}(x)$ at every x so that $\Phi^{(1)}(x)'$ is diagonal. This leaves a residual symmetry corresponding to gauge transformations in the Cartan subgroup $U(1)^{N-1}$. To further reduce the gauge freedom we have to consider a second step where the second eigenvector $\phi^{(2)}$ is taken into account. The gauge transformation that has rotated $\Phi^{(1)}(x)$ to the Cartan subalgebra, maps $\Phi^{(2)}(x)$ to $\Phi^{(2)}(x)'$. The remaining $U(1)^{N-1}$ symmetry can be fixed to Z_N by requiring that some conventionally chosen color components of the twice rotated matrix $\Phi^{(2)}(x)$" vanish. Now we describe explicitly how to perform the presented two–step program for the $SU(2)$ gauge theory[7], where, at each x, we consider the 3–dimensional real vectors $\phi^{(1)}(x)$ and $\phi^{(2)}(x)$.

Step 1. We consider equation (9) for the first eigenvector and we conventionally define $\Omega(x)$ to be the gauge transformation that rotates $\phi^{(1)}(x)$ along the direction 3 in color space: $\dot{\Omega}(x)\phi^{(1)}(x) = \phi^{(1)}(x)' \propto (0,0,1)$. $\Omega(x)$ is unambiguously defined up to gauge transformations $V(x)$ in $U(1)$. So we have reduced the gauge symmetry to the Cartan subgroup of $SU(2)$. This is the Laplacian Abelian gauge[11,12].

Step 2. We apply the gauge transformation $\Omega(x)$ found in **Step 1** to $\phi^{(2)}(x)$: $\phi^{(2)}(x)' = \dot{\Omega}(x)\phi^{(2)}(x)$. $\phi^{(2)}(x)'$ is not invariant under the rotations $V(x) \in U(1)$ that leave $\phi^{(1)}(x)'$ unchanged; then we can fix this symmetry by requiring, for example, that $\phi^{(2)}(x)" = \dot{V}(x)\phi^{(2)}(x)'$ lie in the 1–3 color plane in the positive direction.

We have completely fixed the gauge symmetry in the adjoint representation and, as it is center–blind, we are left with the center symmetry Z_2. The described two–step procedure for $SU(2)$ can be extended[13] to $SU(N)$ with the same pattern of gauge symmetry fixing: $SU(N) \to U(1)^{N-1}$ in the first step and $U(1)^{N-1} \to Z_N$ in the second one.

Two last, important remarks concern the accidental degeneracy of μ_1 and μ_2 in (7) and the arbitrariness in the eigenvectors $\phi^{(j)}$. If it happens that μ_1 or μ_2 is degenerate, the gauge fixing can not be carried out unambiguously and one has a (global) Gribov ambiguity. This case is really exceptional and never occurs in the numerical simulations. The second point is about the scale and sign arbitrariness of the eigenvectors $\phi^{(j)}$. Rescaling can not give rise to any ambiguity in the procedure while the freedom in the choice of the sign does. This global freedom can be eliminated with a conventional choice on $\phi^{(j)}$.

4 Local gauge ambiguities

The adjoint Laplacian gauge fixing procedure has local defects. Now we discuss how these defects can show up and how they can be identified with monopoles and center vortices in $SU(2)$. This discussion can be generalized to $SU(N)$[13].

Step 2 ill–defined: the second step of the Laplacian gauge fixing is not defined at the points x where $\phi^{(1)}(x) \,//\, \phi^{(2)}(x)$. In such a case also $\phi^{(2)}(x)'$ is invariant under rotations $V(x) \in U(1)$. The condition $\phi^{(1)}(x) \,//\, \phi^{(2)}(x)$ sets two constraints and so these points x – where the gauge symmetry is promoted from Z_2 to $U(1)$ – form 2–dimensional surfaces in 4–dimensional space–time.

Step 1 ill–defined: the gauge fixing procedure can not be even started at the points x where $\phi^{(1)}(x) = (0,0,0)$. These defects constitute 1–dimensional strings in the 4–dimensional space–time since 3 constraints must be satisfied. At these points the symmetry is not fixed and the gauge freedom is $SU(2)$.

The 2–dimensional surfaces of **Step 2 ill–defined** can be identified with center vortices. Suppose that at a point x_0 it happens that $\phi^{(1)}(x_0) \,//\, \phi^{(2)}(x_0)$, then moving along a small loop around the singularity point x_0, $\phi^{(2)}$ describes a 2π rotation in color space. As a 2π phase in the adjoint representation of $SU(2)$ corresponds to a π phase in the fundamental one, it follows that a Wilson loop encircling x_0 has a non trivial value with respect to the center Z_2.

The 1–dimensional strings of **Step 1 ill–defined** can be identified with monopole world–lines. In fact, in analogy with the Georgi–Glashow model, the points x where the Higgs field vanishes and the gauge symmetry can not be reduced from $SU(2)$ to $U(1)$, correspond to the monopole world–lines.

In the adjoint Laplacian gauge, monopoles and center vortices turn out to be closely related in a unified description. Consider the 2–dimensional surface $\mathcal{S}$ of the center vortices. At every $x \in \mathcal{S}$, $\phi^{(1)}(x)$ and $\phi^{(2)}(x)$ are parallel or anti–parallel. So $\mathcal{S}$ is divided in parts where $\phi^{(1)}(x)$ is parallel to $\phi^{(2)}(x)$ and parts where it is anti–parallel. By continuity, these patches must be separated by 1–dimensional strings where $\phi^{(1)}(x) = 0$ or $\phi^{(2)}(x) = 0$. Moreover, in the neighbourhood of a monopole, $\phi^{(1)}(x)$ has a hedgehog–like shape and there will be a direction where $\phi^{(1)} \,//\, \phi^{(2)}$. Thus monopole world–lines are embedded within the 2–dimensional surfaces of center vortices.

5 Numerical results and their interpretation

We have performed numerical simulations to investigate the role of the center degrees of freedom in the $SU(2)$ and $SU(3)$ lattice gauge theories. For $SU(2)$ we have collected 1000 configurations at β = 2.3, 2.4, 2.5 on a 16^4 lattice; for $SU(3)$ we have generated 500 configurations on a 16^4 lattice at $\beta = 6.0$. The following figure shows the measurement of the Creutz ratios

$\chi(R) = -\ln(\langle W(R,R)\rangle\langle W(R-1,R-1)\rangle/\langle W(R,R-1)\rangle^2)$ for $SU(2)$ at $\beta = 2.4$ ($W(R,T)$ is the $R \times T$ Wilson loop). Crosses refer to $SU(2)$, circles to center

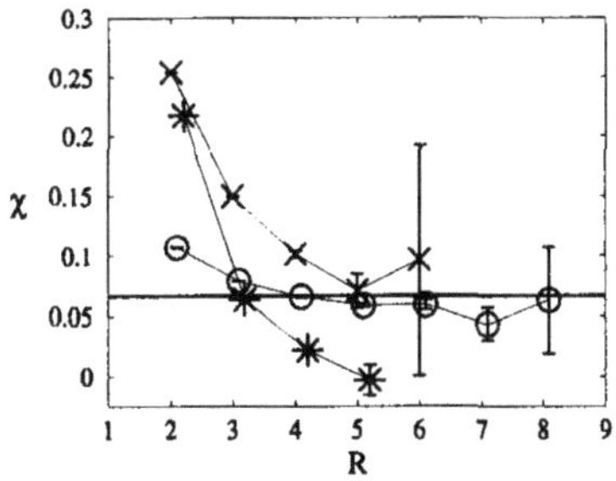

projection after Laplacian gauge fixing and stars to the coset part. The continuous band is the value in the literature[14,15] for the $SU(2)$ string tension at the considered set of parameters. The results show, on one hand, the flattening of the Creutz ratios in the Z_2 sector and, on the other hand, the vanishing of the Creutz ratios computed with the coset links. We have obtained a similar behaviour for $\beta = 2.3$ and 2.5. In the case of $SU(3)$ also, the following figure shows the Creutz ratios in the Z_3 sector after Laplacian gauge fixing. The

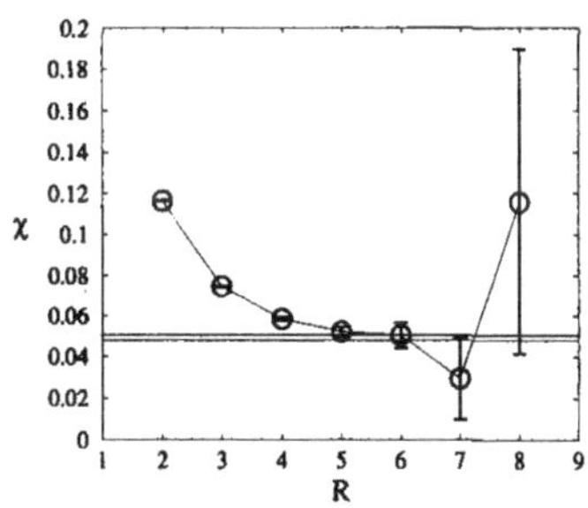

continuous band is the value in literature[16] for the $SU(3)$ string tension at the chosen set of parameters. Also in this case, one can clearly see flattening to a non vanishing value for the Creutz ratios evaluated with center projected links.

The good agreement with the values in the literature for the string tension in $SU(2)$ and $SU(3)$ should not be over–estimated. Numerical simulations are performed at finite lattice spacing and lattice artifacts can give non-negligible effects in the center projected theory. Our conjecture is that even if, at finite lattice spacing, the flattening value of the Creutz ratios in the center sector changes with the particular lattice Laplacian used to fix the gauge, this dependence vanishes in the continuum limit. The observation of such a behaviour would be a robust confirmation of the relevance of the center degrees of freedom in the confinement mechanism. To investigate this issue, we have considered three different lattice Laplacians differing by irrelevant operators: they have been built using smeared links in (6). Every set of 1000 configurations at $\beta = 2.3$, 2.4, 2.5, has been fixed in each one of the three gauges and the Creutz ratios have been measured after center projection. The table summarizes our results: $R_i = \sqrt{\sigma_i/\sigma_{SU(2)}}$ where σ_i is the string tension measured in the Z_2

	$\beta = 2.3$	$\beta = 2.4$	$\beta = 2.5$
R_0	0.813(23)	0.860(20)	0.978(18)
R_1	0.592(12)	0.720(11)	0.804(12)
R_2	0.547(8)	0.653(7)	0.739(11)

sector; $i = 0, 1, 2$ is an index for the three lattice Laplacians and $\sigma_{SU(2)}$ is the value in literature for the $SU(2)$ string tension at the three values of β. Thus, it is in the continuum limit ($\beta \to \infty$) that the string tension measured after center projection correctly reproduces the value of the full gauge theory.

Acknowledgments

We thank C. Alexandrou, M. D'Elia, S. Dürr and J. Fröhlich for useful discussions and G. Bali for providing us with a code for data analysis.

References

1. G. 't Hooft, *Nucl. Phys.* B **190**, 455 (1981).
2. S. Mandelstam, *Phys. Rep.* **23C**, 245 (1976).
3. G. 't Hooft, *Nucl. Phys.* B **138**, 1 (1978).
4. G. Mack and V. B. Petkova, *Ann. Phys. (NY)* **123**, 442 (1979).
5. L. Del Debbio et al., *Phys. Rev.* D **55**, 2298 (1997).
6. Ph. de Forcrand and M. D'Elia, *Phys. Rev. Lett.* **82**, 4582 (1999).
7. C. Alexandrou, Ph. de Forcrand and M. D'Elia, *Nucl. Phys.* B *Proc. Suppl.* **83-84**, 437 (2000).
8. T. Kovacs and E. T. Tomboulis, *Phys. Lett.* B **463**, 104 (1999).
9. V. Bornyakov et al., *JETP Lett.* **71**, 231 (2000).
10. J. C. Vink and U. J. Wiese, *Phys. Lett.* B **289**, 122 (1992).
11. A. van der Sijs, *Nucl. Phys.* B *Proc. Suppl.* **53**, 535 (1997).
12. A. van der Sijs, Prog.Theor.Phys.Suppl. 131 (1998) 149-159.
13. Ph. de Forcrand and M. Pepe, in writing.
14. C. Michael and M. Teper, *Phys. Lett.* B **199**, 95 (1987).

15. G. S. Bali, K. Schilling and C. Schlichter, *Phys. Rev.* D **51**, 5165 (1995).
16. G. S. Bali, K. Schilling, *Phys. Rev.* D **47**, 661 (1993).

FURTHER PROPERTIES OF INSTANTONS AND MONOPOLES IN THE QCD VACUUM

S. THURNER AND H. MARKUM
Institut für Kernphysik, TU Wien, Wiedner Hauptstrasse 8-10, 1040 Vienna, Austria
E-mail: thurner@univie.ac.at

Concerning the instanton sector, we study gauge invariant field strength correlators in pure SU(2) gauge theory by applying renormalization group based smoothing. The correlators are used to extract orientations of clusters of topological charge in configuration and color space. With respect to the fermion sector, we compute the quark condensate, the quark charge and the chiral density in full QCD. Classifying the lattice by elementary 3-cubes being associated to dual links occupied by (or free of) monopoles, the simultaneous occupation by chirality is investigated.

1 Introduction

In this contribution we review two recent directions of our research, one related to the possibility of measuring interactions between topological clusters in the QCD vacuum,[1] the other related to the chirality associated with color magnetic monopoles.[2]

The first part concentrates on pure SU(2) gauge theory, in the spirit of in how far it is possible to describe essential properties of QCD by the existence instantons, and hence reduce the number of integration variables in the path-integral to a feasible number of instanton parameters. In this context interactions of instantons, which are known to exist at least for $T \neq 0$,[3] might play an important role. The motivation for the studies was to get insight into the nature of these interactions, i.e. attraction, repulsion and color orientation, from first principles calculations.

Unfortunately it is not straightforward to identify instantons on realistic Monte Carlo (MC) gauge field configurations. Equilibrium configurations are not classical, nevertheless, a topological charge can be unambiguously assigned to the interpolating field[4] for generic lattice configurations. This has led to the hope to describe the topological (instanton) structure of MC configurations with the help of this new method. Our variant of renormalization-group (RG) smoothing[5] consists of (1) coarse graining a MC configuration (blocking in order to strip off UV fluctuations of scale a) and of (2) inverse blocking to interpolate it by a smooth configuration (SM) on the original lattice.

RG-smoothing as performed in our method does not lead to classical configurations. Clustering of topological charge and action into predominantly (anti)selfdual clusters[6] is a property of SM lattice fields inherited from the MC vacuum. These clusters of topological charge should not be naively identified with instantons. They could very well represent a superposition of a more involved topological structure, such as for example half integer objects, merons. The reason why we here use smoothing as a noise reduction method on the topological content instead of the more frequently used cooling methods, is that SM will not - unlike cooling - move or rotate topological clusters in a more or less uncontrollable way. We have checked that smoothed fields preserve the string tension, its Abelian dominance and contain a topological susceptibility of the right magnitude.

In earlier work we were concentrating on the monopole content[6,7] of SM configurations in certain gauges, relating it to the topological charge distribution and to other gauge-invariant signatures. The second part of this contribution is based on simulations of the full theory, and correlates chirality to paths occupied by monopoles. In this study we work in SU(3) and use ordinary Cabbibo-Marinari cooling as a means of noise reduction.

2 Interactions of topological clusters

The uncorrelated instanton liquid provides a good description of the QCD ground state. There are no indications from the field strength correlator at $T = 0$ [8] for strong correlations between (anti)instantons (I and A). At higher temperatures interactions are expected to be important in the instanton ensemble. Contrary to fermionic interactions, Ansätze for gluonic interactions rely on specific II and IA superpositions. Analyzing the clustered SM configurations across the phase transition we hoped to learn about instantons in the real Yang-Mills vacuum. Instead of supporting the instanton liquid picture our results are more suggestive for a different interpretation of the clusters exposed by smoothing.

The charge density has been measured on a $12^3 \times 4$ lattice by means of Lüscher's prescription which is inexpensive for SM configurations. We have defined topological clusters using a $4D$ site percolation algorithm for assigning neighboring (same sign) lattice sites, marked to have $|q(x)| > q_{th}$ (threshold), to the same cluster. Clusters are characterized by a center identified by q_{max}, a cluster volume V_{cl} and cluster charge Q_{cl}. Spatial distributions of clusters are presented in Fig. 1 by histograms corrected for the lattice geometry. The x-axis represents the relative distance of two clusters. A value of one on the y-axis corresponds to random allocations. Values greater and smaller than one indicate attractive and repulsive forces between the clusters respectively. For equally charged clusters small repulsive interaction is found on the SM

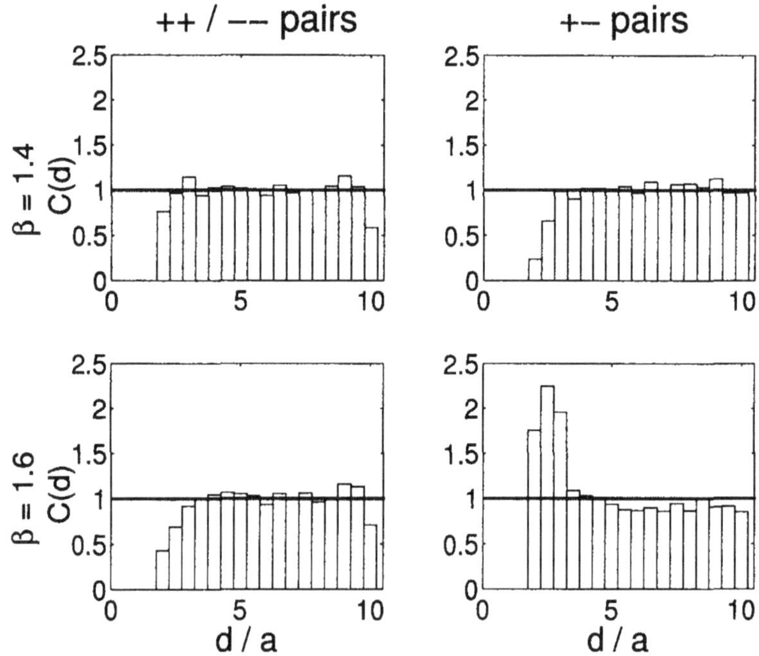

Figure 1: Relative cluster distance-distribution, in confinement (top) and deconfinement (bottom) for same sign and opposite sign clusters.

generated configurations in both phases. For oppositely charged clusters a strong attractive signal is found above T_c.

Color correlations between different clusters can be studied by the normalized *cluster overlap* for a pair of clusters,[1]

$$\mathcal{O} = \frac{\langle \mathrm{tr}\,(G_{\mu\nu}(1)\; S(1,2)\; G_{\mu\nu}(2)\; S(2,1))\rangle_{\mathrm{paths}}}{(\mathrm{tr}\,(G_{\rho\sigma}(1)^2))^{\frac{1}{2}}\; (\mathrm{tr}\,(G_{\tau\lambda}(2)^2))^{\frac{1}{2}}} \quad , \tag{1}$$

with the field strength correlator between the centers inserted. Correlators of specified components would give, for instanton pairs, access to the relative color orientation matrix U_{rel}. The overlap $\mathcal{O}$ can be identified with the adjoint trace $\left((\mathrm{tr}U_{rel})^2 - 1\right)/3$. For random orientations the average of $\mathcal{O}$ over all pairs would vanish. Plotting the average for different pairs as a function of the distance in Fig. 2 shows that the clusters are *not uncorrelated*. This average is sensitive to deviations of the histogram of $\mathrm{tr}U_{rel}$ from the Haar measure. Maximizing the overlap by a gauge rotation one can find U_{rel} for each pair. This

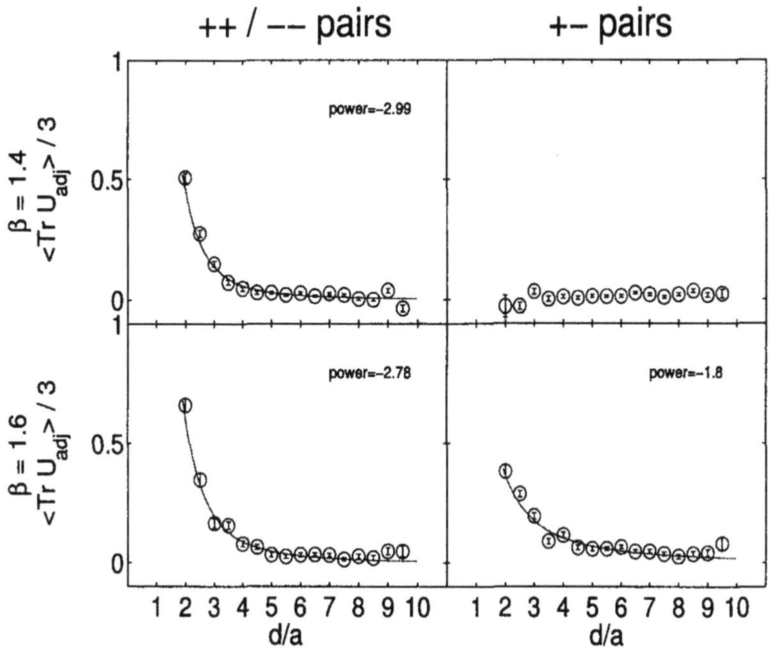

Figure 2: Average cluster overlap $\mathcal{O}$ vs. distance, in confinement (top) and deconfinement (bottom) for same sign and opposite sign clusters.

method can of course be applied independent of the instanton identification method, including the noise reduction and cluster definition. Our analysis has shown for II pairs strong *aligning* correlations in the confined phase (decreasing with distance) but none for IA pairs, and *aligning* correlations *similar* for both types of pairs in the deconfined phase. A recent $T = 0$ study confirms strong II correlations and rather weak ones for IA. These results are difficult to reconcile with any instanton picture.

Therefore we have carefully studied the influence of the cutoff q_{th} (over an interval of space filling fraction between 1 and 10 %) on the cluster composition. The cluster multiplicity goes through a maximum before a cluster percolation transition towards huge and multiply charged clusters is observed. With a cutoff keeping the cluster multiplicity below the maximum we get clustering properties being rather cutoff independent (apart from distance correlations between centers) with an almost Poissonian multiplicity distribution. For instance, on the 12^4 lattice at $\beta = 1.54$ ($\langle Q^2 \rangle \approx 11$) we find for $q_{th} = 0.065$ an average number of clusters of ≈ 35. A strong correlation exists relating charge and volume, $|Q_{cl}| \approx 0.01 V_{cl}$ (with almost all $|Q_{cl}| < 0.5$).

3 Chirality of monopole trajectories

Over the last years one has gained some insight into the mutual interrelations of two distinct excitations of the QCD vacuum: monopoles and instantons.[9] Both of those objects have been used to explain a wide variety of basic QCD properties, such as quark confinement, chiral symmetry breaking[10] and the $U_A(1)$ problem. The first property is usually associated with monopoles, the later ones with instantons. Their integer topological charge Q is related to the chiral zero eigenvalues of the fermionic matrix via the Atiyah-Singer index theorem. Since instantons carry chirality, and on the other hand it has been demonstrated that instantons are predominantly localized at regions where monopoles exist, the question arises whether monopoles carry chirality themselves. For calorons it has been shown that they consist of monopoles[11] which might be a sign that monopoles are indeed carriers of chirality.

Here we discuss this issue by directly looking at the chirality located on monopole loops, and comparing it to the background. We do this by measuring conditional probability distributions of fermionic observables of the form $\bar{\psi}\Gamma\psi$ with $\Gamma = 1, \gamma_4, \gamma_5$ in a standard staggered fermion setting. Those quantities are usually referred to as the quark condensate, quark charge density, and the chiral density. Mathematically and numerically the local quark condensate $\bar{\psi}\psi(x)$ is a diagonal element of the inverse of the fermionic matrix of the QCD action. The other fermionic operators are obtained by inserting the Euclidian γ_4 and γ_5 matrices.

Under a conditional probability distribution we understand the probability of encountering a certain value for a fermionic observable $\bar{\psi}\Gamma\psi$, under the condition that the local position is close to (or away from) a monopole trajectory,

$$P^{(\mathrm{no})\mathrm{mon}}_{s/t}(\bar{\psi}\Gamma\psi, x)|_{x\in(\notin)\mathrm{monopole\,tube}}\,, \tag{2}$$

where x is indicating the local position, and s/t space- or time-like monopole trajectories. The core of the monopole tube is the singular monopole trajectory, living on dual links, as obtained by the standard definition of monopoles in SU(3). We did not distinguish between the two independent colors of monopoles. For each dual link occupied by a monopole trajectory there exists an elementary 3-cube. The 8 sites of such a cube constitute the section of the monopole tube corresponding to that dual link.

Our simulations were performed for full SU(3) QCD on an $8^3 \times 4$ lattice with periodic boundary conditions. Dynamical quarks in Kogut-Susskind dis-

Correlation	cool 0	cool 5	cool 15
$\lvert\bar{\psi}^{\dagger}\psi(0)\rvert\rho(r)$	1.13(02)	1.08(01)	1.16(01)
$\bar{\psi}\psi(0)\rho(r)$	1.14(10)	1.16(05)	1.27(06)
$\lvert\bar{\psi}\gamma_5\psi(0)\rvert^2\rho(r)$	-	0.98(07)	1.30(10)
$q^2(0)\rho(r)$	1.54(47)	1.81(20)	2.41(58)
$\lvert\psi^{\dagger}\psi(0)\rvert q^2(r)$	1.25(66)	1.29(10)	1.42(09)
$\bar{\psi}\psi(0)q^2(r)$	1.32(34)	1.38(16)	1.47(16)
$\bar{\psi}\gamma_5\psi(0)q(r)$	-	0.84(02)	0.48(01)
$q(0)q(r)$	-	1.67(02)	0.84(01)

Table 1: Screening masses in GeV from fits to exponential decays of the various correlators for several cooling steps.

cretization with $n_f = 3$ flavors of degenerate mass $m = 0.1$ were taken into account using the pseudofermionic method. We performed runs in the confinement phase at $\beta = 5.2$. Measurements were taken on 2000 configurations separated by 50 sweeps.

We computed correlation functions between two observables $\mathcal{O}_1(x)$ and $\mathcal{O}_2(y)$,[2]

$$g(y-x) = \langle \mathcal{O}_1(x)\mathcal{O}_2(y)\rangle - \langle \mathcal{O}_1\rangle\langle \mathcal{O}_2\rangle \,. \tag{3}$$

In Fig. 3 we display results for $\mathcal{O}_1$ a local fermionic observable (except in (d)) and $\mathcal{O}_2$ the monopole charge density ρ. All correlations exhibit an extension of several lattice spacings and show an exponential falloff over the whole range. The corresponding screening masses are given in Table 1 in GeV for 3 levels of cooling. They are a coarse measure for the chirality profile of the monopole tube. It is apparent that cooling does not change the screening masses drastically.

For reasons of comparison we included in the table the screening masses for the correlation with the topological charge density (squared). We find that the correlations of the color charge density $\bar{\psi}\psi(x)$ and $|\psi^{\dagger}\psi(x)|$ with the topological charge density are very similar, both in the slopes and the absolute values. This becomes clear because the quark condensate can be interpreted as the absolute value of the quark density. However, cooling is inevitable to obtain nontrivial correlations between the chiral density, $\mathcal{O}_1 = \bar{\psi}\gamma_5\psi(x)$, and the topological charge density. This can be expected since both quantities are connected via the anomaly. The topological charge of a gauge field is related to the chiral density of the associated fermion field by $Q = \int q(x)d^4x = m\int \bar{\psi}\gamma_5\psi(x)d^4x$. We have checked that this relation also holds approximately for the corresponding lattice observables on individual configurations. The

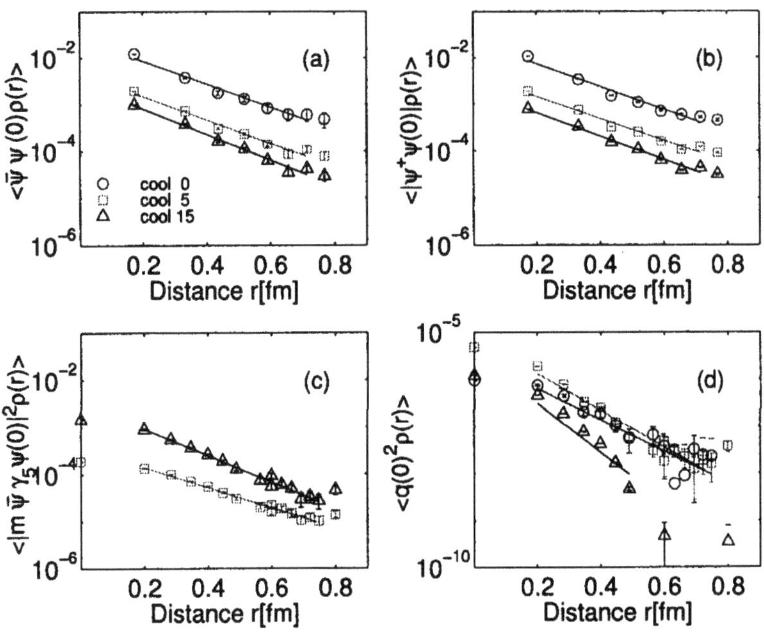

Figure 3: Correlation functions with the monopole density $\rho(r)$.

autocorrelation function of the density of the topological charge $< q(0)q(r) >$ should be compared to $< \bar{\psi}\gamma_5\psi(0)q(r) >$. If the classical t'Hooft instanton with size ρ_I is considered, the topological charge density is

$$q(x) \propto \rho_I^4(x^2+\rho_I^2)^{-4} \quad . \tag{4}$$

On the other hand the corresponding density of fermionic quantities[10]

$$\bar{\psi}\psi(x) \propto \bar{\psi}\gamma_5\psi(x) \propto \rho_I^2(x^2+\rho_I^2)^{-3} \tag{5}$$

is broader. This behavior is reflected in Table 1 and means that the local relation $q(x) = m\bar{\psi}\gamma_5\psi(x)$ does not hold.

Figure 4 shows results for the conditional probability distributions for $\bar{\psi}\psi$ in the case of monopole presence (m=1) or absence (m=0). The m=0 case exhibits a relatively narrow distribution of the fermionic quantity for space- and time-like monopole trajectories respectively. The m=1 case is clearly different and shows a much broader distribution, with both the mean and the variance

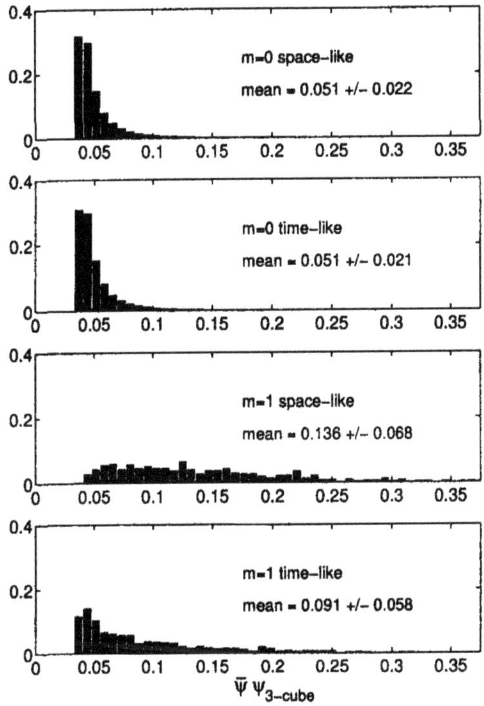

Figure 4: Conditional probability distributions for $\bar{\psi}\psi(x)$ due to monopole appearance.

being about a factor two larger. The time-like trajectories yield distributions which are still peaked on the left, like in the m=0 case, whereas this is not observed for space-like trajectories.

The plot suggests that in the close neighborhood of a monopole it becomes more likely to encounter large values of $\bar{\psi}\psi(x)$. The form of the distributions points towards a picture where monopole tubes carry a space-time dependent density of the fermionic observables. In the confinement, space-like tubes bear more chirality. The same situation is found for various observables $\mathrm{Re}(\psi^\dagger\psi)$ and $|\bar{\psi}\gamma_5\psi|$ (see Fig. 5). The figures depict the outcome after 15 cooling steps. We checked that these observations can be made also after 5 cooling steps.

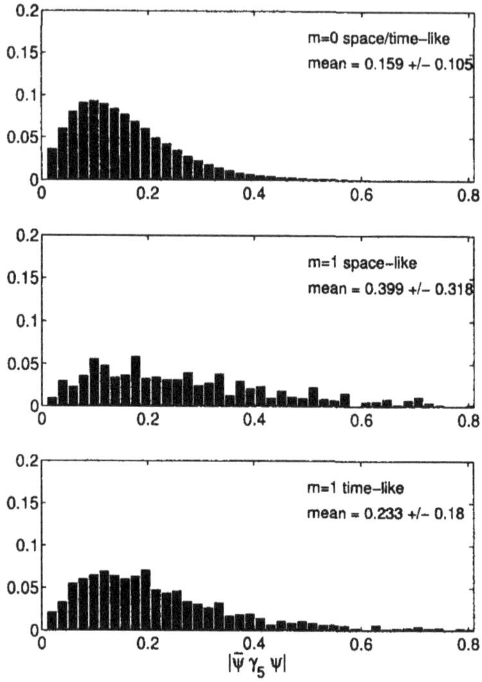

Figure 5: Conditional probability distributions for $|\bar{\psi}\gamma_5\psi(x)|$ due to monopole appearance.

4 Conclusion

Regarding the interactions of topological clusters identified on SM configurations by Lüscher's charge they might be interpreted as sub-clusters of instantons ("instanton quark"= half-instanton for SU(2)). This would explain the unexpected patterns in both the distance and color correlations. Equal sign clusters would be bound into instantons (dipoles) in the confinement phase. They could dissociate in the deconfined phase with opposite charge correlations becoming of equal importance. Further investigations are worthwhile, relating this to the behavior of the field strength correlators above. If this picture can be consolidated this will be a step beyond the present search for instanton structure by various cooling techniques.[12]

The computations of correlation functions between the monopole charge density and the fermionic observables yield an exponential decrease. The screening masses correspond to those of correlators between the topological charge density and the same fermionic observables.[2] Our calculations of conditional distribution functions of fermionic observables point out a significant enhancement for finding large chirality in the neighborhood of monopole trajectories. The same distributions also indicate that the monopoles are not covered by a uniform chirality tube.

Acknowledgments

We thank E.-M. Ilgenfritz and W. Sakuler for collaboration. Support from the Österreichische Forschungsgemeinschaft, project number 06/5984, is greatly acknowledged.

References

1. E.-M. Ilgenfritz and S. Thurner, hep-lat/9810010.
2. H. Markum, W. Sakuler and S. Thurner, *Phys. Lett.* **B 464**, 272 (1999).
3. E.-M. Ilgenfritz and E. V. Shuryak, *Phys. Lett.* **B 325**, 263 (1994); E. V. Shuryak and M. Velkovsky, *Phys. Rev.* **D 50**, 3323 (1994); T. Schäfer, E. V. Shuryak and J .J. M. Verbaarschot, *Phys. Rev.* **D 51**, 1267 (1995); M. Velkovsky and E. V. Shuryak, *Phys. Rev.* **D 56**, 2766 (1997).
4. T. A. DeGrand, A. Hasenfratz and De-cai Zhu, *Nucl. Phys.* **B 475**, 321 (1996); **B 478**, 349 (1996).
5. M. Feurstein, E.-M. Ilgenfritz, M. Müller-Preussker and S. Thurner, *Nucl. Phys.* **B 511**, 421 (1998).
6. E.-M. Ilgenfritz, H. Markum, M. Müller-Preussker and S. Thurner, *Phys. Rev.* **D 58**, 094502 (1998).
7. E.-M. Ilgenfritz, S. Thurner, H. Markum and M. Müller-Preussker, *Phys. Rev.* **D 61**, 054501 (2000).
8. E.-M. Ilgenfritz, B. V. Martemyanov, S. V. Molodtsov, M. Müller-Preussker and Yu. A. Simonov, *Phys. Rev.* **D 58**, 114508 (1998).
9. S. Thurner, H. Markum and W. Sakuler, in *Proceedings of Confinement 1995*, Osaka, World Scientific, 77 (1995); M. N. Chernodub and F. V. Gubarev, *JTEP Lett.* **62**, 100 (1995); H. Reinhardt, *Nucl. Phys.* **B 503**, 505 (1997).
10. E. V. Shuryak, *Nucl. Phys.* **B 302**, 559 (1988); T. Schäfer and E. V. Shuryak, *Rev. Mod. Phys.* **70**, 323 (1998).
11. T. C. Kraan and P. van Baal, *Nucl. Phys.* **B** (Proc. Suppl.) **73**, 554 (1999).
12. M. Teper, *Nucl. Phys.* **B** (Proc. Suppl.) **83-84**, 146 (2000).

6 Quantum and Parallel Computing

Pentium Cluster made by the Zhongshan University lattice group.

WHY QUANTUM COMPUTATION

H. F. CHAU
Department of Physics, University of Hong Kong,
Pokfulam Road, Hong Kong
E-mail: hfchau@hkusua.hku.hk

In this talk, I briefly review the reasons why quantum computation is an interesting subject and why sometimes quantum computation required in order to perform certain tasks efficiently. This talk is addressed to physicists who are new to this area.

1 Introduction

One may question why certain quantum mechanical calculations such as the dynamics of lattice QCD or finding various meta-stable local minima of spin glass so difficult. And perhaps we all agree now that these problems are intrinsically difficult. Nevertheless, it does not necessarily mean that we have to give up. One could look for approximate solutions of these problems using say the Monte Carlo methods. But what if one wants to know the "exact" solution? In this talk, I hope I can convince you that quantum computation may be the way to go. By quantum computation, I mean to use a machine that obeys the laws of quantum physics in performing our computation.

What makes people to think that a quantum computer can be more powerful than a classical one, such as the supercomputers that some of us are using? There are a number of reasons to believe that a quantum computer may be more powerful than a classical one. Let me list some of the reasons below:

- According to the laws of quantum physics, massive parallelism is automatically built in a quantum computer. More precisely, if the computer unitarily evolves from $|i, 0\rangle$ to $|i, f(i)\rangle$ for some function f, then by running the machine once by with the input $\sum_i \alpha_i |i, 0\rangle$, we get $\sum_i \alpha_i |i, f(i)\rangle$.
- The Hilbert space grows exponentially with the number of quantum particles involved.
- The existence of entanglement, such as the well-known EPR pairs in quantum mechanics which has no classical counterpart. Careful use of entanglement may help in computing certain task [1].
- We may be able to simulate a general Hamiltonian to any degree of accuracy using a small set of quantum unitary transformations that we can control [2,3]

2 Some Quantum Algorithms

Backup with all the strong points, does it mean that a quantum computer can perform *any* computation faster than a classical one? The answer turns out to be no. The reason is simple: at the end of the computation, the computer has evolved to a certain, say, pure state. But in order to readout the state of the quantum and hence the result of computation, *measurement* is required. Unfortunately, measurement in general will disturb the state of the computer; and a theorem by Holevo states that the maximum number of bits of classical information that can get out of an unentangled finite dimensional pure quantum state is no greater than the dimension of the state itself [4]. A more down to earth but less precise way to say this theorem is that although at the end of the day we obtain a state that tells us "everything" we want, it does not necessarily mean that we can read them all out by measurements. Thus, one has to devise careful measurement readouts in order to make full use of the computational power of a quantum computer. Although such a measurement is hard and sometimes even impossible to devise, let me mention several examples where this can be done.

The first example is the integer factorization algorithm discovered by Shor in 1994 [5]. To make the long story short, provided that one finds the period of the sequence

$$\{a^k \bmod n : k = 1, 2, 3, \ldots\} \tag{1}$$

efficiently for any given integer a (that is, the number of computational steps is polynomial to the number of input bits $\log n$), then one can make use of this information to find a non-trivial factor of n, a large integer [1,5]. Since we do not know any proper factors of n in advance (or else there is nothing to look for in this problem), classically we have no better way than perhaps to compute the above sequence term by term in order to find its period. Clearly, such an algorithm is very inefficient. But quantum mechanically, we can first prepare a state in the form $\sum_{k=1}^{n^2} |k, 0\rangle$ and then unitarily evolves it to $\sum_{k=1}^{n^2} |k, a^k \bmod n\rangle$. Now, if we measures the second register, the first register will collapse to the form $\sum_k |b + pk\rangle$ for some fixed b depending on the outcome of the measurement and p is the period of the sequence in Eq. (1). Finally, by means of the fast quantum Fourier transform

$$|j\rangle \longmapsto \sum_{k=0}^{m} e^{2\pi ijk/m} |k\rangle \tag{2}$$

where $m = 2^y$ for some integer y, one can easily estimate the period p by measuring the first register. Since all the above operations can be implemented in polynomial time of $\log n$, so the Shor's quantum factorization algorithm is efficient [1,5].

Later on, Grover discovered again quantum algorithm that can search over an unstructured database with n elements in a time that is $O(\sqrt{n})$. In other words, Grover's algorithm achieves a quadratic speedup over all possible classical searching algorithms [6]. This quadratic speedup is later shown to be optimal [7]. But due to the lack of time, I will not discuss the detail of the Grover's algorithm.

At this point, one may wonder if it is possible to simulate a general quantum mechanical system efficiently by a quantum computer. The answer is yes provided that the Hamiltonian is sufficiently local. More precisely, if $H = \sum_j H_j$ where each H_j acts on a Hilbert space of dimension much less than that of the entire Hilbert space dimension. Then, one may approximate the evolution operator e^{iHt} by $\left(e^{iH_1t/n}e^{iH_2t/n}\cdots e^{iH_lt/n}\right)^n$. Most importantly, in order to ensure the simulation accuracy, one has to regulate the time-slicing as

$$e^{iHt} = \left(e^{iH_1t/n}e^{iH_2t/n}\cdots e^{iH_lt/n}\right)^n + \sum_{i>j}[H_i, H_j]t^2/2n + \sum_{k=3}^{\infty} E(k) \qquad (3)$$

where the higher order error terms $E(k)$ are bounded by $\|E(k)\|_{sup} \leq n\|Ht/n\|_{sup}^k/k!$ [3].

3 Fighting with Decoherence

The story I have told you so far is good. But it is a bit too good to be true. Part of the reason is that a quantum computer may interact with the environment in an unwanted fashion, say, due to the presence of noise. We have to somehow control the decoherence so that the *quantum information* stored in the machine can be effectively decoherence free. There are a number of proposals to achieve this. Perhaps the most well-studied one is the use of quantum error correction code [8] and fault-tolerant quantum computation [9].

The idea of quantum error correction code is that by carefully encode the quantum information into a highly entangled state, one is able to protect the quantum information from decoherence. To correct the effect of decoherence, one only needs to measure the *error syndrome* to do the appropriate error correction. But just like the case of classical error correcting code, measurement of the syndrome tells us *nothing* about the encoded quantum information itself[8].

To give you an idea of what a quantum code is like, we consider the nine bit code discovered by Shor [10]. Its encoding is given by the linear operator mapping from the two dimensional Hilbert space H_2 to the 2^9 dimensional Hilbert space H_{2^9}:

$$|x\rangle \longmapsto \sum_{i,j,k=0}^{1} (-1)^{(i+j+k)x} |i,i,i,j,j,j,k,k,k\rangle \tag{4}$$

for $x = 0, 1$.

This code can be used to correct one error out of its nine quantum bits. To do so, we may use auxiliary bits to measure, say, the sum of the first two bits and the sum of the first and the third bits. If any of them are not zero, then we know precisely the location of the bit that is in error. And the error correction procedure in this case is nothing by to apply an σ_x to the corresponding quantum bit. In a similar way, one can test and correct for all the possible spin flip and phase shift occurs in any one of the nine encoded bits [10].

Later on, various researchers extended that error correction code concept so that quantum information is protected and can be manipulated even when the unitary transformation, error detection and correction procedures are all prone to error [9]. This is called the fault-tolerant quantum computation. But unfortunately, time forbids me to say anything more on this beautiful and yet technical topics.

4 Quantum Computing Experiments

After talking so much on the theoretical works, it is time to say something on the experimental realization of a quantum computer. Various proposals including the use of quantum optics / ion trap QED [11], NMR [12] and even solid state implementation say on quantum dots [13] have been proposed. And at this moment, the most impressive experiment is perhaps performed using NMR. Basically, they use the nuclear spins as their quantum bits and the strong magnetic field of the NMR primary magnet defines the z-axis used to identify the two level states of each spin. Without any further interaction, the nuclear spins will evolve according to their interaction with their neighboring spins in the same molecule. And the conventional NMR pulse sequences can then be used to control the evolution of the spins. Unfortunately, the signal of a single molecule is too small to pick up and hence people have to go for an ensemble of molecules in their experiments [12]. So far, the NMR people has performed the simplest possible quantum algorithm on two quantum bits, namely, the Deutsch-Jozsa problem [14]. Nevertheless, the NMR proposal does not scale very well with the number of quantum bits. They can perform experiments perhaps only up to ten quantum bits or so. The solid state implementation is probably the most promising one in the distant future.

5 Summary

In summary, I have talked about why we need quantum computation, what it is good for. I also discussed briefly how to fight with decoherence and the experimental status of the field. There are a number of things that are missed out here in this talk. The possibility of using entangled state to transmit secrets [15], the proof of security of such protocol [16] and the experimental realization of secret transmissions [17] are probably some of the more important things that have been missed out in this talk. But anyway, I hope I have successfully brought the message to you that quantum computing is interesting and useful.

Acknowledgments

This work is supported in part by a the RGC grant number HKU 7143/99P of the Hong Kong SAR Government and the Outstanding Young Researcher Award of the University of Hong Kong.

References

1. A. K. Ekert and R. Jozsa, *Rev. Mod. Phys.* **68**, 733 (1996).
2. R. P. Feynman, *Int. J. Theo. Phys.* **21**, 467 (1982).
3. S. Lloyd, *Science* **273**, 1073 (1996); D. S. Abrams and S. Lloyd, *Phys. Rev. Lett.* **79**, 2586 (1997); B. M. Boghosian and W. Taylor IV, *Physica* D **120**, 30 (1998); S. Somaroo *et al.*, *Phys. Rev. Lett.* **82**, 5381 (1999).
4. A. S. Holevo, *Prob. Pere. Inform.* **9**, 3 (1979) (in Russian).
5. P. W. Shor in *Proc. of the 35th Ann. Symp. on Foundations of Computer Sci.*, (IEEE Press, Los Alamitos, CA, 1994), p. 124.
6. L. K. Grover in *Proc. of the 29th Ann. ACM Symp. of the Theo. of Comp.*, (ACM Press, 1996), p. 212.
7. M. Boyer *et al.* in *Proc. of the Fourth Workshop on Physics and Computation*, p. 36; M. Boyer *et al.*, *Fort. de Phys.* **46**, 493 (1998).
8. See, for example, A. Steane in *Introduction to Quantum Computation and Information*, ed. by H.-K. Lo *et al.*, (World Scientific, Singapore, 1998), p. 184 and the references cited therein.
9. See, for example, J. Preskill in *Inroduction to Quantum Computation and Information*, ed. by H.-K. Lo *et al.*, (World Scientific, Singapore, 1998), p. 213 and the references cited therein.
10. P. W. Shor, *Phys. Rev.* A **52**, 2493 (1995).

11. R. Hughes *et al.*, *Fort. de Phys.* **46**, 329 (1998); D. J. Wineland *et al.*, *Fort. de Phys.* **46**, 363 (1998).
12. see, for example, I. L. Chuang in *Introduction to Quantum Computation and Information*, ed. by H.-K. Lo *et al.*, (World Scientific, Singapore, 1998), p. 311 and the references cited therein.
13. D. Loss and D. P. DiVincenzo, *Phys. Rev.* A **57**, 120 (1998); B. E. Kane, *Nature* **393**, 133 (1998).
14. I. L. Chuang *et al.*, *Nature* **393**, 143 (1998); J. A. Jones *et al.*, *Nature* **393**, 344 (1998).
15. C. H. Bennett and G. Brassard in *Proc. of the IEEE Int. Conf. on Computers, Systems and Signal Proc.* (IEEE Press, New York, 1984), p. 175; A. K. Ekert, *Phys. Rev. Lett.* **67**, 661 (1991).
16. H.-K. Lo and H. F. Chau, *Science* **283**, 2050 (1999); D. Mayers, quant-ph/9802025.
17. P. D. Townsend, *Electronics Lett.* **30**, 809 (1994); G. Ribordy *et al.*, *J. Mod. Opt.* **47**, 517 (2000); T. Jennewein *et al.*, *Phys. Rev. Lett.* **84**, 4729 (2000).

HIGH PERFORMANCE PARALLEL COMPUTER FROM COMMODITY PC COMPONENTS

XIANG-QIAN LUO AND ERIC B. GREGORY
Physics Department, Zhongshan University, Guangzhou 510275, China

JIECHAO YANG, YULI WANG, DI CHANG, AND YIN LIN,
G.D. National Communication Network Science and Technology Stock Co., Ltd, Guangzhou 510080, China

We describe the construction and configuration of a parallel computer composed of a cluster of personal computers. Furthermore, we show that such a cluster is an extremely inexpensive way of building computational power.

1 Introduction

The last two decades have ushered in the computer revolution for the consumer. In this period computers have moved from the domain of large companies, universities, and governments, to private homes and small businesses. As computational power has become more accessible, our demands and expectations for this power have increased accordingly. We demand an ever-increasing amount of computational ability for business, communication, entertainment, and scientific research. This rapid rise in both the demand for computational ability as well as the increase of that capability itself have forced a continual redefinition of the concept of a "super computer." The computational speed and ability of household computers now surpasses that of computers which helped guide men to the moon. The demarcation between super computers and personal computers has been further blurred in recent years by the high speed and low price of modern CPUs and networking technology and the availability of low cost or free software. By combining these three elements - all readily available to the consumer - one can assemble a true super computer that is within the budget of small research labs and businesses. This type of cluster is generally termed a Beowulf class computer. The idea was originally developed as a project at the US National Aeronautics and Space Administration[1]. We document the construction and performance of a cluster of PCs, configured to be capable of parallel processing. We believe this to be the first such installation at an academic institution in China.

2 Construction of a Parallel Cluster

2.1 Overview

As we stated above, we can broadly classify the components of a parallel computer cluster as computational hardware, communications hardware, and software. Our particular cluster consists of ten dual processor computers linked with fast ethernet technology.

2.2 Computational Hardware

We built a cluster of 10 PC type computers, all the components of which we purchased at normal consumer outlets for computer equipment. The major difference in our computers from one likely to be found in a home or business is that each is equipped with two CPUs. This allows us to roughly double our processing power without the extra overhead cost for extra cases, power supplies, network cards, etc. Specifically, we have installed two 500MHz Pentium III Processors in each motherboard. For the purposes of this report we will describe each computer as one "node" in the cluster; i.e., a node has two processors. Each node has its own local EIDE hard disk, in our case each has 10GB. This amount of space is not necessary, as the operating system requires less than one gigabyte per node, however the price of IDE hard disks has dropped so rapidly that it seems a reasonable way to add supplementary storage space to the cluster. Furthermore, each node is equipped with memory (at least 128MB), a display card, a 100Mbit/s capable fast ethernet card, a CDROM drive and a floppy drive. These last two items are not an absolute necessity as installation can be done over the network, but they add a great deal of convenience and versatility for a very modest cost.

One node is special and equipped with extra or enhanced components. The first node acts as a file server and has a larger (20GB) hard disk. This disk is the location of all the home directories associated with user accounts. The first node also has a SCSI adaptor, for connecting external backup devices such as a tape drive.

What each computer does not have is a monitor, keyboard, and mouse. Monitors can easily be one of the most expensive components of any home computer system. For a cluster such as this one, the individual nodes are not intended for use as separate workstations. Most users access the cluster through network connections. We use a single console (one small monitor, a keyboard and mouse) for administrative tasks. It is handy when installing the operating system on a new node. In this situation we move the console cables to the particular node requiring configuration. Once we installed communications programs such as telnet and ssh, it is almost never necessary to move the monitor and cables to the subordinate nodes.

2.3 Communications Hardware

There are many options for networking a cluster of computers, including various types of switches and hubs, cables of different types and communication protocol. We chose to use fast ethernet technology, as a compromise between budget and performance demands. We have already stated that we equipped each node with a 100Mbit/s capable fast ethernet card. A standard ethernet hub has the limitation on not being able to accommodate simultaneous communications between two separate pairs of computers, so we use a fast ethernet switch. This is significantly more expensive than a hub, but necessary for parallel computation involving any amount of inter-node communication. We found a good choice to be a Cisco Systems 2900 series switch. For ten nodes a bare minimum is a 12 port switch: one port for each node plus two spare ports for connecting either workstations or a connection to an internet router. We have in fact opted for a 24 port switch to leave room for future expansion of the cluster as our budget permits.

100Mbit per second communication requires higher quality "Catagory-5" ethernet cable, so we use this as the connection between the nodes and the switch. It should be noted that while a connection can be made from one of the switch ports to an external internet router, this cable must be "crossover" cable with the input and output wire strands switched.

2.4 Software

For our cluster we use the Linux open source UNIX-like operating system. Specifically, we have installed a Red Hat Linux distribution, due to the ease of installation. The most recent Linux kernel versions automatically support dual CPU computers. Linux is also able to support a Network File System (NFS), allowing all of the nodes in the cluster to share hard disks, and a Network Information System (NIS), which standardizes the usernames and passwords across the cluster.

The one precaution one must take before constructing such a cluster is that the hardware components are compatible with Linux. The vast majority of PC type personal computers in the world are running a Windows operating system, and hardware manufacturers usually write only Windows device drivers. Drivers for Linux are usually in the form of kernel modules and are written by Linux developers. As this is a distributed effort, shared by thousands of programmers worldwide, often working as volunteers, every PC

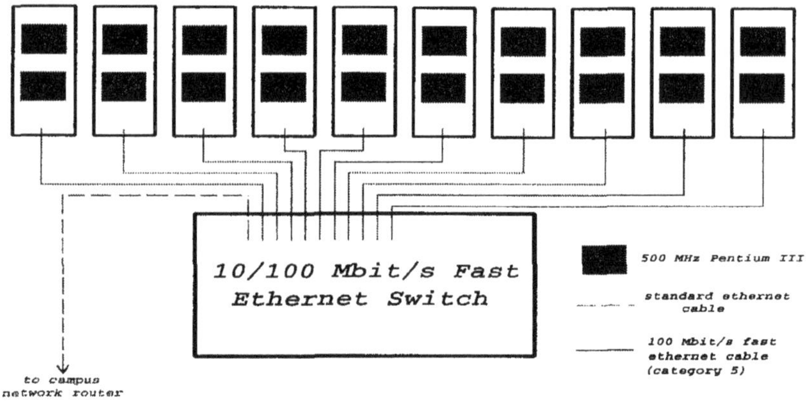

Figure 1. The layout of a 10 dual-CPU node cluster.

hardware component available is not necessarily immediately compatible with Linux. Some distributions, such as Redhat have the ability to probe the hardware specifications during the installation procedure. It is rather important to check online lists of compatible hardware — particularly graphics cards and network cards — before purchasing hardware. We began by purchasing one node first and checking the compatibility with the operating system first before purchasing the rest of the nodes.

To provide parallel computing capability, we use an Message Passing Interface (MPI) implementation. MPI is a standard specification for message passing libraries[2]. Specifically we use the *mpich* implementation, which is available for free download over the world wide web[3]. An MPI implementation is a collection of software that allows communication between programs running on separate computers. It includes a library of supplemental *C* and Fortran functions to facilitate passing data between the different processors.

3 Performance vs. Cost

We ran a standard LINPACK benchmark test and determined the peak speed of a single 500MHz Pentium III processor. The results of this test are shown in Table 1 to be about 100 million floating point operations a second (Mflops). With this in mind, we can say that the theoretical upper limit for the aggregate speed of the whole cluster (20 CPUs) approaches 2 Gflops. Of course this

Table 1. Results of LINPACK benchmark test on a single CPU.

single precision	86 - 114 Mflop
double precision	62 - 68 Mflop

is possible only in a computational task that is extremely parallelizable with minimum inter-node communications, no cache misses, etc. In the year 2000, the cost for our entire cluster was about US$15,000, including the switch. This means that the cost for computational speed was about US$7.50/Mflop. It is instructive to compare this to other high performance computers. One example is a Cray T3E-1200. Starting at US$630,000 for six 1200 Mflop processors[4], the cost is about US$87.50 per Mflop. The Cray is more expensive by an order of magnitude. Clearly there are advantages in communication speed and other performance factors in the Cray that may make it more suitable for some types of problems. However, this simple calculation shows that PC clusters are an affordable way for smaller research groups or commercial interests to obtain a high performance computer.

4 Physics Applications

The motivation for building the cluster at ZSU was to provide a platform for lattice QCD simulations. In such a simulation, a discrete lattice represents space-time and the computer must generate lattice gluon field configurations $\{U\}$ obeying the probability distribution:

$$\mathcal{P}\{U\} \sim \exp\left(-S\{U\}\right), \tag{1}$$

where $S\{U\}$ is a Euclidean lattice action for QCD with the quark fields integrated out. Generating appropriately weighted configurations is a time consuming task involving many local calculations. As such, it is well suited for parallelization [5]. A parallel lattice QCD algorithm divides the lattice into sections and assigns the calculations relevant to each section to a different processor. Near the boundaries of the lattice sections, information must be exchanged between processors. However, since the calculations are generally quite local, the inter-processor communication is not extremely large.

We have tested the performance of our cluster in actual lattice QCD simulations. Hioki and Nakamura [6] provide comparison performance data on SX-4 (NEC), SR2201(Hitachi), Cenju-3 (NEC) and Paragon (Intel) machines. Specifically, we compare the computing time per link update in microseconds per link and the inter-node communication speed in MB/s. The link update is

Table 2. Comparison of performance of MPI QCD benchmark. Comparison data from Hioki and Nakamura.

Machine	μ-sec/link	MB/s
SX-4	4.50	45
SR2201	31.4	28
Cenju-3	57.42	8.1
Paragon	149	9.0
ZSU's Pentium cluster	**7.3**	**11.5**

a fundamental computational task within the QCD simulation and is therefore a useful standard. The test was a simulation of improved pure guage lattice action (1×1 plaquet and 1×2 rectangle terms) on a 16^4 lattice. In each case the simulation was run on 16 processors. We used the QCDimMPI [7] Fortran code. Table 2 shows the results of this testing.

This type of cluster has a wide variety of scientific applications beyond lattice physics. Many sorts of problems can be parallelized. Notable examples are numerical general relativity, climate modeling, and fluid mechanics.

5 Conclusions

A parallel cluster of PC type computers is an economical way to build a powerful computing resource for academic purposes. On an MPI QCD benchmark simulation it compares favorably with other MPI platforms. It is also drastically cheaper than commercial supercomputers for the same amount of processing speed. PC clusters such as this one have applications in both academia and in commercial enterprises. It is particularly suitable for developing research groups in countries where funding for pure research is more scarce. We believe that our cluster may be the first such facility at an academic physics department in mainland China.

6 Acknowledgements

This work is supported by the National Science Fund for Distinguished Young Scholars (19825117), National Science Foundation, Guangdong Provincial Natural Science Foundation (990212) and Ministry of Education of China.

We are grateful for generous additional support from Guoxun (Guangdong National Communication Network) Ltd.. We would also like to thank Shinji Hioki of Tezukayama University for the use of the QCDimMPI code. The C version of the LINPACK benchmark was written by Bonnie Toy.

References

1. http://cesdis.gsfc.nasa.gov/linux/beowulf/beowulf.html
2. William Gropp and Ewing Lusk *Users Guide for mpich* (Argonne National Laboratory:ANL/MCS-TM-ANL-96).
3. http://www-unix.mcs.anl.gov/mpi/mpich/
4. Cray, Inc.
   http://www.cray.com/products/systems/crayt3e/1200data.html
5. R. Gupta, hep-lat/9905027.
6. S. Hioki, A. Nakamura. *Nucl. Phys.* **B(Proc. Suppl.)73**, 895,(1999).
7. http://tupc3472.tezukayama-u.ac.jp/QCDMPI/

7 Statistical Mechanics

A workshop session.

FIRST ORDER PHASE TRANSITION OF THE Q-STATE POTTS MODEL IN TWO DIMENSIONS

H. ARISUE, K. TABATA
Osaka Prefectural College of Technology, Saiwai-cho, Neyagawa
Osaka 572, Japan
E-mail: arisue@las.osaka-pct.ac.jp

We have calculated the large-q series of the energy cumulants, the magnetization cumulants and the correlation length at the first order phase transition point both in the ordered and disordered phases for the q-state Potts model in two dimensions. The series enables us to estimate the numerical values of the quantities more precisely by a factor of $10^2 - 10^4$ than the Monte Carlo simulations. From the large-q series of the eigenvalues of the transfer matrix, we also find that the excited states form a continuum spectrum and there is no particle state at the first order phase transition point.

1 Introduction

In many of the physical systems that exhibit the first order phase transition, the order of the transition changes to the second order by changing the parameter of the system. It is important to know how the quantities at the first order phase transition point diverge when the parameter approaches the point at which the order of the transition changes. The q-state Potts model[1,2] in two dimensions gives a good place to investigate this subject. It exhibits the first order phase transition for $q > 4$ and the second order one for $q \leq 4$. Many quantities of this model are known exactly at the phase transition point $\beta = \beta_t$ for $q > 4$, including the latent heat[3] and the correlation length in the disordered phase,[4,5,6] which vanishes and diverges, respectively, in the limit $q \to 4_+$. One the other hand other physically important quantities such as the specific heat, the magnetic susceptibility, and the correlation length (in the ordered phase) at the transition point, which also diverge as $q \to 4_+$, are not solved exactly.

Here we calculate the large-q expansion series of the energy cumulants including the specific heat and the magnetization cumulants including the magnetic susceptibility in both the phases and the correlation length in the ordered phase at the transition point using the finite lattice method.[7,8,9] Obtained long series for the energy and magnetization cumulants give the estimates of the quantities that are more precise by a factor of $10^2 - 10^4$ than the Monte Carlo simulations. Especially its estimates are within an accuracy of 0.1% at $q = 5$, where the correlation length is as large as a few thousands of the lattice spacing. Bhattacharya *et al.*[10] made a stimulating conjecture on the asymptotic behavior of the energy cumulants at the first order transition point; the relation between the energy cumulants and the correlation length in their asymptotic behavior at the first order transition point for $q \to 4_+$ will be the same as their relation in the second order phase transition for $q = 4$ and $\beta \to \beta_t$, which is well known from their critical exponents. The obtained series enables us to confirm the correctness of the conjecture.

As for the correlation length at the first order phase transition point, the results of the Monte Carlo simulation[11] and the density matrix renormalization group[12] indicate that the correlation lengths are very close to each other in the ordered and disordered phases for $q \gtrsim 10$. On the other hand, at the second order phase transition point($q \leq 4$) their ratio is known to be 1/2. It is interesting whether the ratio is exactly equal to unity, remains close to unity, or approaches 1/2 when $q \to 4_+$. To investigate it we calculate the first few terms of the large-q expansion for the eigenvalues of the transfer matrix and find that from the second largest to the N-th largest eigenvalues with N the one-dimensional size of the lattice make a continuum spectrum in the thermodynamic limit both in the ordered and disordered phases. We also calculate the long series of the second moment correlation length in both the phases, which serve to investigate the behavior of the spectrum of the eigenvalues of the transfer matrix in the region of q close to 4.

2 Finite lattice method

Here we use the finite lattice method,[7,8,9] to generate the large-q series for the Potts model, instead of the standard diagrammatic method used by Bhattacharya *et al.*[13] The finite lattice method can in general give longer series than those generated by the diagrammatic method especially in lower spatial dimensions. In the diagrammatic method, one has to list up all the relevant diagrams and count the number they appear. In the finite lattice method we can skip this job and reduce the main work to the calculation of the expansion of the partition function for a series of finite size lattices, which can be done using the straightforward site-by-site integration[14,15] without the diagrammatic technique. This method has been used mainly to generate the low- and high-temperature series in statistical systems and the strong coupling series in lattice gauge theory. We note that this method is applicable to the series expansion with respect to any parameter other than the temperature or the coupling constant. Using this method we calculated[17,21] the series for the n-th energy cumulants ($n = 0-6$) to z^{23}, n-th magnetization cumulants ($n = 1-3$) to z^{21}, and second moment correlation length to z^{19} with $z \equiv 1/\sqrt{q}$.

Figure 1: Padé approximants of $F_d^{(2)}\mathcal{L}^p$ at $q = 4$ plotted versus p.

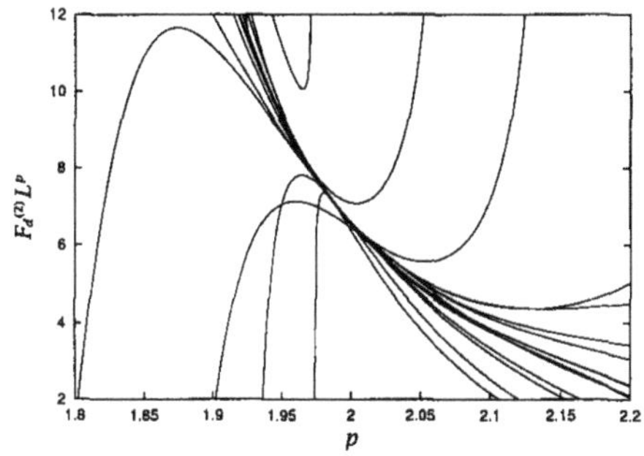

3 Energy cumulants

The latent heat $\mathcal{L}$ at the transition point are known to vanish at $q \to 4_+$ as

$$\mathcal{L} \sim 3\pi x^{-1/2} ,$$

with $x = \exp(\pi^2/2\theta)$ and $2\cosh\theta = \sqrt{q}$. Bhattacharya *et al.*'s conjecture says that the n-th energy cumulants $F_{d,o}^{(n)}$ at the first order transition point $\beta = \beta_t$ will diverge for $q \to 4_+$ as

$$F_d^{(n)}, (-1)^n F_o^{(n)} \sim \alpha B^{n-2} \frac{\Gamma\left(n - \frac{4}{3}\right)}{\Gamma\left(\frac{2}{3}\right)} x^{3n/2-2} . \tag{1}$$

The constants α and B in Eq.(1) should be common to the ordered and disordered phases from the duality relation for each n-th cumulants.

If this conjecture is true, the product $F^{(n)}\mathcal{L}^{3n-4}$ is a smooth function of θ, so we can expect that the Padé approximation of $F^{(n)}\mathcal{L}^p$ will give convergent result at $p = 3n - 4$. It has been examined for the large-q series obtained by the finite lattice methods for $n = 2, \cdots, 6$ both in the ordered and disordered phases, which in fact give quite convergent Padé approximants for $p = 3n - 4$ and as p leaves from this value the convergence of the approximants becomes bad rapidly. An example can be seen in Fig. 1 for n=2 in the disordered phase. We give in Table 1 the values of the specific heat $C = \beta_t^2 F^{(2)}$ evaluated from these Padé approximants for some values of q. These estimates are three or four orders of magnitude more precise than (and consistent with) the previous result for $q \geq 7$ from the large-q expansion to order z^{10} by Bhattacharya *et al.*[13] and the result of the Monte Carlo simulations for $q = 10, 15, 20$ carefully done by Janke and Kappler[16] as in Table 2. What should be emphasized is that we obtained the values of the specific heat in the accuracy of about 0.1 percent at

Table 1: The specific heat for some values of q. The exact correlation length is also listed.

q	C_d	C_o	ξ_d(exact)
5	2889(2)	2886(3)	2512.2
6	205.93(3)	205.78(3)	158.9
7	68.738(2)	68.513(2)	48.1
8	36.9335(3)	36.6235(3)	23.9
9	24.58761(8)	24.20344(7)	14.9
10	18.38543(2)	17.93780(2)	10.6
15	8.6540358(4)	7.9964587(2)	4.2
20	6.13215967(2)	5.36076877(1)	2.7

Table 2: Comparison with the Monte Carlo simulations by Janke and Kappler(1997).

		$q=10$	$q=15$	$q=20$
C_d	large-q	18.38543(2)	8.6540358(4)	6.13215967(2)
	Monte Carlo	18.44(4)	8.651(6)	6.133(4)
C_o	large-q	17.93780(2)	7.9964587(2)	5.36076877(1)
	Monte Carlo	18.0(1)	7.99(2)	5.361(9)

$q = 5$ where the correlation length is as large as 2500. As for the asymptotic behavior of $F^{(n)}$ at $q \to 4_+$, the Padé data of $F_d^{(2)}/x$ and $F_o^{(2)}/x$ have the errors of a few percent around $q = 4$ and their behaviors are enough to convince us that the conjecture (1) is true for $n = 2$ with $\alpha = 0.073 \pm 0.002$. Furthermore from the conjecture (1) the combination $\left\{\Gamma\left(n-\frac{4}{3}\right)|F^{(n)}|/\Gamma\left(\frac{2}{3}\right)F^{(2)}\right\}^{\frac{1}{n-2}}x^{-\frac{3}{2}}$ is expected to approach the constant B for each $n(\geq 3)$, and in fact the Padé data for every $n(=3,\cdots,6)$ gives $B = 0.38 \pm 0.05$, which also gives strong support to the conjecture for $n \geq 3$.

4 Magnetization cumulants

The behavior of the n-th magnetization cumulants $M^{(n)}$ for $\beta \to \beta_t$ at $q = 4$ is well known as $M_{d,o}^{(n)} \simeq A_{d,o}^{(n)}(\xi)^{\frac{15}{8}n-2}$ and parallel to the conjecture for the energy cumulants by Bhattacharya *et al.* we can make a conjecture that

$$M_{d,o}^{(n)} \sim \mu_{d,o}^{(n)} x^{\frac{15}{8}n-2} . \tag{2}$$

in the limit $q \to 4_+$ with $\beta = \beta_t$. We have examined the Padé approximation of $M^{(n)}\mathcal{L}^p$ for the large-q series generated by the finite lattice method for $n = 2$ and 3 both in the ordered and disordered phases, which in fact gives

Table 3: The magnetic susceptibility for some values of q.

q	χ_d	χ_o
5	$9.13(3) \times 10^4$	$9.01(3) \times 10^4$
6	$6.585(4) \times 10^2$	$6.665(4) \times 10^2$
7	70.54(1)	77.31(1)
8	19.359(1)	21.525(1)
9	8.0579(1)	9.0106(2)
10	4.23276(2)	4.73823(4)
15	0.7304214(1)	0.8056969(2)
20	0.309365682(1)	0.33556421(1)

Table 4: Comparison with the Monte Carlo simulations by Janke and Kappler(1997).

		$q=10$	$q=15$	$q=20$
$M_d^{(2)}$	large-q	4.23276(2)	0.7304214(1)	0.309365682(1)
	M.C.	4.233(2)	0.73039(8)	0.30936(4)
$M_o^{(2)}$	large-q	4.73823(4)	0.8056969(2)	0.33556421(1)
	M.C.	4.74(4)	0.805(3)	0.3355(8)

quite convergent Padé approximants for $p = 15n/4 - 4$ and as p leaves from this value the convergence of the approximants becomes bad rapidly again. In Table 3 we present the resulting estimates of the magnetic susceptibility $\chi_{d,o} = M_{d,o}^{(2)}$. Our result is much more precise than the Monte Carlo simulation[16] at least by a factor of 100 as in Table 4. From the behavior of $M_{d,o}^{(n)}/x^{(15n/8-2)}$ we obtain the coefficients in Eq.(2) as $\mu_d^{(2)} = 0.0020(2)$, $\mu_o^{(2)} = 0.0016(1)$ and $\mu_d^{(3)} = 7.4(5) \times 10^{-5}$, $\mu_o^{(3)} = 7.9(2) \times 10^{-5}$. These convince us that the conjecture made for the magnetization cumulants is also true.

5 Exponential correlation length

Next we investigate the exponential correlation length $\xi_{1,o}$ in the ordered phase. Here the exponential correlation length is defined by $\xi_1 = \log(\Lambda_1/\Lambda_0)$ with Λ_0 and Λ_1 the largest and the second largest eigenvalues of the transfer matrix, respectively. (We have omitted the subscript '1' in the previous sections) There is an obstacle to extract the correction to the leading term of the large-q expansion for the correlation length at the phase transition point from the correlation function $< \mathcal{O}(t)\mathcal{O}(0) >_c$, since we know from the graphical expansion that it behaves like

$$< \mathcal{O}(t)\mathcal{O}(0) >_c \propto z^t(1 + 2zt^2 + \cdots) \qquad (3)$$
$$\neq \exp(-mt).$$

for a large distance t. Thus we will diagonalize the transfer matrix T directly for large-q. The eigenfunction for the largest eigenvalue Λ_0 in the leading order of z is

$$|0> \equiv |\underbrace{0\ 0\ 0\ .....0\ 0\ 0}_{N}>$$

where all of the N spin variables (each of which can take the value of $s = 0, 1, \cdots, q-1$) are fixed to be zero, with the element of the transfer matrix $< 0|T|0 > = 1 + O(z^2)$ and the corresponding eigenvalue is $\Lambda_0 = 1 + O(z)$. The eigenfunctions for the second largest eigenvalue are

$$|I> \equiv \frac{1}{\sqrt{N-I+1}} \sum |0\ ...0\ \underbrace{e\ e\ e\ ...e\ e}_{I} 0\ ...0>$$
$$(I = 1, \cdots, N),$$

$$|e> \equiv \frac{1}{\sqrt{q-1}} \sum_{s=1}^{q-1} |s>$$

with the diagonal matrix elements

$$< I|T|I > = z + O(z^2),$$

and the off-diagonal matrix elements starting from higher orders in z. The second largest eigenvalues of T degenerate in the leading order with $\Lambda_i = z + O(z^{3/2})$. The degeneracy of the eigenvalues of the first N 'excited states' is the reason why the expansion series (3) of the correlation function cannot be exponentiated into a single exponential term. The off-diagonal matrix elements resolve the degeneracy with

$$\Lambda_1/\Lambda_0 = z + 4z^{3/2} + 6z^2 + O(z^3), \qquad (4)$$
$$\cdots$$
$$\Lambda_N/\Lambda_0 = z - 4z^{3/2} + 6z^2 + O(z^3).$$

for $N \to \infty$. These eigenvalues constitute a continuum spectrum. It appears to be kept in any higher order of z. From Eq.(4) we obtain

$$1/\xi_{1,o} = -\log z - 4z^{1/2} + 2z + 8/3z^{3/2} + O(z^2).$$

This is the same as the large-q expansion of $1/\xi_{1,d}$ given by Buffenoir and Wallon[5] to this order.

In the disordered phase the situation is quite similar. The eigenvalues of the transfer matrix for the first N excited states constitute a continuum spectrum with their values exactly the same as in the ordered phase at least to the order of $z^{5/2}$.

The eigenvalues form the continuum spectrum just on the first order phase transition point. Off the transition point, we can see that the spectrum is discrete with $\Lambda_i - \Lambda_{i+1} \sim O(\epsilon)$ for $\sqrt{z} \ll \epsilon \ll 1$ and $\Lambda_i - \Lambda_{i+1} \sim O(\epsilon^{2/3})$ for $\epsilon \ll \sqrt{z} \ll 1$ where $\epsilon \equiv \beta/\beta_t - 1$.

6 Second moment correlation length

Here we give the results of the large-q expansion of the second moment correlation lengths $\xi_{2nd,o}$ in the ordered phase and $\xi_{2nd,d}$ in the disordered phase defined by

$$\xi_{2nd}^2 = \frac{\mu_2}{2d\mu_0},$$

where μ_2 and μ_0 are the second moment of the correlation function and the magnetic susceptibility, respectively. The obtained expansion coefficients[21] for the ordered and disordered phases coincide with each other to order z^3 and differ from each other in higher orders. The ratio of the second moment correlation length in the ordered phase to that in the disordered phase is estimated by the Padé analysis to be not far from unity even in the limit of $q \to 4$ ($\xi_{2nd,o}/\xi_{2nd,d} = 0.930(3)$). Another point is that the ratio $\xi_{2nd,d}/\xi_{1,d}$ of the second moment correlation length to the exponential correlation length is much less than unity in the region of q where the correlation length is large enough. It approaches 0.51(2) for $q \to 4$. It is known that in the limit of the large correlation length,

$$\frac{\xi_{2nd}^2}{\xi_1^2} \to \frac{\sum_{i=1}^{\infty} c_i^2 (\xi_i/\xi_1)^3}{\sum_{i=1}^{\infty} c_i^2 (\xi_i/\xi_1)} < 1,$$

with $\xi_i \equiv -\log(\Lambda_i/\Lambda_0)$. If the 'higher excited states' ($i = 2, 3, \cdots$) did not contribute so much, this ratio would be close to unity, as in the case of the Ising model on the simple cubic lattice, where $\xi_{2nd}/\xi_1 = 0.970(5)$ at the critical point.[18,19] Our result implies that the contribution of the 'higher excited states' is large in the disordered phase of the Potts model in two dimensions even when q is close to 4. This strongly suggests that the eigenvalues of the transfer matrix for the first N excited states in the disordered phase form the continuum spectrum not only in the large-q region but also when q approaches 4.

As for the exponential correlation length $\xi_{1,o}$ in the ordered phase for $q \to 4$, it is difficult to calculate the eigenvalues of the transfer matrix in much higher orders. It is quite natural, however, to expect that the ratio $\xi_{1,o}/\xi_{1,d}$ would be close to unity even in the limit of $q \to 4$. The reason is the following. If the ratio $\xi_{1,o}/\xi_{1,d}$ would be $1/2$ in the limit of $q \to 4$, which is the known ratio in the second order phase transition point($q \leq 4$), then the ratio $\xi_{2nd,o}/\xi_{1,o}$ should be close to unity, which would imply that the higher excited states would not contribute so much to $\xi_{2nd,o}$ and the eigenvalue of the transfer matrix for the first excited state would be separated from the higher excited states. This scenario is not plausible, since as already mentioned in section 5 the continuum spectrum of the eigenvalues of the transfer matrix appears to be maintained in any high order in z. In this case, we can expect that the ratio $\xi_{2nd,o}/\xi_{1,o}$ would be around $1/2$ as is the case in the disordered phase, resulting that $\xi_{1,o}/\xi_{1,d}$ is close to unity.

7 Summary

We generated the large-q series for the energy and magnetization cumulants at the first order phase transition point of the two-dimensional q-state Potts model in high orders using the finite lattice method. They gave very precise estimates of the cumulants for $q > 4$ and confirmed the correctness of the Bhattacharya *et al.*'s conjecture that the relation between the cumulants and the correlation length for $q = 4$ and $\beta \to \beta_t$ (the second order phase transition) is kept in their asymptotic behavior for $q \to 4_+$ at $\beta = \beta_t$ (the first order transition point). If this kind of relation is satisfied as the asymptotic behavior for the quantities at the first order phase transition point in more general systems when the parameter of the system is varied to make the system close to the second order phase transition point, it would serve as a good guide in investigating the system.

Further the large-q expansion of the eigenvalues of the transfer matrix was calculated in the first 4 terms. We found that they have the same spectra of the eigenvalues of the transfer matrix in the ordered and disordered phases giving the same exponential correlation length ($\xi_{1,o} = \xi_{1,d}$) to the order of $z^{3/2}$ and that the spectra are continuous in the thermodynamic limit. We also calculated the large-q expansion of the second moment correlation length in the ordered and disordered phases in high orders and found that they differ from each other in higher orders than z^3, but that the ratio $\xi_{2nd,d}/\xi_{1,d}$ is not far from unity for all region of $q > 4$. We also found that $\xi_{2nd,d}/\xi_{d,1}$ is far from unity even in the limit of $q \to 4$. It receives significant contributions not only from the 'first excited state' but also 'higher excited states' and this suggest strongly that the continuum spectrum would be maintained (i.e. there would be no particle state) in the disordered phase. From these results it is quite natural to expect that the exponential correlation length $\xi_{1,o}$ in the ordered phase is not far from that in the disordered phase even in the limit of $q \to 4$ and it is not plausible that their ratio approaches 1/2 that is their ratio in the second order phase transition point($q \leq 4$).

References

1. R. B. Potts, *Proc. Camb. Phil. Soc.* **48**, 106 (1952).
2. F. Y. Wu, *Rev. Mod. Phys.* **54**, 235 (1982).
3. R. J. Baxter, *J. Phys.* C **6**, L445 (1973); *J. Stat. Phys.* **9**, 145 (1973).
4. A. Klümper, A. Schadschneider and J. Zittartz, *Z. Phys.* B **76**, 247 (1989).
5. E. Buffenoir and S. Wallon, *J. Phys.* A **26**, 3045 (1993).
6. C. Borgs and W. Janke, *J. Phys.* I (France) **2**, 649 (1992).
7. T. de Neef and I. G. Enting, *J. Phys.* A **10**, 801 (1977); I. G. Enting, *J. Phys.* A **11**, 563 (1978); *Nucl. Phys.* B (Proc. Suppl.) **47**, 180 (1996).
8. M. Creutz, *Phys. Rev.* B **43**, 10659 (1991).
9. H. Arisue and T. Fujiwara, *Prog. Theor. Phys.* **72**, 1176 (1984); H. Arisue, *Nucl. Phys.* B (Proc. Suppl.) **34**, 240 (1994).
10. T. Bhattacharya, R. Lacaze and A. Morel, *Nucl. Phys.* **B435**, 526 (1995).
11. W. Janke and S. Kappler, *Nucl. Phys.* B (Proc. Suppl.) **34**, 1155 (1994).
12. F. Iglói and E. Carlon *Phys. Rev.* B **59**, 3783 (1999).
13. T. Bhattacharya, R. Lacaze and A. Morel, *J. Phys. I* (France) **7**, 1155 (1997).
14. I. G. Enting, *J. Phys. A* **13**, 3713 (1980).
15. G. Bhanot, *J. Stat. Phys.* **60**, 55 (1990).
16. W. Janke and S. Kappler, *J. Phys. I* (France) **7**, 1155 (1997).
17. H. Arisue and K. Tabata, *Phys. Rev.* E **59**, 186 (1999), *Nucl. Phys.* B **546[FS]**, 558 (1999).
18. M. Caselle, M. Hasenbusch and P. Provero, hep-lat/9903011.
19. M. Campostrini, A. Pelissetto P. Rosi and E. Vicari, *Phys. Rev.* E **57**, 184 (1998).
20. H. Arisue and K. Tabata *Phys. Rev.* E **59**, 186 (1999), H. Arisue and K. Tabata *Nucl. Phys.* B **546[FS]**, 558 (1999).
21. H. Arisue, hep-lat/0001014.

MULTI-OVERLAP MONTE CARLO STUDIES OF SPIN GLASSES

WOLFHARD JANKE
Institut für Theoretische Physik, Universität Leipzig, 04109 Leipzig, Germany
E-mail: wolfhard.janke@itp.uni-leipzig.de

BERND A. BERG
Department of Physics, The Florida State University, Tallahassee, FL 32306, USA
E-mail: berg@scri.fsu.edu

ALAIN BILLOIRE
CEA/Saclay, Service de Physique Théorique, 91191 Gif-sur-Yvette, France
E-mail: billoir@spht.saclay.cea.fr

We discuss recent high statistics Monte Carlo simulations of the Edwards-Anderson Ising spin-glass model in three and four dimensions based on a non-Boltzmann sampling technique. Particular emphasis is placed on those properties of the probability density $P_{\mathcal{J}}(q)$ of the Parisi overlap parameter q which are difficult to obtain with canonical simulations relying on the standard Boltzmann distribution. This comprises the free-energy barriers F_B^q which are visible in $P_{\mathcal{J}}(q)$ and the behavior of the tails of the probability density.

1 Introduction

Typical examples of spin glasses are dilute solutions of transition metal magnetic impurities in noble hosts[1]. Experimentally well studied systems are, for instance, Cu-0.9% Mn and $(Fe_{0.15}Ni_{0.85})_{75}P_{16}B_6Al_3$[2,3]. The randomly distributed impurities induce a magnetic polarization of the host metal electrons which is positive at some distances and negative at others. This in turn is responsible for competing interactions between the magnetic moments and hence frustration in the spin configurations which is the characteristic feature of these materials. Measurements of the remanent magnetization in the spin-glass phase below the freezing point exhibit the phenomenon of aging and thus suggest the presence of many equilibrium or metastable configurations with a distribution of free-energy barriers separating them. This is one of the reasons why up today spin glasses are among the most challenging systems of statistical physics[1,4,5,6].

Since the middle of the 1970's the peculiar behavior of the spin-glass phase has attracted the interest of many workers employing a variety of different approaches. Despite the large amount of experimental, mathematical, theoretical and simulational work done, the physical mechanisms underlying these phenomena are not yet fully understood. Due to the enormous complexity even for simplified models, analytical approaches so far focussed mainly on approximations. Mean-field treatments[7] are expected to become accurate in high dimensions, but their status is unclear in physical dimensions, where the less controlled droplet approximation[8] has been proposed as an alternative. Numerical approaches such as Monte Carlo simulations can, in principle, provide arbitrarily precise results in physical dimensions. In practice, however, the simulational approach is severely hampered by the extremely slow (pseudo-) dynamics which numerically reflects the experimental findings of free-energy barriers separating metastable configurations.

A quantitative characterization of these free-energy barriers would be a clue for a better understanding of spin glasses. This was one of the main motivations for our series of Monte Carlo simulations[9,10,11] of the Edwards-Anderson[12] Ising (EAI) spin-glass model whose main results obtained so far will be briefly summarized in the following.

2 The Model and Simulation Methods

The energy of the EAI spin-glass model is defined by

$$E = -\sum_{\langle ik \rangle} J_{ik}\, s_i s_k \, , \tag{1}$$

where the sum is over nearest-neighbor pairs of a d-dimensional (hyper-) cubic lattice of size $N = L^d$ with periodic boundary conditions, and the fluctuating spins s_i as well as the quenched exchange coupling constants J_{ik} take on the values ± 1, with equal probabilities. As order parameter one usually takes the Parisi overlap parameter[7]

$$q = \frac{1}{N} \sum_{i=1}^{N} s_i^{(1)}\, s_i^{(2)} \, , \tag{2}$$

where the spin superscripts label two independent (real) replicas for the same realization of randomly chosen exchange coupling constants $\mathcal{J} = \{J_{ik}\}$. For given $\mathcal{J}$ the probability density of q is denoted by $P_{\mathcal{J}}(q)$ and Parisi's function $x_{\mathcal{J}}(q) = \int_{-q}^{q} dq' P_{\mathcal{J}}(q')$ is essentially its cumulative distribution function as, for instance, defined in Ref. 13. By averaging over the quenched disorder, one arrives at the functions

$$P(q) = [P_{\mathcal{J}}(q)]_{\rm av} = \frac{1}{\#J} \sum_{\mathcal{J}} P_{\mathcal{J}}(q) \, ; \quad x(q) = [x_{\mathcal{J}}(q)]_{\rm av} = \frac{1}{\#J} \sum_{\mathcal{J}} x_{\mathcal{J}}(q) \, , \tag{3}$$

Table 1. Number of realizations $\#\mathcal{J}$, number of requested flips $n_{\rm flip}$ during weight construction, number of equilibration sweeps $n \times 65536$, and average number of sweeps per realization $n_{\rm sw}$.

	3d, $\beta = 1.0$				4d, $\beta = 0.6$			
L	$\#\mathcal{J}$	$n_{\rm flip}$	n	$n_{\rm sw}$	$\#\mathcal{J}$	$n_{\rm flip}$	n	$n_{\rm sw}$
4	8 192	10	1	0.2×10^6	4 096	10	2	0.4×10^6
6	8 192	10	4	1.0×10^6	4 096	20	8	3.7×10^6
8	8 192	10	16	7.6×10^6	1 024	20-30	16	49.3×10^6
12	640	20	32	154.0×10^6				

where $\#J$ ($\rightarrow \infty$) is the number of realizations considered. Below the freezing temperature, in the infinite-volume limit $N \rightarrow \infty$, an increasing *continuous* part of $x(q)$ characterizes the mean-field picture[7] of spin glasses, whereas in ferromagnets as well as in the droplet picture[8] of spin glasses $x(q)$ is a step function. Most studies so far have focussed on this averaged quantity.

The information on free-energy barriers, however, is encoded in the distribution functions for individual realizations $\mathcal{J}$. In our work we concentrated on those free-energy barriers F_B^q which are reflected by the $P_{\mathcal{J}}(q)$ minima. Conventional, canonical Monte Carlo (MC) simulations are not suited for such a study because the likelihood to generate the corresponding configurations in the Gibbs canonical ensemble is very small. This problem is overcome by the multi-overlap MC algorithm[9] which samples with the non-Boltzmann weight

$$w_{\mathcal{J}}(q) = \exp\left[-\beta(E_{\mathcal{J}}^{(1)} + E_{\mathcal{J}}^{(2)}) + S_{\mathcal{J}}(q)\right] , \tag{4}$$

where $\beta = J_0/k_B T$ is the inverse temperature in natural units. $E_{\mathcal{J}}^{(1)}$ and $E_{\mathcal{J}}^{(2)}$ are the energies of the two replicas, coupled by $S_{\mathcal{J}}(q)$ in such a way that a broad histogram in q over the entire accessible range $-1 \leq q \leq 1$ is obtained. The iterative construction of the weight function $S_{\mathcal{J}}(q)$ is the first step of the algorithm. Once $S_{\mathcal{J}}(q)$ is determined and kept fixed, the system is equilibrated and then the data production is performed. Finally any canonical quantity can be computed by reweighting.

We assess the (pseudo-) dynamics of the multi-overlap algorithm by measuring the autocorrelation time $\tau_{\mathcal{J}}^{\rm muq}$ which counts the average number of sweeps it takes to create a "tunneling" event $q = 0 \rightarrow q = \pm 1$ and back. The iteration for the weight construction was stopped after $n_{\rm flip}$ "tunneling" events, cf. Table 1. Each production run of data taking was concluded after at least 20 "tunneling" events. To allow for standard reweighting in tempera-

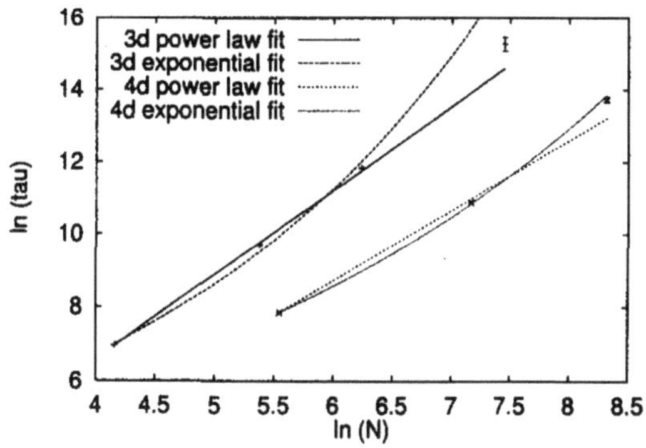

Figure 1: Average autocorrelation times $[\tau_{\mathcal{J}}^{\mathrm{muq}}]_{\mathrm{av}}$ of the multi-overlap algorithm.

ture we stored besides $P_{\mathcal{J}}(q)$ also a time series of measurements for the order parameter, energies and magnetizations of the two replicas. The number of sweeps between two successive points in a time series was adjusted by an adaptive data compression routine to ensure that each time series consists of $2^{16} = 65536$ measurements separated by approximately $\tau_{\mathcal{J}}^{\mathrm{muq}}$ sweeps.

3 Results

The simulations were performed in the spin-glass phase at $\beta = 1.0 > \beta_c = 0.90 \pm 0.03$ (Ref. 14) in three dimensions (3*d*) and at $\beta = 0.6 > \beta_c = 0.485 \pm 0.005$ (Ref. 15) in four dimensions (4*d*). The number of simulated realizations and the average length of the runs are compiled in Table 1. The J_{ik} realizations were drawn using the pseudo random number generators RANMAR [16] and RANLUX [17] (luxury level 4). In the simulations themselves we always employed the RANMAR generator.

3.1 Autocorrelation times of the Multi-Overlap Algorithm

For each realization we measured the autocorrelation time $\tau_{\mathcal{J}}^{\mathrm{muq}}$. Fitting the averages over the realizations $\mathcal{J}$ to the power-law ansatz $\ln([\tau_{\mathcal{J}}^{\mathrm{muq}}]_{\mathrm{av}}) = a + z\ln(N)$ we obtained $z = 2.32 \pm 0.07$ in 3*d* and $z = 1.94 \pm 0.02$ in 4*d*. The quality of the fits is bad; see Fig. 1 where for comparison also exponential fits are shown. But they clearly show that the slowing down is quite off from the theoretical optimum $z = 1$ one would expect if the free-energy barriers visible

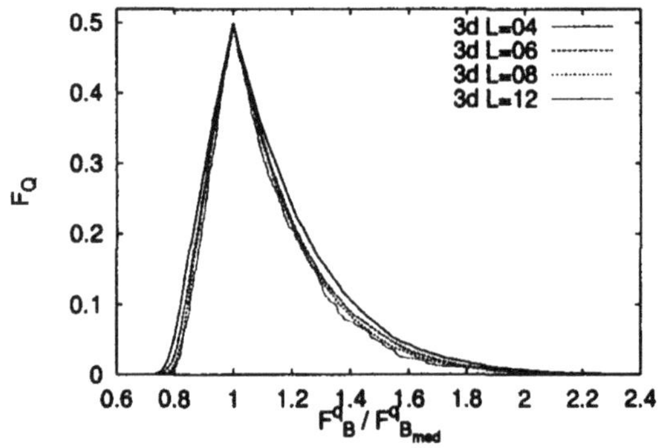

Figure 2. Distribution function F_Q (7) for the $3d$ overlap barriers (6) in units of their median value.

in the overlap parameter q were the only cause for the slowing down of the canonical dynamics. In this case the multi-overlap autocorrelation time $\tau_{\mathcal{J}}^{\mathrm{muq}}$ should be dominated by a random walk behavior between $q = -1$ and $q = +1$ and scale proportional to N in units of sweeps in the limit of perfect flatening. The large values of z imply that the canonical overlap barriers cannot be the exclusive cause for the slowing down of spin-glass dynamics below the freezing point, i.e., the projection of the multi-dimensional phase space onto the q direction hides important features of the free-energy landscape of the model.

3.2 Finite-Size Scaling Behavior of Free-Energy Barriers

Our effective free-energy barriers F_B^q are defined through an auxiliary $1d$ Metropolis-Markov chain[18] which has the canonical $P_{\mathcal{J}}(q)$ probability density as its equilibrium distribution. The tridiagonal transition matrix T is given in terms of the probabilities $w_{i,j} = \frac{1}{2}\min\Big(1, P_{\mathcal{J}}(q_i)/P_{\mathcal{J}}(q_j)\Big)$ $(i \neq j)$ to jump from state $q = q_j$ to $q = q_i = i/N$ in steps of $\Delta q = \pm 2$ or 0. Since T fulfills the detailed balance condition (with $P_{\mathcal{J}}$) it has only real eigenvalues. The largest eigenvalue (equal to one) is non-degenerate, and the second largest eigenvalue λ_1 determines the autocorrelation time of the chain (in units of sweeps),

$$\tau_B^q = -\frac{1}{N \ln \lambda_1} \approx \frac{1}{N(1-\lambda_1)}\,, \tag{5}$$

which in turn defines the associated effective free-energy barrier

$$F_B^q = \ln(\tau_B^q)\,. \tag{6}$$

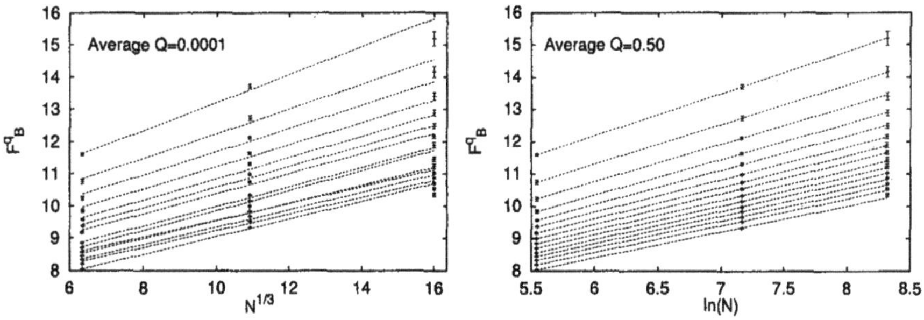

Figure 3. FSS fits at $F = i/16$, $i = 1, \ldots, 15$ (from down to up) of the $4d$ free-energy barriers F^q_B to the mean-field behavior (9) (left) and the logarithmic ansatz (10) (right).

This definition generalizes the standard formula $F^q_B = \ln[P^{\max}_{\mathcal{J}}/P^{\min}_{\mathcal{J}}] + const.$. for the simple double-peak situation of first-order phase transitions.

For the analysis of the free-energy barriers we employed the peaked distribution function[19]

$$F_Q(x) = \begin{cases} F(x) \text{ for } F(x) \le 0.5 \ ; \\ 1 - F(x) \text{ for } F(x) \ge 0.5 \ , \end{cases} \tag{7}$$

where $F(x)$ is the standard cumulative distribution function defined by[13]

$$\frac{i}{n} - \frac{1}{2n} \le F(x) \le \frac{i}{n} + \frac{1}{2n} \quad \text{for} \quad x_i \le x \le x_{i+1} \ , \tag{8}$$

with straight-line interpolations in-between. For self-averaging data the function F_Q collapses in the infinite-volume limit to $F_Q(x) = 0.5$ for $x = [x]_{\text{av}}$ and 0 otherwise. For non-self-averaging quantities the width of F_Q stays finite. The concept carries over to quantities which diverge in the infinite-volume limit, when for each lattice size scaled variables x/x_{med} are used.

The behavior of $F_Q(F^q_B/F^q_{B,\text{med}})$ shown in Fig. 2 for the $3d$ case clearly suggests that F^q_B is a non-self-averaging quantity. In $4d$ the evidence is even stronger than in $3d$. Non-self-averaging was also observed for the autocorrelation times $\tau^{\text{muq}}_{\mathcal{J}}$ of our algorithm while the energy is an example for a self-averaging quantity. For non-self-averaging quantities one has to investigate many samples and should report the finite-size scaling (FSS) behavior for fixed values of the cumulative distribution function F. In Ref. 11 we performed for $F = i/16$, $i = 1, \ldots, 15$, fits to the ansatz

$$F^q_B = a_1 + a_2\, N^{1/3} \ , \tag{9}$$

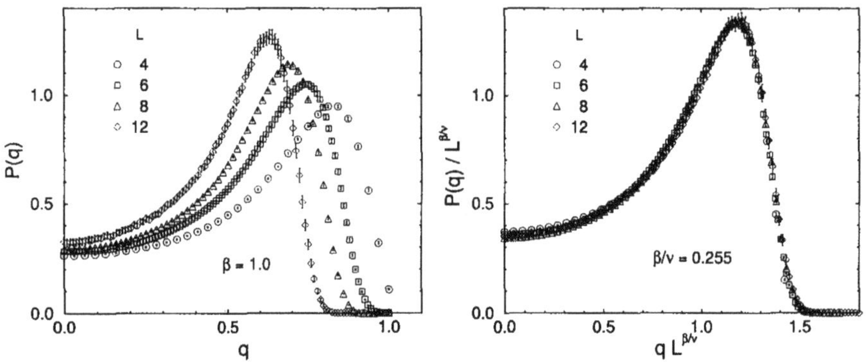

Figure 4. Averaged $3d$ probability densities $P(q)$ in raw (left) and scaled (right) form.

corresponding to $\tau_B^q = e^{a_1}\, e^{a_2\, N^{1/3}}$, as suggested by investigations of barriers and autocorrelation times in the mean-field limit[20]. In both dimensions the goodness-of-fit parameter[13] Q turned out to be unacceptably small. The $4d$ fits are depicted on the l.h.s. of Fig. 3. We therefore tried fits to the ansatz

$$F_B^q = \ln(c) + \alpha\, \ln(N)\ , \tag{10}$$

corresponding to $\tau_B^q = c\, N^\alpha$; cf. the r.h.s. of Fig. 3. Since in $4d$ as well as in $3d$ the average Q-value is now within the statistical expectation, the ansatz (10) is strongly favored over (9). As a function of F ($= 1/16 - 15/16$) the exponent $\alpha = \alpha(F)$ in the power law (10) varies smoothly from 0.8 to 1.1 in $3d$ and from 0.8 to 1.3 in $4d$. A similar analysis for the autocorrelation times $\tau_{\mathcal{J}}^{\rm muq}$ of the multi-overlap algorithm gives exponents $\alpha(F)$ which are larger, $\alpha^{\rm muq}(F) \approx \alpha_B^q(F) + 1$. This re-iterates our previous observation that relevant barriers exist, which are invisible in the overlap variable q.

3.3 Finite-Size Scaling Behavior of Averaged Probability Densities

The averaged densities $P(q)$ of the $3d$ model at $\beta = 1.0$ are shown on the l.h.s. of Fig. 4. By reweighting we estimated $\beta_c \approx 0.88$ and found a FSS of the spin-glass susceptibility $\chi_{\rm SG} = N[\langle q^2\rangle]_{\rm av} \propto L^{\gamma/\nu}$ with $\gamma/\nu = 2.37(4)$. Close to the transition point one expects that $P(q)$ satisfies the FSS relation $P(q) = L^{\beta/\nu}\hat{P}(L^{\beta/\nu}q)$. Using the scaling relation $\beta/\nu = (d - \gamma/\nu)/2 \approx 0.317$ this is indeed the case at β_c. More surprisingly is the fact that also in the spin-glass phase at the simulation point $\beta = 1.0$ the data for different lattice sizes fall onto a common master curve; cf. the r.h.s. of Fig. 4, where

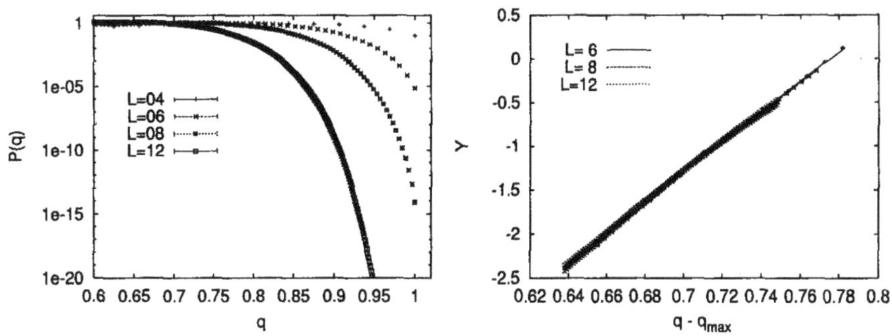

Figure 5. Tails of the $P(q)$ densities in $3d$ (left) and fits to the ansatz (12) (right).

an exponent of $\beta/\nu = 0.255$ is used. This implies that the spin-glass phase is either governed by massless modes or that at least the correlation length ξ is large ($\xi > L_{\max} = 12$).

3.4 Finite-Size Scaling Behavior of the Tails of $P(q)$

The multi-overlap algorithm becomes particularly powerful when studying the tails of the distributions which are highly suppressed compared to the peak values; see the l.h.s. of Fig. 5. Based on the replica mean-field approach, theoretical predictions for the scaling behavior of the tails have been made by Parisi and collaborators[21]. They show that $P(q) = P_{\max} f(N(q - q^{\infty}_{\max})^x)$ for $q > q_{\max}$ and predict more quantitatively that

$$P(q) \sim \exp\left[-c_1 N (q - q^{\infty}_{\max})^x\right] \quad \text{for} \quad N(q - q_{\max})^x \quad \text{large}, \qquad (11)$$

with a mean-field exponent of $x = 3$. The additional condition $q \le q_{\text{cut}} < 1$ ensures that q stays away from the lattice artefacts close to $q = 1$. By allowing for an overall normalization factor $c_0^{(N)}$ and taking the logarithm twice we have performed fits of the form

$$Y \equiv \ln\left[-\ln(P/c_0^{(N)})\right] - \ln N = \ln c_1 + x \ln(q - q^{\infty}_{\max}) \ . \qquad (12)$$

To enforce the conditions $N(q - q^{\infty}_{\max})^x$ large as well as $q \le q_{\text{cut}} < 1$ we restrict our fits to the q-range defined by $2^{-1} P_{\max} \ge P(q) \ge 2^{-10} P_{\max}$ which excluded the use of the $L = 4$ data in $3d$. The fits of the data from the remaining lattices in $3d$ are depicted on the r.h.s. of Fig. 5. Leaving the exponent x as a free parameter, we arrived at the estimates

$$x = 12 \pm 2 \text{ in } 3d \ (\beta = 1.0) \quad \text{and} \quad x = 5.3 \pm 0.3 \text{ in } 4d \ (\beta = 0.6) \qquad (13)$$

for the infinite-volume exponents.

4 Summary and Conclusions

We have investigated the probability densities $P_{\mathcal{J}}(q)$ of the Parisi order parameter q.

The free-energy barriers F_B^q as defined in (6) are found to be non-self-averaging. The logarithmic scaling ansatz (10) for the barriers at fixed values of F is favored over the mean-field ansatz (9). Further, relevant barriers are still found in the autocorrelations of the multi-overlap algorithm.

The averaged densities exhibit a pronounced FSS collapse onto a common master curve even in the spin-glass phase. For the scaling of their tails towards $q = \pm 1$ we find qualitative agreement with the decay law predicted by mean-field theory, but with an exponent x that is, in particular in $3d$, much larger than theoretically expected.

Acknowledgements

W.J. would like to thank A. Aharony, K. Binder and E. Domany for useful discussions on non-self-averaging quantities. We acknowledge partial financial support by the German-Israel-Foundation (GIF-I-0438-145.07/95) and by the US Department of Energy (DOE-DE-FG02-97ER41022) as well as the computer-time grants p526, bvpl01, and hmz091 on the T3E computers of CEA (Grenoble), ZIB (Berlin), and NIC (Jülich), respectively.

References

1. K.H. Fischer and J.A. Hertz, *Spin Glasses* (Cambridge University Press, Cambridge, England, 1991).
2. C.A.M. Mulder, A.J. van Duyneveldt, and J.A. Mydosh, *Phys. Rev.* B **23**, 1384 (1981); *Phys. Rev.* B **25**, 515 (1982).
3. P. Granberg, P. Svedlindh, P. Nordblad, L. Lundgren, and H.S. Chen, *Phys. Rev.* B **35**, 2075 (1987); E. Vincent, J. Hammann, M. Ocio, J.-P. Bouchaud, and L.F. Cugliandolo, in: *Complex Behaviour of Glassy Systems*, ed. M. Rubi, Lecture Notes in Physics, Vol. 492 (Springer-Verlag, Berlin, 1997), pp. 184-219 [cond-mat/9607224], and references given therein.
4. A.P. Young (ed.), *Spin Glasses and Random Fields* (World Scientific, Singapore, 1997).

5. M. Mézard, G. Parisi, and M.A. Virasoro, *Spin Glass Theory and Beyond* (World Scientific, Singapore, 1987).
6. K. Binder and A.P. Young, *Rev. Mod. Phys.* **56**, 801 (1986).
7. G. Parisi, *Phys. Rev. Lett.* **43**, 1754 (1979).
8. D.S. Fisher and D.A. Huse, *Phys. Rev.* B **38**, 386 (1988).
9. B.A. Berg and W. Janke, *Phys. Rev. Lett.* **80**, 4771 (1998).
10. W. Janke, B.A. Berg, and A. Billoire, *Ann. Phys. (Leipzig)* **7**, 544 (1998); *Comput. Phys. Commun.* **121-122**, 176 (1999).
11. B.A. Berg, A. Billoire, and W. Janke, *Phys. Rev.* B **61**, 12143 (2000); and in preparation.
12. S.F. Edwards and P.W. Anderson, *J. Phys.* F **5**, 965 (1975).
13. W.H. Press, B.P. Flannery, S.A. Teukolsky, and W.T. Vetterling, *Numerical Recipes* (Cambridge University Press, Cambridge, England, 1986).
14. N. Kawashima and A.P. Young, *Phys. Rev.* B **53**, R484 (1996).
15. D. Badoni, J.C. Ciria, G. Parisi, F. Ritort, J. Pech, and J.J. Ruiz-Lorenzo, *Europhys. Lett.* **21**, 495 (1993).
16. G. Marsaglia, A. Zaman, and W.W. Tsang, *Stat. Prob. Lett.* **8**, 35 (1990).
17. M.L. Lüscher, *Comput. Phys. Commun.* **79**, 100 (1994).
18. N. Metropolis, A.W. Rosenbluth, M.N. Rosenbluth, A.H. Teller, and E. Teller, *J. Chem. Phys.* **21**, 1087 (1953).
19. B.A. Berg, *Introduction to Monte Carlo Simulations and Their Statistical Analysis*, in preparation.
20. N.D. Mackenzie and A.P. Young, *Phys. Rev. Lett.* **42**, 301 (1982); *J. Phys.* C **16**, 5321 (1983); K. Nemoto, *J. Phys.* A **21**, L287 (1988); G.J. Rodgers and M.A. Moore, *J. Phys.* A **22**, 1085 (1989); D. Vertechi and M.A. Virasoro, *J. Phys. (France)* **50**, 2325 (1989); R.N. Bhatt, S.G.W. Colborne, M.A. Moore, and A.P. Young, unpublished, as quoted by Rodgers and Moore.
21. S. Franz, G. Parisi, and M.A. Virasoro, *J. Phys. I (France)* **2**, 1869 (1992); G. Parisi, F. Ritort, and F. Slanina, *J. Phys.* A **26**, 3775 (1993); J.C. Ciria, G. Parisi, and F. Ritort, *J. Phys.* A **26**, 6731 (1993).

SHORT-TIME CRITICAL DYNAMICS

L. SCHÜLKE

Fachbereich Physik, Universität Siegen, D-57068 Siegen, Germany
E-mail: schuelke@pollux.physik.uni-siegen.de

An introductory review to short-time critical dynamics is given. From the scaling relation valid already in the early stage of the evolution of a system at or near the critical point, one derives power law behaviour for various quantities. By a numerical simulation of the system one can measure the critical exponents and, by searching for the best power law behaviour, one can determine the critical point. Critical slowing down as well as finite size corrections are nearly absent, since the correlation length is still small for times far before equilibrium is reached. By measuring the (pseudo) critical points it is also possible to distinguish (weak) first-order from second-order phase transitions.

In this talk I like to give an introductory report about the main features of short-time critical dynamics. The topic exists since about ten years ago. It will be shown that it is possible to measure the critical temperature as well as all the static and dynamic critical exponents already in the short-time regime of the evolution of statistical systems, far before the equilibrium is reached. This can be done by starting either from an initial system of very high temperature, suddenly quenched to the critical temperature, or by starting with a completely ordered state, leading to the same results. For a comprehensive review we refer to references[1] and[2]. It will also be shown that short-time critical dynamics can in addition be used as a tool to distinguish first-order from second order phase transitions by comparing the critical temperature obtained from the different starting conditions.

It has for long been known that statistical system for particular values of their coupling show critical behaviour. This behaviour is mainly characterised by the fact that the correlation length ξ_x becomes infinite as well as the corresponding correlation time ξ_t. If we define the difference between the coupling and the critical one by τ, we have

$$\xi_x \to \tau^{-\nu}, \qquad \xi_t \to \tau^{-\nu z}; \qquad \tau = \frac{K - K_c}{K_c}. \tag{1}$$

Other quantities approach zero or infinity when $\tau \to 0$, e.g., the magnetisation of a spin system behaves as

$$M(\tau) \to (\tau)^{\beta}. \tag{2}$$

The exponents ν, β and z are static rsp. dynamic critical exponents. Mainly due to the large correlation length and -time there exists a dynamical scaling form. We do not explicitly quote the scaling form here, but refer to it in the subsequent discussion.

In 1989 Janssen, Schaub and Schmittmann[3] have shown, that this scenario is not only valid in equilibrium, but *a scaling relation is already valid in the short-time regime* of the evaluation of a critical system. The system should be prepared at a very high temperature, with a small magnetisation m_0 remaining. It is then suddenly quenched to the critical temperature and released to the dynamical evolution of model A. The authors show, using renormalisation group analysis, that after a macroscopic small time t_{mic} a scaling form is valid. For the $k-th$ moment of the magnetisation it reads

$$M^{(k)}(t,\tau,L,m_0) = b^{-k\beta/\nu}\, M^{(k)}(\, t/b^z,\; b^{1/\nu}\tau,\; L/b,\; b^{x_0}m_0). \tag{3}$$

In this equation b is an arbitrary scaling factor, t is the time, τ is defined in Eq.(1), L is the linear lattice size, and $m_0 \gtrsim 0$ the initial magnetisation. Except of this additional last argument the scaling form looks exactly like that in equilibrium. This argument m_0 gives rise to a new, independent critical exponent x_0, the scaling dimension of the initial magnetisation.

The questions arises how to exploit the scaling relation (3) in order to get the desired information about the critical exponents. We investigate the system numerically with Monte-Carlo simulation at a temperature at or near the critical temperature T_c, measuring the magnetisation and its second moment, $M(t)$ and $M^{(2)}(t)$, as well as the autocorrelation $A(t) = \langle \sum_i S_i(t)\; S_i(0)\rangle$, where a time unit is defined as a complete update of the whole lattice. Since the system is in the early stage of the evolution the correlation length is still small and finite size problems are nearly absent. Therefore we generally consider L large enough and skip this argument. An occasional check is made to prove this assumption. We now choose the scaling factor $b = t^{1/z}$ so that the main t-dependence on the right is cancelled. Expanding the scaling form (3) for $k = 1$ with respect to the small quantity $t^{x_0/z}\, m_0$, one obtains

$$M(t) \sim m_0\, t^{\theta}\, F(t^{1/\nu z}\tau) = m_0\, t^{\theta}\, \left(1 + at^{1/\nu z}\tau + O(\tau^2)\right), \tag{4}$$

where $\theta = (x_0 - \beta/\nu)/z$ has been introduced. In most cases $\theta > 0$, i.e., for $\tau = 0$ the small initial magnetisation *increases* in the short-time region. Figure 1 shows the initial increase for the 3-dimensional Ising model as an example[4]. For small $\tau \neq 0$ we have expanded the function $F(t^{1/\nu z}\tau)$. There appear corrections to the simple power law dependent on the sign of τ. Therefore simulating the system for values of the coupling K in the neighbourhood of the critical point one obtains $M_K(t)$ with non-perfect power law behaviour, and the critical point K_c can be obtained by interpolation. Figure 2 shows a plot of the magnetisation for the 2-dimensional 3-state Potts model for three different values of the coupling[5]. The curve with the best power law behaviour is found for $K_c = 1.0055(8)$, while the exact value is 1.00505.

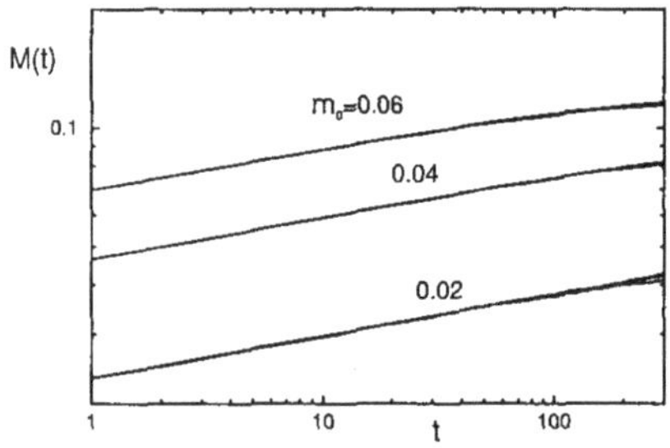

Figure 1. Three-dimensional Ising model: Initial increase of the magnetisation for different initial magnetisations m_0, plotted vs. time on log-log scale. $L = 128$. Result is $\theta = 0.108(2)$

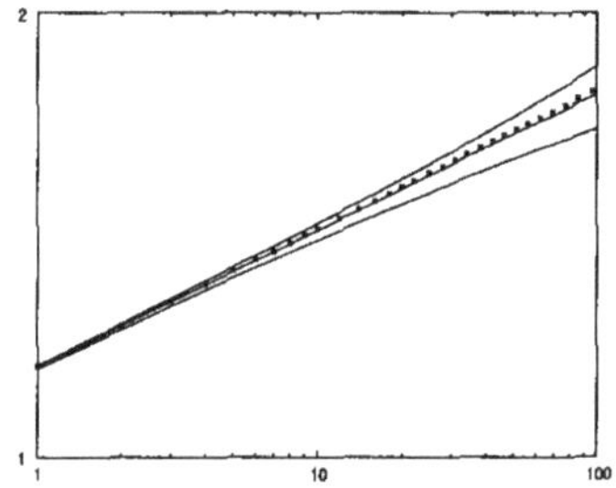

Figure 2. Two-dimensional 3-state Potts model: Magnetisation for three different values of the coupling near the critical point. From above $J = 1.00750$, 1.00505, and 1.0025. Dotted curve $J_c = 1.0055(8)$

For the second moment of the magnetisation one can deduce $M^{(2)}(t) \sim L^{-d}$ (d = dimension of the system), because the correlation length is still small in the early time region even at the critical point. Combining this with the result of the scaling form for $\tau = 0$ and $b = t^{1/z}$, $M^{(2)}(t) \sim t^{-2\beta/\nu z} M^{(2)}(1, t^{-1/z}L)$, one obtains the power law

$$M^{(2)}(t) \sim t^{c_2} \qquad c_2 = \left(d - 2\frac{\beta}{\nu}\right)\frac{1}{z}. \tag{5}$$

An example for the power law behaviour of this quantity is given[4] in Fig. 3.

From the scaling form (3), setting again $b = t^{1/z}$, one derives for $\partial_\tau \ln M(t,\tau)|_{\tau=0}$ the power law

$$\partial_\tau \ln M(t,\tau)|_{\tau=0} \sim t^{\frac{1}{\nu z}}. \tag{6}$$

Approximating numerically the derivative of $M(t)$ involves the difference of small quantities and is therefore affected more by uncertainties. However, Fig. 4 shows for the 2-dimensional 3-state Potts model as example that the data here also, for $t > 20$, show perfect power law behaviour.

Another interesting quantity is the autocorrelation

$$A(t) = \frac{1}{L^d}\langle \sum_i S_i(t) S_i(0) \rangle. \tag{7}$$

An analysis[6] (for $m_0 = 0$) leads to

$$A(t) \sim t^{c_a} \qquad c_a = \frac{d}{z} - \theta. \tag{8}$$

The plot in Fig. 5 shows a nearly perfect power law. Again the 3-dimensional Ising model has been taken as an example[4].

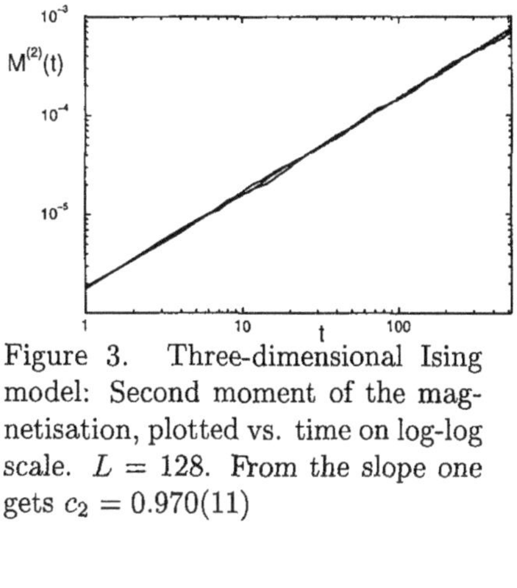

Figure 3. Three-dimensional Ising model: Second moment of the magnetisation, plotted vs. time on log-log scale. $L = 128$. From the slope one gets $c_2 = 0.970(11)$

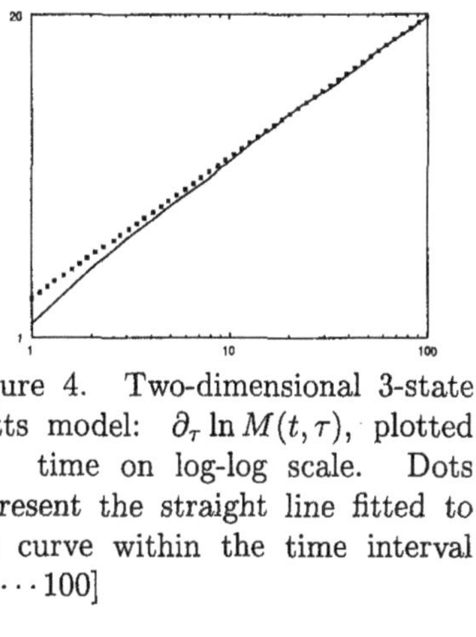

Figure 4. Two-dimensional 3-state Potts model: $\partial_\tau \ln M(t,\tau)$, plotted vs. time on log-log scale. Dots represent the straight line fitted to the curve within the time interval $[20 \cdots 100]$

Care has to be taken to exclude the time region $t < t_{mic}$, as well as the large time region where deviations of the the power law behaviour occur due to finite size effects or due to too large fluctuations. An example for a more detailed analysis of the straight line (on log-log plot) in Fig. 5 is given in Fig. 6. The individual slope in bins between t and $1.5\,t$ is plotted versus t. It is seen that the slope for $t \lesssim 10$ is not yet constant and therefore the region $t < 10$ should be excluded. It is also seen that the fluctuations within the higher bins increase although much more points enter.

Till now a completely disordered initial state has been considered as starting point, i.e., a state of very high temperature. The question arises how a *completely ordered initial state* evolves, when heated up suddenly to the critical temperature. In the scaling form (3) one can skip besides L, also the argument $m_0 = 1$:

$$M^{(k)}(t,\tau) = b^{-k\beta/\nu}\, M^{(k)}(\, t/b^z,\ b^{1/\nu}\tau) \qquad (m_0 = 1). \tag{9}$$

The system is simulated numerically by starting with a completely ordered state, whose evaluation is measured at or near the critical temperature. The quantities measured are $M(t)$ and $M^{(2)}(t)$. With $b = t^{1/z}$ one avoids the main t-dependence on the right of eq.(9), and for $k = 1$ one has

$$M(t,\tau) = t^{-\beta/\nu z}\, M(\,1,\ t^{1/\nu z}\tau) \ = t^{-\beta/\nu z}\left(1 + a\, t^{1/\nu z}\, \tau + O(\tau^2)\right). \tag{10}$$

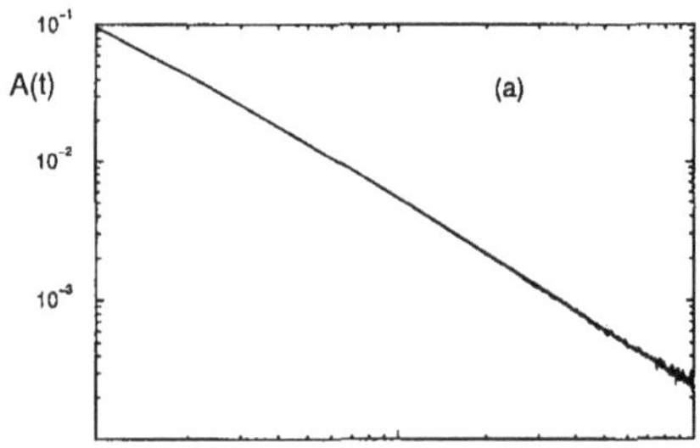

Figure 5. Three-dimensional Ising model: Autocorrelation plotted vs. time on log-log scale. $L = 128$. From the slope one gets $c_a = 1.36(1)$

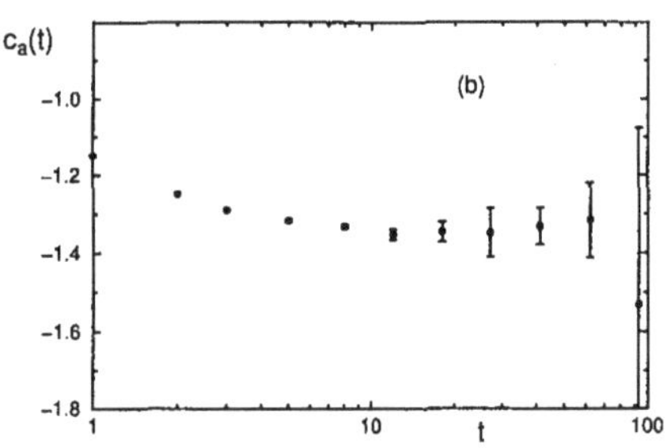

Figure 6. Three-dimensional Ising model: slope c_a measured between t and $1.5t$ plotted vs. time on log-log scale.

Similarly as in case of the a completely disordered start, see eq.(4), one can choose couplings in the neighbourhood of the critical point and fix from these measurements the critical point by looking for the best power law behaviour. We would like to mention that measurements starting from $m_0 = 1$ are much less affected by fluctuations, because the quantities measured are rather big in contrast to those from a random start. Figure 7 is an example of the 3-dimensional Ising model[4]. Both the critical point as well as the slope at that point can be determined from the measurement. Figure 8 shows the decay of the magnetisation at the critical point for different lattice sizes from $L = 32$ to 256. The data points for all lattice sizes agree completely up to $t = 1000$, only the data for $L = 32$ show some deviation for $t \gtrsim 300$. This justifies our statement that finite size effects are unimportant in short-time critical dynamics.

Also for the cold start one derives from (9)

$$\partial_\tau \ln M(t,\tau)|_{\tau=0} \ \sim \ t^{\,1/\nu z} \tag{11}$$

which allows to measure the ratio $1/\nu z$. With the magnetisation and its second moment the time dependent Binder cumulant

$$U(t) = \frac{M^{(2)}}{(M)^2} - 1 \quad \sim \quad t^{d/z} \tag{12}$$

is defined. From its slope one can directly measure the dynamic exponent z. For the last two quantities we give examples in Figs. 9 and 10, using the 3-dimensional Ising model.

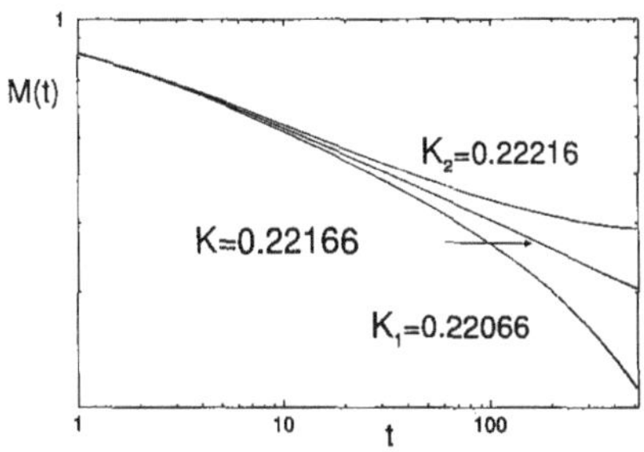

Figure 7. Three-dimensional Ising model: Decay of the magnetisation for initial magnetisations $m_0 = 1$, for three different values of the coupling near the critical point.

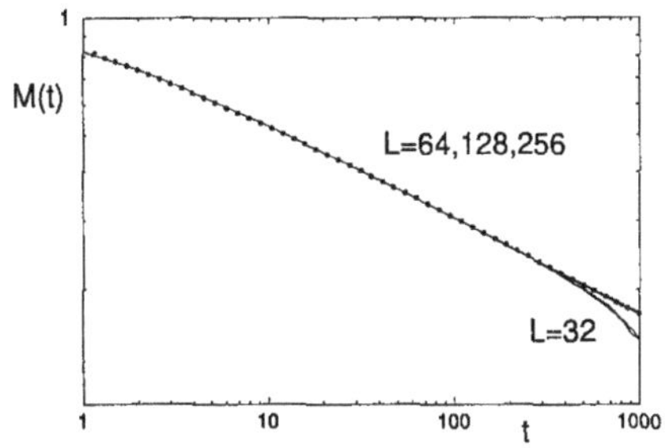

Figure 8. Three-dimensional Ising model: Decay of the magnetisation for different lattice sizes at the critical point.

In summarising the topics discussed up to now, we would like to make the following remarks:

- From an investigation of the system from a high-temperature initial state, a new independent critical exponent θ can be determined.
- A determination of the critical point and of all the critical exponents is possible starting from a high-temperature or from a zero-temperature initial state. The results are the same for second-order transitions.
- In contrast to investigations in equilibrium, the correlation length is still finite in the short-time regime, therefore finite size effects are strongly reduced.
- Also the correlation time is still finite, therefore critical slowing down is absent.
- We work directly with sample averages, not with time averages.

As is well known, critical slowing down can be overcome in some cases by the successful non-local cluster algorithm[7]. However, the cluster algorithm is not applicable, e.g., for systems with randomness or frustration. With short-time dynamic simulations we have, e.g., successfully investigated the chiral degree of freedom in the 2-dimensional fully frustrated XY model, where critical slowing down makes measurements in equilibrium extremely difficult[8,9].

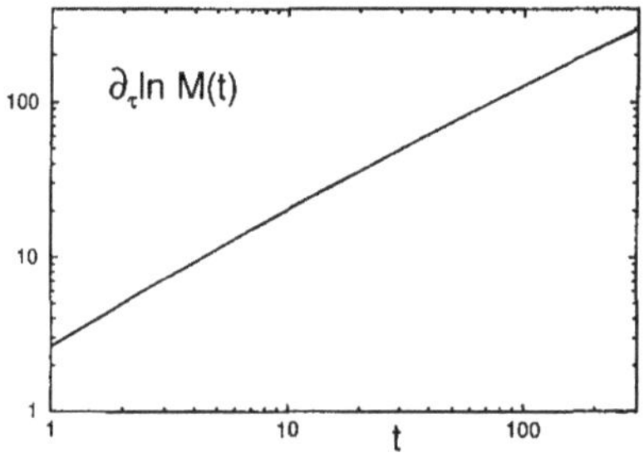

Figure 9. Three-dimensional Ising model: Logarithmic derivative of the magnetisation with respect to τ, obtained from an ordered start.

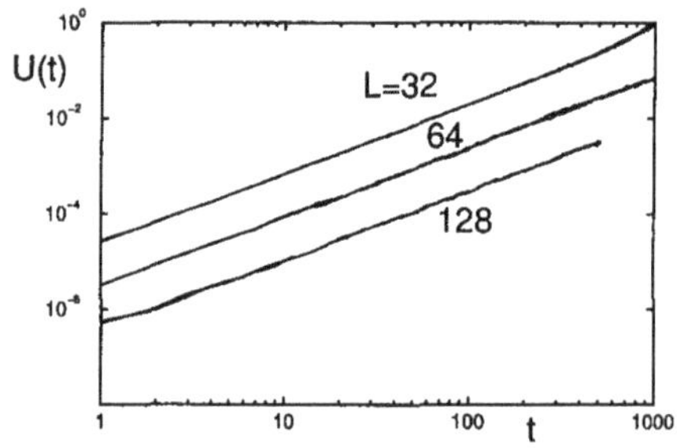

Figure 10. Three-dimensional Ising model: Time evolution of the Binder cumulant $U(t)$ for different lattice sizes.

Till now we have investigated critical systems which undergo a second order phase transition. The relaxation of initial states with very high temperature suddenly quenched to the critical temperature, or of initial states with zero temperature, suddenly heated up to the critical temperature, is measured in the short-time regime. Various quantities exhibit a power law behaviour. From measurements with couplings in the neighbourhood of the critical point we can find by interpolation the optimal power law behaviour and determine in this way the critical point. The result is the same for both initial conditions within the statistical error.

If the phase transition is first order, at least if it is weak, the behaviour should be similar to second order. By cooling down a very hot initial state, one should find a *pseudo critical point* K^*, and by heating up a cold initial state *another pseudo critical point* K^{**} is found. We expect

$$K^{**} < K_c < K^* .$$

A difference between K^{**} and K^* is a clear signal for a first-order transition.

The p-state Potts model undergoes a second order phase transition for $p \leq 4$ at the critical point $K_c = \ln(1+\sqrt{p})$. For $p \geq 5$ the transition is of first order, but still weak for small p. We have investigated this model[10] for $p = 5$ and $p = 7$. For the initial condition $m_0 = 0$ the second moment $M^{(2)}(t)$ of the magnetisation has been measured, and for initial condition $m_0 = 1$ the quantities $M(t)$ and $M^{(2)}(t)$. In case of $p = 7$ the time interval extended up to $t = 6\ 000$ for $L = 280$ and 560, while in case of $p = 5$ a more careful analysis is necessary. Here we have chosen time intervals up to 40 000 for $L = 560$. In both cases we find a clear difference between K^* and K^{**}. In case of K^{**}, both the magnetisation and the Binder cumulant (12) have been used. Figure 11 shows the result for $p = 7$. The results for $p = 5$ are similar. This example shows that short-time critical dynamics can be successfully used as a tool to distinguish between first- and second order phase transitions.

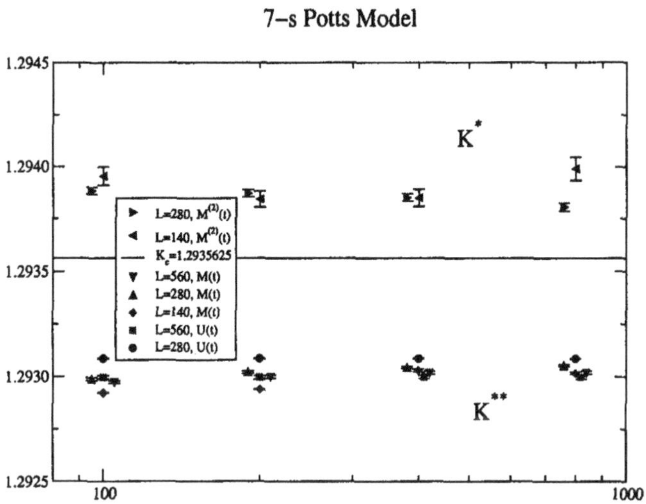

Figure 11. 7-state Potts model: Determination of the pseudo-critical point K^* determined from $M^{(2)}(t)$ and of K^{**} determined from $M(t)$ and $U(t)$. For the time interval $[t \cdots t_{max}]$ used for the fit various starting points between $t = 100$ and 800 and the maximum time have been used.

References

1. B. Zheng, Int. J. Mod. Phys. **B 12**, 1419 (1998).
2. H. Luo, *Short-time critical dynamics of statistical systems with quenched disorder*, *Berichte aus der Physik* (Shaker Verlag, Aachen, 2000).
3. H. K. Janssen, B. Schaub and B. Schmittmann, Z. Phys. **B 73**, 539 (1989).

4. A. Jaster, J. Mainville, L. Schülke and B. Zheng, J. Phys. A: Math. Gen. **32**, 1395 (1999).
5. L. Schülke and B. Zheng, Phys. Lett **A 215**, 81 (1996).
6. H. K. Janssen, in *From Phase Transition to Chaos*, edited by G. Györgyi, I. Kondor, L. Sasvári and T. Tél, Topics in Modern Statistical Physics (World Scientific, Singapore, 1992).
7. R. H. Swendsen and J. S. Wang, Phys. Rev. Lett. **58**, 86 (1987).
8. H.J. Luo, L. Schülke and B. Zheng, Phys. Rev. Lett. **81**, 180 (1998).
9. H.J. Luo, L. Schülke and B. Zheng, Phys. Rev. **E 57**, 1327 (1998).
10. L. Schülke and B. Zheng, *Dynamic Approach to Weak First Order Phase Transitions*, Univ. Siegen, Univ. Halle, 2000, cond-mat/0003161.

NONPERTURBATIVE APPROACH TO FRUSTRATED MAGNETS

M. TISSIER, B. DELAMOTTE, D. MOUHANNA
Laboratoire de Physique Théorique et des Hautes Energies, Université Paris VI-Pierre et Marie Curie - Paris VII-Denis Diderot, 2 Place Jussieu, F-75252 Paris Cedex 05, France
E-mail: tissier@lpthe.jussieu.fr

Frustrated magnets are a notorious example where the usual perturbative methods are in conflict. We show that a nonperturbative approach, based on the concept of effective average action enables one to get a coherent picture of the physics of Heisenberg frustrated magnets everywhere between $d = 2$ and $d = 4$. We recover all known perturbative results in a single framework and find the transition to be weakly first order in $d = 3$. The effective critical exponents found by this method are in good agreement with numerical and experimental data.

1 Introduction

Among the physical properties of a material, universal quantities are of particular interest since they depend on very few features of the material, such as the symmetry of the order parameter, the space dimensionality, the range of interaction, but are insensitive to most of its microscopic details. As a consequence, two systems may have very different physical origins, but still share similar behaviours: in the critical domain of a second order phase transition certain quantities exhibit power-law behaviours characterized by universal critical exponents. Among the second order phase transitions which have been studied so far, many belong to the well-known vectorial $SO(N)/SO(N-1)$ universality class. A famous example is the ferromagnetic-paramagnetic phase transition. But some other are characterized by order parameters with different symmetries, and ask for a new treatment. This is the case of the Antiferromagnetic Heisenberg model on a Triangular lattice (AFHT model) for which the occurence of competing interactions tends to modify the ground state: the minimum of the energy is reached for a configuration in which spins on a triangular plaquette lie 120° one from another. As a consequence the order parameter is matrix-like[1] (a 2×3 matrix) with a corresponding symmetry-breaking scheme $SO(3) \otimes SO(2)/SO(2)$.

Although many experimental, numerical and theoretical works have been devoted to this system, a full understanding of the underlying physics has not been reached so far. As we shall see, the situation is quite awkward. From the experimental point of view, people agree that materials supposed to belong to the AFHT universality class exhibit a critical behaviour, but no universality seem to occur, and critical exponents differ from one material to another: for VCl_2[2]: $\beta = 0.20(2), \gamma = 1.05(3), \nu = 0.62(5)$, for VBr_2[3]: $\alpha = 0.30(5)$, for CuFUD[4]: $\beta = 0.22(2)$ and for $Fe[S_2CN(C_2H_5)_2]_2Cl$[5,6,7]: $\beta = 0.24(1), \gamma = 1.16(3)$. We should note also that the usual scaling laws are violated. The Monte Carlo (MC) simulations (see[8] for a review and references therein) show that the critical exponents are affected by the microscopic realization of the symmetry-breaking scheme: for AFHT model $\nu \simeq 0.59(1), \gamma \simeq 1.17(2), \beta \simeq 0.29(1), \alpha \simeq 0.24(2)$, and for models supposed to belong to the same universality class, AFHT with rigid constraints $\nu = 0.504(10), \gamma = 1.074(29), \beta = 0.221(9), \alpha = 0.488(30)$, $V_{3,2}$ Stiefel models $\nu \simeq 0.51(1), \gamma \simeq 1.13(2), \beta \simeq 0.193(4), \alpha \simeq 0.47(3)$. Moreover numerical predictions of the critical exponents differ significantly from those found by experiments. These observations are not compatible with the phenomenon of universality. To cope with this discrepancy, it has been proposed that the system does not experience a second order phase transition but a weak first order one, characterized by a large (though not infinite) correlation length[9,10]. This would explain the weak universality displayed by the systems supposed to be described by the AFHT model, and would be consistent with the critical behaviour observed, since it is very difficult to discriminate a truly second order phase transition from a situation where the correlation length is very large.

As we will see in the next section, the situation in not well understood from the theoretical point of view neither. The main problem being that different perturbative approaches lead to qualitatively different results. We will conclude that this discrepancy can only be overcome by a nonperturbative approach. In the third section, we give a brief introduction to the renormalization group in the framework of the effective average action. We present our results in the fourth section and give our conclusions in the last one.

2 Perturbative results

We are now going to focus on the two usual perturbative approaches, namely the Non-Linear σ (NLσ) model and the Landau-Ginzburg-Wilson (LGW) model, whose respective domain of validity are the vicinity of the lower critical dimension (d=2), and that of the upper critical dimension (d=4).

2.1 Non-Linear σ model

Around $d = 2$, the AFHT model is expected to experience a second order phase transition at very low temperature (in dimension d=2+ϵ, the critical temperature is of order ϵ). As a consequence, only the less energetic excitations contribute to the partition function. Therefore the field configurations relevant for the critical behaviour are small fluctuations around a common direction, varying slowly in space, that can be decomposed in plane waves labeled by a momentum in Fourier space. This type of excitations are called spin-wave excitations. The contribution of other field configurations are suppressed exponentially with their energy. The analysis of the critical properties is most conveniently carried out in the framework of the NLσ model. The action for the spin-wave excitations is expressed in terms of a geometric object: the metric tensor on the manifold of the order parameter space. The flow equations of the coupling constants can then be extracted from other geometric objects such as Riemann and Ricci tensors. Notice finally that, since the field configurations taken into account by the perturbative NLσ model correspond to small fluctuations, the critical behaviour of the model is, in this approach, only sensitive to the local properties of the manifold of the order parameter (the Lie algebras of the symmetry groups in the high and low temperature phases), but not on the global properties (such as the topology of the manifold): we cannot see any difference between two order parameters having the same local structure, but differing nonetheless by global aspects.

The result of this study is threefold[11]:

- a stable fixed point is found, which is the signature of a second order phase transition.
- At the fixed point, the symmetry breaking scheme is enlarged from $SO(3) \otimes SO(2)/SO(2)$ to $SO(3) \otimes SO(3)/SO(3)$. At large distances, the theory is equivalent to that of a 3×3 matrix. It is non trivial to know if this phenomenon of enlarged symmetry persists beyond perturbation theory, and in particular in $d = 3$;
- it is known that the manifold $SO(3) \otimes SO(3)/SO(3)$ has the same local structure as the 4-component vector model $SO(4)/SO(3)$. As seen from the NLσ model calculations, the two above theories are equivalent to all orders in ϵ.

The natural conclusion of this study is that a fixed point of symmetry $SO(4)/SO(3)$ exists everywhere between two and four dimensions, associated with the AFTH model. We now see that this conflicts with the results of the LGW model.

2.2 Landau-Ginzburg-Wilson model

By analogy with the $SO(N)/SO(N-1)$ vectorial model one can treat the AFHT model by means of a LGW approach, where interactions are treated perturbatively. We may expect that just below four dimensions a new fixed point appears very close to the gaussian fixed point. However, in the case of AFHT model, no fixed point is found by this approach[1]. The flow drives the system toward a region of instability, where the ϕ^4-like coupling constant becomes negative, and therefore the potential is not bounded by below anymore. The ϕ^6-like terms become then relevant as they are necessary to stabilize the potential, and the phase transition is expected to be first order.

2.3 Discussion

These results call for few remarks. One of the most striking features is that NLσ and LGW models predict different qualitative behaviours when extrapolated to three dimensions. There has to be a qualitative change in the order of the transition for some critical dimension d_c, $2<d_c<4$. Determining wether d_c is lower or greater than three is one of the most important question. The discrepancy between perturbative results around upper and lower critical dimensions is actually quite common, and has been observed in many models: the dipole-locked phase of superfluid ^{3}He, superconductors, Josephson junction arrays, electroweak phase transition, smectic liquid crystal, etc... In this sense, the simpler situation observed in the vectorial model, where a fixed point exists both around two and four dimensions, seem to be the exception more than the rule.

Another point is to be understood: as seen from the NLσ model, the $SO(4)/SO(3)$ vectorial model and the $SO(3)\otimes SO(2)/SO(2)$ model are identical at large distances to all orders in $\epsilon = d-2$. However in four dimensions, the former exhibits a fixed point whereas the latter does not. Where does the difference come from? The matrix model has a non-trivial topology and displays stable field configurations that are not taken into account by spin-wave expansion. Such excitations are called topological excitations or vortices, and are known to play á crucial role in other models (they are responsible for the Kosterlitz-Thouless phase transition in the two-component vector model). Those configurations are too energetic to influence the critical physics in $d = 2+\epsilon$ that takes place at $T_c \simeq \epsilon$. However as the dimension increases, T_c rises also and vortices start to be excited and tend to disorder the system. On the other hand, the four-component vectorial model has no vortices. Thus the difference between the two models observed in four dimensions may be caused by vortices: comparing the two theories would then give informations on the role of vortices.

As we can see, there are at least two reasons why we must use nonperturbative methods. First of all, we want to study the theory in arbitrary dimension, in particular in $d = 3$, where no small parameter exists. Second we think that topological defects play a role that cannot be obtained from the NLσ model, and that are not obviously taken into account by the perturbative approach of the LGW model.

3 Exact renormalization group equation

We have used a nonperturbative approach, based on the concept of the effective average action $\Gamma_k[\phi]$[12]. It corresponds to the functional generating the one-particle irreducible vertices, except that only modes with momentum greater than k – the running cutoff – have been integrated out. In the limit $k \to 0$, all the modes have been taken into account and Γ_k identifies with the usual vertex functional. At the scale $k = \Lambda$ – the overall cutoff – no fluctuation has been taken into account, and Γ_k then identifies with the microscopic action. Therefore, the effective average action connects the microscopic to the macroscopic physics. The evolution of Γ_k as k is varied is governed by a concise equation, which reads:

$$\frac{\partial \Gamma_k}{\partial t} = \frac{1}{2}\mathrm{Tr}\left\{\frac{\partial R_k}{\partial t}\left(\Gamma_k^{(2)}(q) + R_k(q)\right)^{-1}\right\}, \tag{1}$$

where $t = \ln(k/\Lambda)$ with Λ, and $\Gamma_k^{(2)}$ the second derivative of the effective average action with respect to the fields. The trace has to be understood as a sum over internal degrees of freedom as well as an integral over momenta. Finally, $R_k(q)$ is the infrared cutoff, which governs the integration of energetic modes as the running cutoff is lowered. This function can be chosen at will, but must decrease sufficiently fast for large momenta, to ensure an efficient integration of modes. The equation (1) is exact, and no approximation is needed to derive it. It is a nonlinear functional differential equation that cannot be solved analytically, however it happens to be a good starting point to perform approximations: we can truncate Γ_k to keep only a manageable number of coupling constants, and derive their flow equations using Eq.(1). For the sake of concreteness, let us consider the case of the $SO(N)/SO(N-1)$ vectorial model. A simple truncation consists in taking an ansatz for Γ_k composed of a potential quartic in fields, and a kinetic term which take into account the field renormalization:

$$\Gamma_k = \int d^d x \left\{ \frac{Z_k}{2} \nabla\phi_i \nabla\phi_i + \frac{\lambda_k}{2} \left(\frac{\rho}{2} - \kappa_k \right)^2 \right\} \tag{2}$$

where $\rho = \phi_i \phi_i / 2$. We obtain in this way nonperturbative flow equations, and no small parameter is needed to derive them. The flow equations are easily found and, for $N \geq 3$ display the following properties:

- Around two dimensions, a fixed point is found, with critical exponents identifying with those of the NLσ model at one loop.
- We also retrieve the flow equations of the LGW model by expanding the nonperturbative results in powers of the ϕ^4 coupling constant. As a consequence, the different perturbative results are recovered in a single framework.
- As the dimension is lowered from four, the coordinates of the fixed point found in the LGW model change continuously and finally identify with those of the NLσ model around $d = 2$. Critical exponents can be evaluated for any dimension by linearizing the flow equations around the fixed point. In particular, for $d = 3$, the critical exponents extracted in this way differ only by 10% from those found by numerical simulations.
- by extending the truncation, one can get more accurate values of the critical exponents in 3 dimensions.

The main drawback of this approach is that we have no estimate of the bias introduced by such a truncation. However, this method has been tested on many theories, and give good results in various situations. We cannot cite here all of them, and refer the reader to other publications[13].

4 Nonperturbative treatment

4.1 AFHT model

Previous observations are very encouraging, and led us to use the same approach to study the phase diagram of the AFHT model[14]. We have written an ansatz for the effective average action in terms of the 2×3 matrix field ϕ_{ab}, by keeping a potential and a kinetic term, both having the symmetry of the high temperature phase, i.e. SO(3)⊗SO(2) (SO(3) group is acting on the left of the field ϕ_{ab} and SO(2) acts on the right. One can view the field ϕ_{ab} as being composed of two three-component vectors $\boldsymbol{e_1}$ and $\boldsymbol{e_2}$. Then

SO(3) corresponds to rotating both vectors in the usual way, while SO(2) turns e_1 and e_2 into each other). We have considered in the potential all the terms up to fourth power in fields. For the kinetic term, apart from the conventional term, we considered a "current" term which is responsible for the phenomenon of enlarged symmetry in two dimensions[11]. This term is irrelevant around four dimensions and is usually discarded of the perturbative analysis by power-counting arguments, but can be handled in our approach. Our approximation for the effective average action reads:

$$\Gamma_k = \int d^d x \left\{ \frac{Z_k}{2} \nabla\phi_{ab}\nabla\phi_{ab} + \frac{\omega_k}{4}\left(\epsilon_{ab}\phi_{ca}\nabla\phi_{cb}\right)^2 + \frac{\lambda_k}{4}\left(\frac{\rho}{2} - \kappa_k\right)^2 + \frac{\mu_k}{4}\tau \right\} \tag{3}$$

where $\rho = \mathrm{Tr}^t\phi\phi$ and $\tau = \frac{1}{2}\mathrm{Tr}(^t\phi\phi)^2 - \frac{1}{4}(\mathrm{Tr}^t\phi\phi)^2$ are the two independent $SO(3) \otimes SO(2)$ invariants built out of the matrix field ϕ_{ab}. The model is parametrized by the coupling constants $\{\omega_k, \lambda_k, \kappa_k, \mu_k, Z_k\}$. For $\lambda_k \geq 0$ and $\mu_k \leq 0$, the ground state corresponds to a configuration of fields $\phi_{ab} = \sqrt{\kappa_k} R_{ab}$, where R_{ab} is a matrix built out of two orthonormal three-component vectors. The residual symmetry of the ground state is a diagonal $SO(2)$ and we retrieve the right symmetry-breaking scheme in this region of the parameter space. We have derived the flow equations for the five retained coupling constants. Those are too lengthy to be written here, but can be downloaded from our web site[15]. It is then quite easy to look for the fixed points of the flow equations.

Around $d = 2$ we retrieve the one loop β functions found in the NLσ model by expanding our nonperturbative flow equations in powers of the temperature (in our parametrization, the temperature is κ_k^{-1}). We therefore extract the same critical exponents, and observe the phenomenon of enlarged symmetry at the fixed point. We can follow the coordinates of this fixed point as the dimension is increased. The fixed point survives up to a critical dimension $d_c \simeq 2.87$ above which no fixed point is found. For d=3, the flow drives the system to a range of parameters where the potential is not bounded by below anymore. As discussed above, the transition is therefore expected to be first order. However, as the critical dimension d_c is close to the physical dimension $d = 3$, there exist a range of parameters for which the flow is very slow. As a consequence, the system exhibits a large correlation length (a rough estimate gives few thousands of lattice spacing), which corroborates the scenario of a weakly first order phase transition discussed in the introduction. We can also extract effective critical exponents by diagonalizing the flow equations in the region where the flow is slow. To get a more accurate result, we considered a ϕ^6 truncation, and get: $\nu = 0.53$, $\gamma = 1.03$ and $\beta = 0.28$, which lie in between the various sets of exponents found experimentally and numerically (see in the introduction for comparison).

To test the precision of our method, we have considered a generalization of the previous model for 2×6 matrix. The situation is simpler in this case because we find a fixed point of the flow equations in $d = 3$, and can extract the following critical exponents: $\nu = 0.74$ and $\gamma = 1.45$ which compare well with the Monte Carlo data $\nu = 0.700(11)$ and $\gamma = 1.383(36)$[16]. Our results appear to be better than those found at three-loop in $d = 3$ which provides $\nu = 0.575$ and $\gamma = 1.121$[16]. We have checked that our results are stable when higher order monomials in fields are added to the anstaz.

5 Conclusion

Using a nonperturbative method, we have reached a global understanding of frustrated Heisenberg magnets including a matching between previous perturbative predictions and a good agreement with experimental and numerical data. It remains to understand the very origin of the disappearance of the NLσ model fixed point. The role of non trivial topological configurations can be invoked and will be studied in a forthcoming paper.

LPTHE is a laboratoire associé au CNRS UMR 7589.

References

1. T. Garel and P. Pfeuty, J. Phys. C L **9**, 245, (1976).
2. H. Kadowaki, K. Ubukoshi, K. Hirakawa, J.L. Martinez, and G. Shirane, J. Phys. Soc. Japan **56**, 4027, (1987).
3. J. Wosnitza, R. Deutschmann, H.v. Löhneysen, and R.K. Kremer, J. Phys.: Condens. Matter **6**, 8045 , (1994).
4. K. Koyama and M. Matsuura, J. Phys. Soc. Japan **54**, 4085, (1985).
5. G.C. DeFotis, F. Palacio, and R.L. Carlin, Physica B **95**, 380, (1978).
6. G.C. DeFotis and S.A. Pugh, Phys. Rev. B **24**, 6497, (1981).
7. G.C. DeFotis and J.R. Laughlin, J. Magn. Magn. Matter **54-57**, 713, (1986).
8. D. Loison and K.D. Schotte, cond-mat/0001135.
9. G. Zumbach, Phys. Rev. Lett. **71**, 2421, (1993).
10. G. Zumbach, Nucl. Phys. **B413**, 771, (1994).
11. P. Azaria, B. Delamotte, F. Delduc, and Th. Jolicoeur, Nucl. Phys. **B408**, 485, (1993).
12. C. Wetterich, Nucl. Phys. **B352**, 529, (1991).
13. J. Berges, N. Tetradis and C. Wetterich hep-ph/0005122.

14. M. Tissier, B. Delamotte, and D. Mouhanna Phys. Rev. Lett. **84**, 5208 (2000).
15. M. Tissier, B. Delamotte, and D. Mouhanna, in preparation. Equations available at
http://www.lpthe.jussieu.fr/~tissier.
16. D. Loison, A.I. Sokolov, B. Delamotte, S.A. Antonenko, K.D. Schotte, and H.T. Diep, cond-mat/0001105.

UNIVERSAL SHORT-TIME CRITICAL BEHAVIOR ON THE TWO-DIMENSIONAL TRIANGULAR LATTICES

L. WANG, H.-P. YING, Z.-G. PAN

Zhejiang Institute of Modern Physics, Zhejiang University, Hangzhou 310027, P.R. China

The universal behavior of the short-time dynamics for spin models on a two-dimensional triangular lattice are investigated by using a dynamic Monte Carlo simulation. Our simulation results of the dynamic evolutions from fully ordered initial states show that the universal scaling exists already in the short-time regime by observing the power-law behavior of the magnetization and Binder cumulant. The values estimated for the dynamic and static critical exponents, θ, β and ν, confirm explicitly that the Potts models on the triangular lattices and square lattices are belong to the same universality class. Also our work strongly suggests that the simulation for the dynamic relaxations can be used to determine the universality.

1 Introduction

The study of the phase transition and critical phenomena has been an abstractive and important topic in statistical physics for a long time [1]. As well known, there appear critical scaling and universality due to the infinite spatial and time correlations at the second-order phase transition points, and the universal behavior of a critical system is characterized by a number of critical exponents such that the models in the same universality class have the same values of the critical exponents. Therefore, the estimation of critical exponents to determine the universality class for different statistical models is an interesting challenge. However such research is difficult to be carried out analytically by the reason that only a few statistical models can be exactly resolved[2]. As a result, numerical simulations supply a powerful method for estimating critical exponents which are measured by generating equilibrium states and updating the configurations in the Monte Carlo (MC) studies. For example, the dynamical exponents z can be measured from the exponential decay of time correlation for finite systems in the long-time regime. Unfortunately, the simulations near the critical point in the equilibrium suffer from critical slowing down [3], because the auto-correlation length τ divergence as $\tau = L^z$, where L denotes the linear size of a lattice and $z \geq 2$.

In the last decade, the exploration of critical phenomena have been greatly broadened and much progress has been made in critical dynamics, and the most important discovery is that there exist critical scaling and universality for systems far from equilibrium states[4]. Traditionally, it is believed that universal behavior exists only in the equilibrium or in the long-time regime of a dynamical evolution. However, recent researches in critical dynamics for many statistical models have shown that universal scaling behavior also emerges within the macroscopic short-time regime of the dynamic process after microscopic time scale t_{mic} [5,6]. The significance of the ***short-time critical dynamics*** (SCD) not only discovers the existence of universal dynamic scaling behavior within the short-time regime, but also prefers a very efficient method to determine the critical exponents. In Ref.[4], the recent progress in the short-time dynamical studies of two-dimensional (2d) statistical models has been reviewed, where several simulation methods and the corresponding numerical results have been presented. Up to now, the values of critical exponents for a variety of statistical models, such as the Ising and Potts models on square lattices [7,8,9], the quantum XY and the random-bond Potts models in two-dimensions [10,11], the 2d FFXY model and (2+1)d SU(2) lattice-gauge models [12,13], have been calculated precisely. But only very few attention has been concentrated on the 2d triangular lattices [14,16].

By the SCD method, we can estimate not only the dynamic critical exponents θ and z, but also the static critical exponents β and ν. More important, the results for both the dynamic critical exponents and static critical exponents (z, β and ν) are consistent with these obtained by the traditional MC simulations performed in the long-time regime. Furthermore, similar to the measurements of the critical exponents, the determination of a critical temperature is also very difficult in equilibrium states. Now we have had the SCD by which the critical temperature can be also extracted from the power-law scaling behavior in the critical regime [12]. Therefore, we can take a comprehensive investigation to a statistical model by the SCD approach independent of the study in equilibrium states.

In this paper we show our numerical study for the q-state Potts models on a 2d triangular lattice by using the SCD method. Our attention is specially paid to the universal short-time evolutions from the fully ordered initial states. the calculations give evidence that there exists universal scaling already in the short-time regime by observing the power-law behavior of the magnetizations and auto-correlations, as well as their finite size scaling collapse. Those characters can be used to estimate the critical exponents and then to check the corresponding results each other. Our numerical results confirm explicitly the universality proposal in the short-time regime [15,16].

2 Model and SCD Scaling

The Hamiltonian for the q-state Potts model with ferromagnetic coupling ($J > 0$) defined on a 2d triangular lattice is given by

$$H = -\frac{J}{2}\sum_{i=1}^{N}\sum_{l_i=1}^{6}\delta_{\sigma_i,\sigma_{l_i}}, \qquad \sigma = 1,\cdots,q, \tag{1}$$

where $\sum_i$ represents the sum over all N lattice sites , and l_i denotes the nearest-neighbors of site i, whose co-ordinate is six on a triangular lattice. σ_i is a spin variable, taking q values, on each site i. It is known that in the equilibrium the Potts model is exactly solvable, and the critical points are $J_c = 0.5493$ and $J_c = 0.6310$ for the $q = 2$ and $q = 3$ respectively [2]. In our work we consider both the $q = 2$ and $q = 3$ cases and calculate the critical exponents, θ, z, β and ν, to investigate their universality explicitly.

General we consider a $O(n)$ vector model ($n = 1$ for the Ising model) with the dynamics of model A [17] and let it suddenly quenched from a very high temperature with small initial magnetization m_0 to the critical temperature T_c, Janssen, Schaub and Schmittmann showed [5] dynamic scaling behavior already at critical temperature within the short-time regime,

$$M^{(k)}(t,\tau,L,m_0) = b^{-k\beta/\nu}M^{(k)}(b^{-z}t, b^{1/\nu}\tau, b^{-1}L, b^{x_0}m_0), \tag{2}$$

where $M^{(k)}$ is the kth moment of the magnetization, $\tau = (T - T_c)/T_c$ is the reduced temperature, β and ν are the well known static critical exponents and b is a scaling factor. While x_0, a *new independent* exponent, is the scaling dimension of initial magnetization m_0. Here a MC sweep over all sites on the lattice is defined as the unit of the MC updating time t.

For a sufficiently large lattice ($L = \infty$), and setting $\tau = 0$, $b = t^{1/z}$, from the scaling form Eq.(2), the power-law of the time evolution of the magnetization at the critical temperature can be deduced

$$M(t) \sim m_0 t^{\theta}, \tag{3}$$

where θ is a new dynamic exponent which characterize the universality in the short-time regime. θ is related to x_0 by $\theta = (x_0 - \beta/\nu)/z$. The relation shows that, after the microscopic time t_{mic} the magnetization undergoes an initial increase at critical point, and we can easily obtain the exponent θ based on this power-law form.

Then, we focus our contention on the dynamic process starting from an ordered initial state. Although no detailed analytical study has been made about this process, the MC simulations on this evolution to some statistical spin models has showed there also exist a similar scaling relation [18],

$$M^{(k)}(t,\tau,L) = b^{-k\beta/\nu} M^{(k)}(b^{-z}t, b^{1/\nu}\tau . b^{-1}L) \tag{4}$$

At the exact critical temperature($\tau = 0$) and setting $b = t^{1/z}$, The Eq. (4) leads a power-law decay for the magnetization,

$$M(t) \sim m_0 t^{-\beta/\nu z}, \tag{5}$$

where $M(t)$ is defined by

$$M(t) = \frac{1}{N} < \sum_i S_i(t) >, \tag{6}$$

for the $q = 2$, and

$$M(t) = \frac{3}{2N} < \sum_i (\delta_{\sigma_i(t),1} - \frac{1}{3}) >, \tag{7}$$

for the $q = 3$. Where $N = L \times L$ is also the total spins defined on each size of a lattice with the periodic boundary conditions in our simulations, and $< \cdots >$ denotes the averages with respect to all independent initial configurations and the random numbers sequences.

However, the independent determination of $1/\nu$ seems slightly more complicated than z and $2\beta/\nu$. We should differentiate $\ln M(t,\tau)$ with respect to τ at the critical point

$$\partial_\tau \ln M(t,\tau)|_{\tau=0} = t^{1/(\nu z)} \partial_{\tau'} \ln M(t',\tau')|_{\tau'=0} \tag{8}$$

the exponent $1/\nu$ can be determined from this power-law behavior by input of the z value. In the practical numerical calculation, the differential to τ is substituted with a reasonable small difference $\Delta\tau$.

The dynamic exponent z can also be determined independently. In order to do it, we introduce a Binder cumulant

$$U(t,L) = \frac{M^{(2)}(t)}{(M(t))^2} - 1. \tag{9}$$

Here the second moment magnetization $M^{(2)}(t,L)$ is defined as

$$M^{(2)}(t) = \frac{1}{N^2} < (\sum_i S_i(t))^2 >, \tag{10}$$

for the Ising model ($q = 2$), and

$$M^{(2)}(t) = \frac{9}{4N^2} < (\sum_i (\delta_{\sigma_i(t),1} - \frac{1}{3}))^2 >, \tag{11}$$

for the Potts model ($q = 3$). For sufficiently large lattices, the finite size scaling analysis shows a power-law increase the Binde cumulant $U(t, L)$ obeys,

$$U(t, L) \sim t^{d/z}. \tag{12}$$

Based on this feature the dynamic exponent z can be estimated and then be used to calculate the exponent $1/\nu$ through the Eq.(8).

As done by many authors[7,11,19], the critical exponents can be measured both from disordered initial states and from the ordered initial states, and we make models involved from those two different initial states for comparisons to obtain more creditable results. We mostly perform our simulations on the scaling form which describe the dynamic process from an ordered initial state instead of a disordered initial state to avoid the finite m_0 effect on the measured results. There is an advantage to reduce the statistical fluctuations for the determination of the critical temperature from a dynamic process from ordered initial states, especially when t_{mic} is not so small.

3 Simulations and Results

It has been demonstrated in the previous works [8,20,21] that the results obtained by the Metropolis and Heat-bath algorithms are consistent each other and the latter is more efficient than the former. Therefore, in the present paper, the MC simulation is only performed by the Heat-bath algorithm. Samples for average are taken over 150,000 independent initial configurations on the L^2 triangular lattices with $L = 32, 64$ and 128. Statistical errors are simply estimated by performing three groups of the averages selecting different random seeds for the initial configurations. Our simulation is taken at the critical temperature points $J_c = 0.5493$ ($q = 2$) and $J_c = 0.6310$ ($q = 3$).

We begin by considering the short-time critical behavior in a dynamic process starting from random initial states with small m_0 to determine the critical exponent θ. We prepare the disordered initial configurations to the given values of m_0 by a sharp preparation method [8], and simulate the evolution of magnetization $M(t)$ versus the MC updating time t for different small $m_0 = 0.04, 0.02$ and 0.01 on the $N = 128^2$ lattice. The curves of $M(t)$ relaxation in a double-log scale are showed in Figs. 1 and 2 for the $q = 2$ and $q = 3$ models respectively. Obviously, there are very nice power-law increases of $M(t)$ after $t_{mic} \simeq 10$ and all the curves are almost parallel each other. Thus the θ can be estimated from the slopes of the curves in the regime of t=[10, 200]. In Table 1, the values of θ at different initial magnetizations m_0 and in the limit of $m_0 \to 0$ are presented for the $q = 2$ and $q = 3$ models respectively. In Table 2 the final results of θ, after an extrapolation to $m_0 \to 0$ limit, are given for the q=2 and q=3 models on the triangular and square lattices, and it can be found that the values of θ on the triangular lattices are the same as those on the square lattices. The calculations show an evidence of the universality in the short-time regime.

Secondly we study the evolution of the magnetization in the initial stage of the dynamic relaxation starting from fully ordered initial states. In Figs.3 and 4, the power-law behavior of the Binder cumulant $U(t)$ for the q=2 and q=3 models on the $N = 128^2$ is displayed in a double-log scale and the nice power-law of $U(t)$ is clearly shown. Then the exponent z can be estimated from the slopes of the curves. In Table 1, the values of z for two models on the lattice L=128 are also presented. The values of z is 2.145(3) for q=2 and 2.148(4) for q=3. As a comparison, we also list in the Table 2 the z values of the corresponding results on the square lattices. Next, we investigate the short-time behavior of magnetization of $M(t)$ starting from the fully ordered initial states. The time evolutions of $M(t)$ for the two models are displayed in Fig.5 and Fig.6 respectively. Those two figures show that the finite size effect is small as there is no difference within the scope of statistical errors on the lattice sizes of $L = 32, 64$ and 128. We can determine the value of exponent $\beta/\nu z$ from the slopes of the power-law decay curves. Then the values of β/ν are measured on the L=128 lattice by input of z, as listed in the Table 2, for both the q=2 and q=3 cases. Obviously, our measured results are more reasonable than that obtained in the previous works, because the exact value of β/ν is 1/8 for the Ising model and 2/15 for the q=3 Potts model.

The MC simulations of $\partial_\tau \ln M(t)$ are carried out by taking the difference $\Delta\tau$ =0.02. The simulations are take on L=128, 64 and 32 lattice. Fig.7 and Fig.8 plot the curves of time emulation $\partial_\tau \ln M(t)$ of two models respectively. The overlap of curves of three different size lattice indicates that the the effect of finite size can be ignored. We also notice that the power-law is not showed between $t = [1, 20]$, which is just the microscopic time scale t_{mic}. Therefore, we measure the exponent $1/\nu z$ from the interval [50,500]. The slopes yield the exponent $1/\nu z$=1.027(6) for the q=2 model and $1/\nu z$=1.223(5) for the q=3 model. The results of $1/\nu$ are also listed in Table 2.

Table 1. The measured values of θ versus the initial m_0 for the Ising model ($q = 2$) and Potts model ($q = 3$) on a $N = 128^2$ lattice. The last column gives the results of θ after extrapolation to $m_0 = 0$.

m_0	0.04	0.02	0.01	0.00
$q = 2$	0.183(1)	0.185(1)	0.189(2)	0.191(2)
$q = 3$	0.101(1)	0.089(1)	0.082(1)	0.076(2)

Table 2. The results of critical exponents θ, $2\beta/\nu$ and z for q=2 and q=3 models on the 2d triangular lattices. For comparison, the values on the square lattice are also listed.

	θ	$2\beta/\nu$	$1/\nu$	z
q=2(Triangular)	0.191(2)	0.250(5)	1.027(6)	2.143(3)
(Sq., Ref.[4])	0.191(1)	0.240(15)	1.03(2)	2.155(3)
exact		1/4	1	
q=3(Triangular)	0.076(2)	0.256(6)	1.223(8)	2.146(5)
(Sq., Ref.[4])	0.075(3)	0.269(7)		2.196(8)
exact		4/15	1.2	

4 Conclusion and Summary

We have systematically investigated the short-time critical dynamics of $q = 2$(Ising) and $q = 3$ Potts models on the 2d triangular lattices at the critical temperatures by using the dynamic MC simulation method, starting from both the random and ordered initial states. By observing the power-law increase of magnetization $M(t)$ from the random initial states, the new dynamic exponents θ has been calculated and its values are the same as those for the $q = 2$ and $q = 3$ Potts models on the square lattices. Mainly, the power-law behavior of the Binder cumulant $U(t)$, magnetization $M(t)$ and the $\partial_\tau \ln M(t)$ from the fully ordered initial states are studied to determine the dynamic exponents z, and static critical exponents β and ν for the $q = 2$ and $q = 3$ models.

By comparison to the results of critical exponents β and ν for the corresponding models on the square lattices, it has been confirmed numerically that they are belong to the same universality class, from the short-time regime of the dynamical relaxations, for the two spin models ($q = 2$ and $q = 3$) on the 2d triangular and square lattices, respectively. The great advantage of the SCD method is that the critical slowing down is eliminated because the measurements are carried out at the beginning of the evolution rather than in the equilibrium, so that the finite size effect is small and the method is efficient to study the critical phenomena. The further application of SCD is interesting to investigate the quantum phase transitions for their universal behavior.

Acknowledgments
H.P.Y. would like to thank the Heinrich Hertz - Stiftung Foundation for the financial support. This work was supported in part by the Center for Simulation and Scientific Computing at the College of Science, Zhenjiang University. This work was also supported by the National Natural Science Foundation of China under Grant No. 19975041.

References

1. J. Zinn-Justin, *Quantum Field theory and Critical Phenomena*, (Clarendon Press, Oxford, 1989); K. Binder and D.W. Heermann, *Monte Carlo Simulation in Statistical Physics* (Springer, Berlin, 1992).
2. R.J. Baxter, *Exactly solved models in statistical mechanics*, (Academic Press, New York, 1982).
3. R.H. Swendsen and J.S. Wang, Phys. Rev. Lett. **58**(1987)86; U. Wolff, Phys. Rev. Lett. **62**(1989)361.
4. B. Zheng, Int. J. Mod. Phys. **B12**(1998)1419.
5. H.K. Janssen, B. Schaub, and B. Schmittmann, Z. Phys. **B73**(1989)539.
6. D.A. Huse, Phys. Rev. **B53**(1989)304.
7. Z.B. Li, L. Schülke, and B. Zheng, Phys. Rev. Lett. **74**(1995)3396; Phys. Rev. **E53**(1996)2940.
8. K. Okano, L. Schülke, K. Yamagishi and B. Zheng, Nucl. Phys. **B485** [**FS**] (1997)727.
9. B. Zheng, M. Schulz and S. Trimper, Phys. Rev. Lett. **82**(1999)1891.
10. H.P. Ying, H.J. Luo, L. Schülke, and B. Zheng, Mod. Phys. Lett. **B12**(1998)1237.
11. H.P. Ying and Kenji Harada, Phys. Rev. **E62**(2000)174.
12. H.J. Luo, L. Schülke and B. Zheng, Phys. Rev. Lett. **81**(1998)180.
13. K. Okano, L. Schülke, and B. Zheng, Phys. Rev. **D57**(1998)1411.
14. U. Ritsche and H.W. Diehl, Nucl. Phys. **B464**(1996)512.
15. U. Ritschel and P. Czerner, Phys. Rev. Lett. **75**(1995)3882.
16. U. Ritschel and P. Czerner, Phys. Rev. **E55**(1997)3958.

17. P.C. Hohenberg and B.I. Halperin, Rev. Mod. Phys. **49**(1977)435.
18. Nobuyasa Ito, Physics A **192**(1993)604.
19. L. Schülke, B. Zheng, Phys. Lett.**A204**(1995)295.
20. H.J. Luo, L. Schülke and B. Zheng, Phys. Rev. **E57**(1998)1327.
21. J.B. Zhang, L. Wang, D.W. Gu, H.P. Ying and D.R. Ji, Phys. Lett. **A262**(1999)226.

Figure 1. The plot of $M(t)$ versus t in a double-log scale for the Ising model on a $N = 128^2$ lattice. The lines are presented for different initial magnetizations m_0=0.04,0.02,0.01 (from up to low).

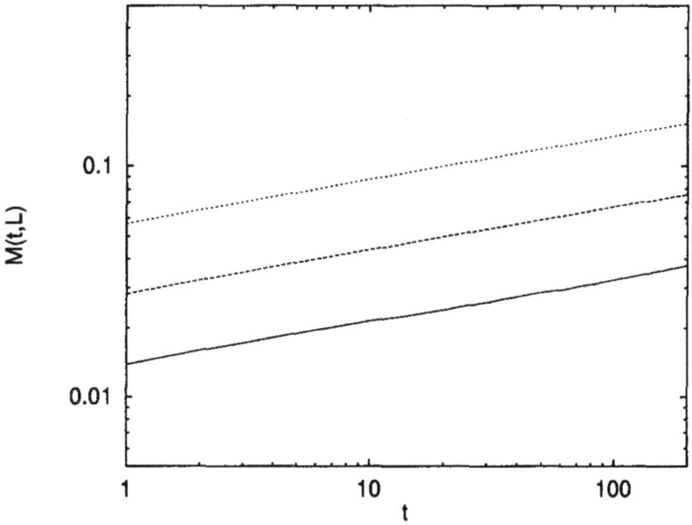

Figure 2. The same as Fig.1 but for the q=3 Potts model.

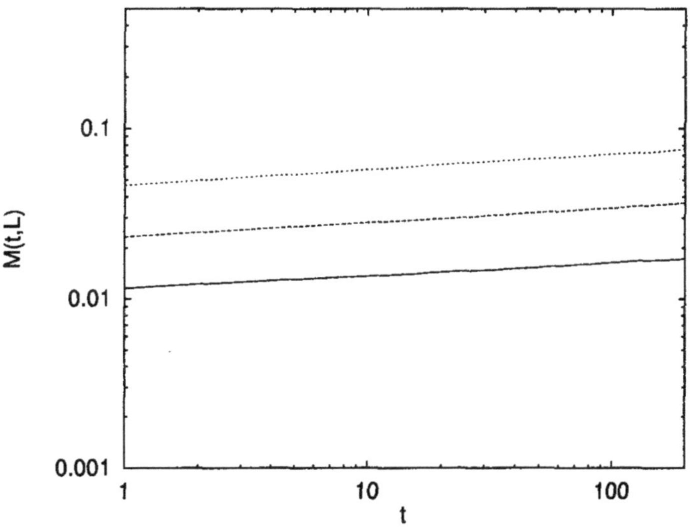

Figure 3. The time evolution of Binder cumulant $U(t)$ in a double-log scale for the Ising model on a $N = 128^2$ lattice.

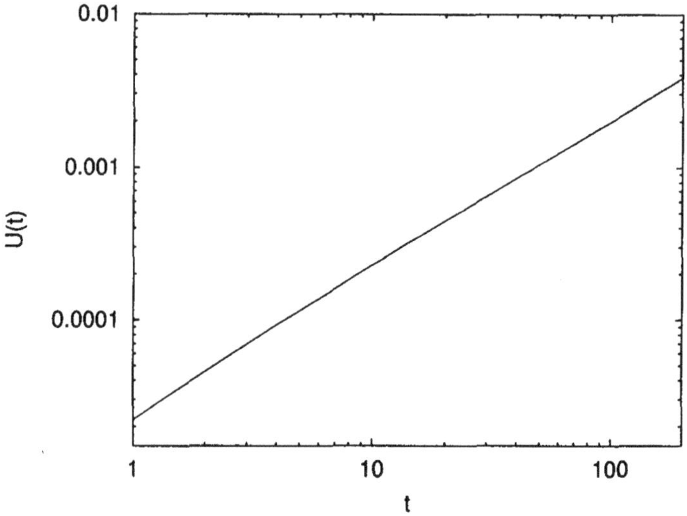

Figure 4. The same as Fig.3 but for the q=3 Potts model.

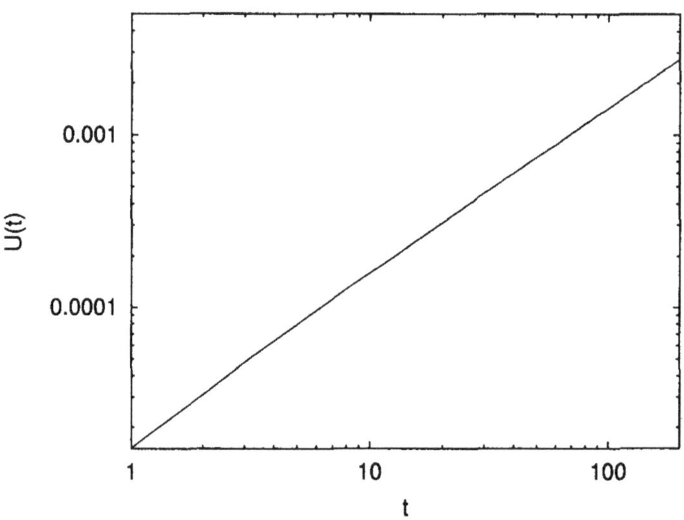

Figure 5. The power-law decay of $M(t)$ versus t in a double-log scale for the Ising model with $m_0 = 1$. The lines for different lattices of L=128,64,32 are almost overlapped.

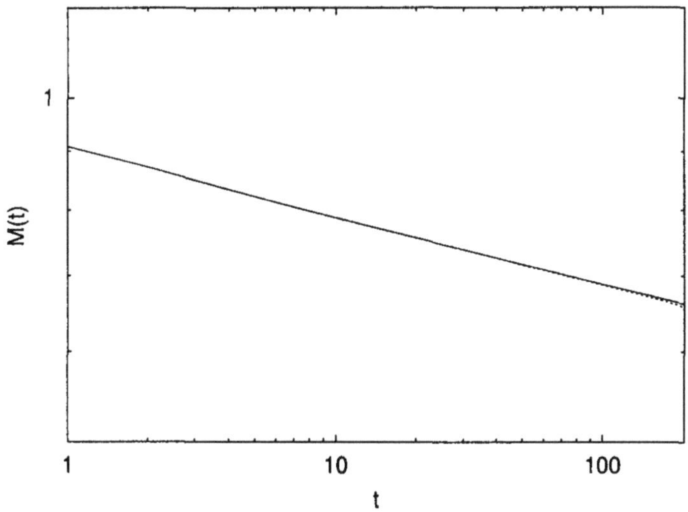

Figure 6. The same as Fig.5 but for the q=3 Potts model.

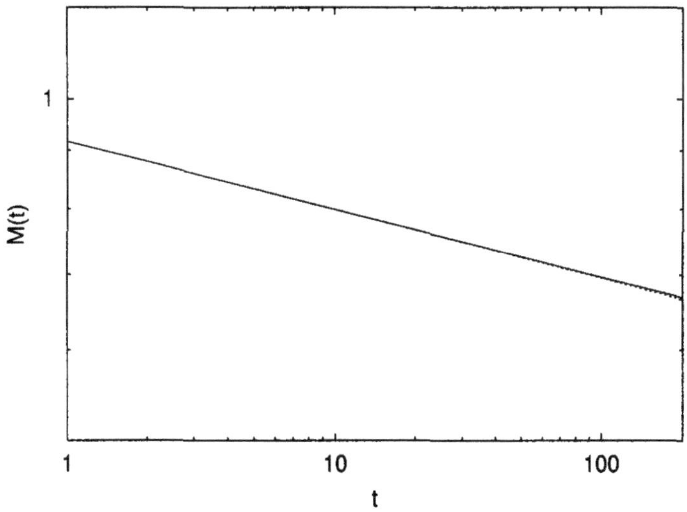

Figure 7. The curves of $\partial_\tau \ln M(t)$ in a double-log scale for the Ising model with $m_0 = 1$ for different lattices of L=128,64,32. It is obvious that the finite size effect seems small, and the power-law behavior exists after $t_{mic} \simeq 20$.

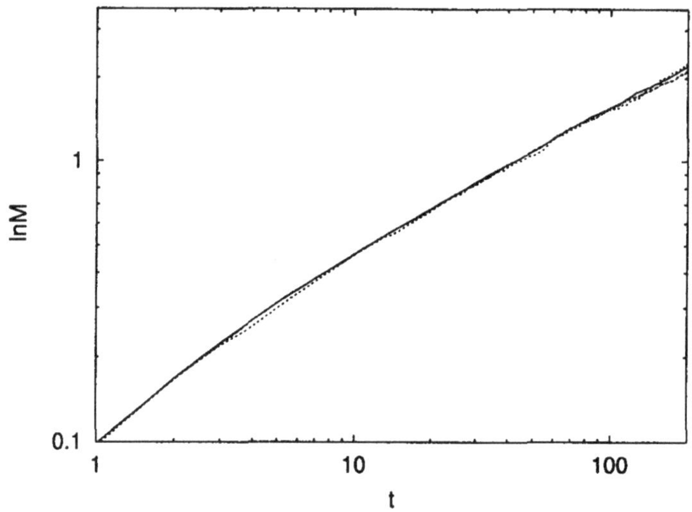

Figure 8. The same as Fig.7 but for the q=3 Potts model.

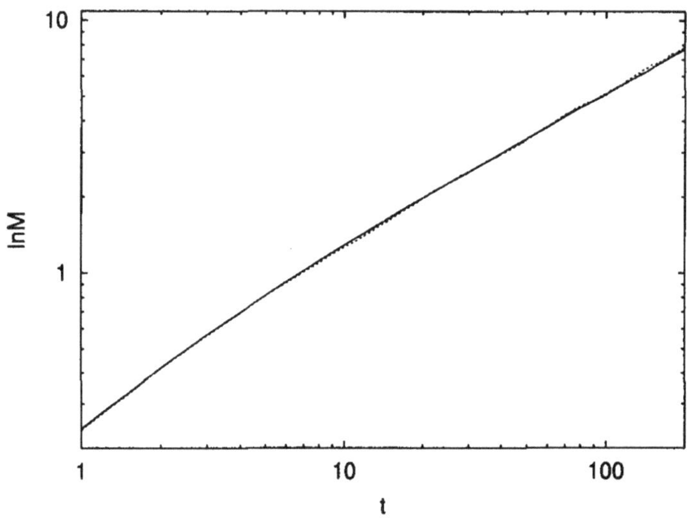

8 Supersymmetry, and Beyond the Standard Model

The Xing Pavillion, a Zhongshan University landmark.

SUPERSYMMETRY ON THE LATTICE

SIMON M. CATTERALL
Physics Department, Syracuse University, Syracuse, NY, USA 13244
E-mail smc@phy.syr.edu

ERIC B. GREGORY
Physics Department, Zhongshan University, Guangzhou, China 510275
E-mail: steric@zsu.edu.cn

We report on a simulation of the supersymmetric anharmonic oscillator computed using lattice path integral techniques[1]. Our numerical work utilizes a Fourier accelerated hybrid Monte Carlo scheme to sample the path integral. Combining this with the one-dimensional nature of the problem we find that we can generate high statistics data on large lattices for very modest computational cost.

1 Introduction

Supersymmetry (SUSY) is a scheme commonly proposed to unify the interactions of the standard model. It calls for a symmetric correspondence between bosonic and fermionic quantum fields. Such a symmetric correspondence is not manifest in the physics of everyday energy scales, so if SUSY exists it must be unbroken only at sufficiently high energies.

A theoretical investigation of such a spontaneously broken symmetry usually requires non-perturbative methods. Lattice field theory is such a non-perturbative method that has proven useful in disciplines such as QCD and quantum gravity, so it is natural to want to apply these techniques to supersymmetry. Unfortunately this proves to be a non-trivial task. The discrete nature of a lattice introduces problems. Many lattice field theory studies are plagued by complications arising from the fermion doubling problem. The naive lattice discretization of the fermion field in D dimensions produces produces 2^D fermion modes per single mode found in the naive boson discretization. Hence, in the simplest discretization of boson and fermion fields, SUSY is impossible because the boson modes are outnumbered by the fermion modes. Adding a Wilson mass term decouples the fermion doubles, and one can restore SUSY in the continuum limit in this manner[2], but only for the free case.

One encounters a further complication because supersymmetry is a space-time symmetry. As continuous translational symmetry is broken by lattice discretization, supersymmetry is explicitly broken. The challenge, then is to develop a prescription which allows supersymmetry to be regained as the continuum limit is taken.

With this in mind we performed a simulation of a one-dimensional model to test if it is possible, in a simple interacting case, to find a restored SUSY in the continuum limit. Using a non-standard lattice action[3] we find a supersymmetric continuum limit not just for the free action, but for an interacting case as well (similar to the action proposed in [4].

2 Model

We approached the problem of doubled fermions by using the doubled boson action and adding a Wilson mass term for the bosons as well as for the fermions. Recall that the naive lattice discretization of fermions

$$S = \frac{1}{2}\sum_{ij} \overline{\psi}_i D_{ij} \psi_j \tag{1}$$

has the momentum space form

$$\tilde{S} = \frac{i}{2\pi}\sum_{k} \overline{\psi}_{-k} \psi_k \sin(k), \tag{2}$$

which has zeros at $k = 0$ and at $k = \pi$. Here we define $D_{ij} = \frac{1}{2}(\delta_{j,i+1} - \delta_{j,i-1})$ The "naive" boson treatment uses the free action

$$S = \frac{1}{2}\sum_{ij} x_i \Box_{ij} x_j = \frac{1}{2}\sum_{ij} (x_{i+1} - xi)^2, \tag{3}$$

the momentum space expression

$$\tilde{S} = \frac{i}{2(2\pi)}\sum_{k} 4\sin^2(k/2) x_k x_{-k} \tag{4}$$

having only the zero at the origin. Note that

$$\begin{aligned} \Box_{ij} &= D^{+}_{ik} D^{-}_{kj} \\ &= \sum_{k} (\delta_{i+1,k} - \delta i, k)(\delta_{k,j} - \delta k - 1, j) . \end{aligned}$$

However, we can employ "doubled bosons"

$$S = -\sum_{ij} x_i D^2_{ij} x_j \tag{5}$$

with

$$D^2_{ij} = \frac{1}{4}(x_{k,i+1} - x_{k,i-1})(x_{j,k+1} - x_{j,k-1}). \tag{6}$$

Now

$$\tilde{S} = \frac{i}{2\pi}\sum_k 4sin^2(k)x_k x_{-k}, \tag{7}$$

which also has a zero at the zone boundary. We can at this point add appropriate Wilson terms to remove the doubles for both the fermions and bosons in a parallel fashion.

We simulate a bosonic field x and a fermionic field ψ in a one dimensional lattice with L sites and periodic boundary conditions. We use the action

$$S = \sum_{ij}\frac{1}{2}\left[-x_i D^2_{ij} x_j + \overline{\psi}_i\left(D_{ij} + P'\right)_{ij}\psi_j\right] + \sum_i \frac{1}{2}P_i^2, \tag{8}$$

where

$$P_i = \sum_j K_{ij}x_j + gx_i^3 \tag{9}$$

and

$$P'_{ij} = K_{ij} + 3gx_i^2\delta_{ij}. \tag{10}$$

The choice of a cubic interaction term in $P(x)$ guarantees unbroken supersymmetry [5]. The Wilson mass matrix is

$$K_{ij} = m\delta_{ij} - \frac{r}{2}\left(\delta_{i,j+1} + \delta_{i,j-1} - 2\delta_{i,j}\right). \tag{11}$$

We use dimensionless lattice units where $m = a_{\text{phys}}a$, $g = g_{\text{phys}}a^2$, and $x = x_{\text{phys}}a^{-\frac{1}{2}}$. The lattice spacing is $a = 1/L$

Now, the fermion matrix is

$$M = D + P' \tag{12}$$

$$= \left[m - r + 3gx_i^2\right]\delta_{ij} - \left[\frac{1}{2} + \frac{r}{2}\right]\delta_{j,i-1} + \left[\frac{1}{2} - \frac{r}{2}\right]\delta_{j,i+1}. \tag{13}$$

Note that when $r = 1$, M is almost lower triangular and therefore

$$\det(M) = \prod_{i=1}^{L}\left(1 + m + 3gx_i^2\right) + (-1)^{L-2}. \tag{14}$$

So M is positive definite for the choices $g > 0$ and $m > 0$. Furthermore, this choice completely eliminates the doubles for the free case.

3 Simulation

The observation that the choice of $r = 1$ can produce a positive definite fermion matrix is useful in that it allows us to employ Monte Carlo numerical simulation. We use the standard trick of replacing the fermionic term in the action,

$$\sum_{ij} \overline{\psi}_i M_{ij} \psi_j \tag{15}$$

with a psuedofermionic term

$$\frac{1}{2} \sum_{ij} \phi_i \left(M^T M\right)_{ij} \phi_j, \tag{16}$$

where the psuedofermionic fields ϕ are simulated as bosonic variables, hence producing the same $\det(M)$ in the partition function.

We simulate the whole system by using a hybrid Monte Carlo [6] (HMC) scheme. We simulate not S but $H = S + \Delta S$, where ΔS is given by

$$\Delta S = \frac{1}{2} \sum_i \left(p_i^2 + \pi_i^2\right). \tag{17}$$

The p and π act as classical momenta conjugate to the x and ϕ variables in an artificial time. We generate moves by reversible classical evolution of the variables, then submitting the whole system to a Metropolis test. This classical "leapfrog" evolution follows from Hamilton's equations:

$$\dot{x} = p \tag{18}$$

$$\dot{\phi} = \pi \tag{19}$$

$$\dot{p} = -\frac{\partial H}{\partial x} = -F \tag{20}$$

$$\dot{\pi} = \frac{\partial H}{\partial \phi} = -\mathcal{F} \tag{21}$$

So, in artificial time, t,

$$\begin{aligned} x_i(t + \Delta t) &= x_i(t) + \Delta t p_i(t) - \frac{(\Delta t)^2}{2} F_i \\ \phi_i(t + \Delta t) &= \phi_i(t) + \Delta t \pi_i(t) - \frac{(\Delta t)^2}{2} \mathcal{F}_i \end{aligned} \tag{22}$$

$$p_i(t+\Delta t) = p_i(t) + \frac{\Delta t}{2}\left(x_i(t) + x_i(t+\Delta t)\right)$$
$$\pi_i(t+\Delta t) = \pi_i(t) + \frac{\Delta t}{2}\left(\phi_i(t) + \phi_i(t+\Delta t)\right) \tag{23}$$

We determine the bosonic forces F_i and the fermionic forces $\mathcal{F}_i$ by solving the linear system

$$\left(M^T M\right)_{ij} s_j = \phi_i. \tag{24}$$

One can show that

$$F = D_{ij}^2 x_j + P'_{ij} P_j + 6g x_i s_i M_{ij} s_i \tag{25}$$

$$\mathcal{F}_i = s_i \tag{26}$$

Furthermore, we combat critical slowing down by performing the updates in momentum space. By using a momentum dependent timestep, slow modes can be updated at the same rate as the faster ones. Specifically, for bosons we use $\Delta t = \tau_B$ with

$$\tau_B = epsilon \frac{(m_{\text{eff}} + 2r)}{\sqrt{\sin^2 n + \left(m_{\text{eff}} + 2r\sin^2 \frac{2\pi n}{2L}\right)^2}} \tag{27}$$

and for the psuedofermions, $\Delta t = \tau_F = 1/\tau_B$.

To summarize, the simulation proceeds as follows. At the beginning of an update we Fourier transform the fields and choose starting conjugate momenta p_k and π_k from a Gaussian distribution. Using the current values of the fields, we determine the forces F_k and $\mathcal{F}_k$ and evolve the fields using equations (22) in momentum space using the appropriate timestep. Using the new fields we determine an updated set of forces, and finally update the momenta using (23). The repeated application of this step causes the system to move along an almost constant energy trajectory in phase space. We repeat the process N_{leap} times, then apply a Metropolis test to the entire update. The combination of drawing new momenta each update and the integration errors introduced during the classical evolution provides ergodicity. The reversibility of (22) and (23) together with the Metropolis test enforce detailed balance. In practice we set $\epsilon \sim 0.1$ and the number of leapfrog integrations per trajectory at $N_{\text{leap}} = 10$.

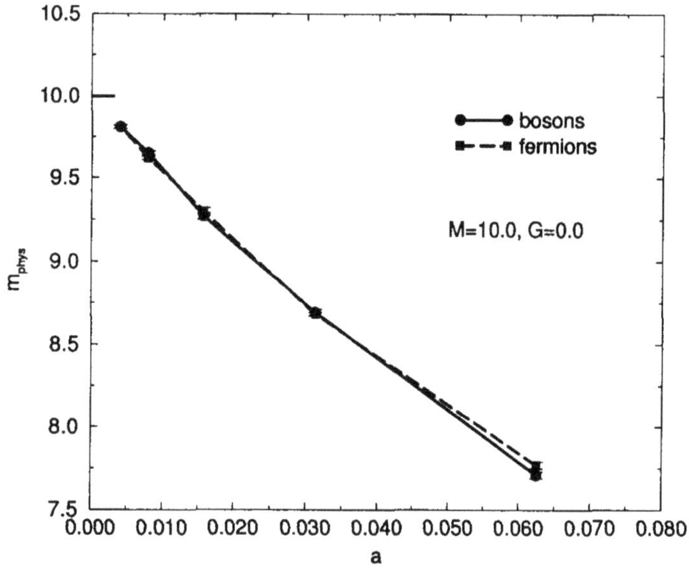

Figure 1. Boson and fermion masses vs. lattice spacing a at $m_{\text{phys}} = 10.0$, $g_{\text{phys}} = 0.0$, $r = 1$.

4 Results

We measured correlators to determine the physical mass of both the bosons and the fermions. For the bosons we measure

$$G_i j^B = \langle x_i x_j \rangle \tag{28}$$

and for the fermions

$$G_i j^F = \langle s_i M_i k s_j \rangle, \tag{29}$$

which one can show is an estimator for $\langle \overline{\psi}_i \psi \rangle$.

First we examined the simple non-interacting case ($g = 0$). It can be shown analytically that in this case there is an exact degeneracy between boson and fermion masses. We performed a million HMC trajectories at $m_{\text{phys}} = 10.0$ for lattice sizes $L = 256$, 128, 64, 32, and 16 and measured the correlators described above. Then we extracted the mass from an exponential fit to the first $L/4$ points. The results are displayed in Figure 1.

For finite lattice sizes, the mass is far from the inputted value of $m_{\text{phys}} = 10.0$ due to the influence of the Wilson mass term, but it is clear that for both the bosons and fermions the mass gap approaches 10.0 as the a goes toward zero. Furthermore, the boson and fermion masses are indistinguishable within the statistical accuracy, even at finite lattice spacing. As noted above, this is expected from analytical study.

In Figure 2 we show the results of a a simulation of the interacting case $g_{\text{phys}} = 100$, $m_{\text{phys}} = 10.0$ and $r = 1$. Here again the bosonic and fermionic mass gaps are converging on the same limit as the lattice spacing goes to zero. Analytic calculations show that this limit should be $m_{\text{cont}} = 16.87$ and the simulation results appear consistent with this. What is even more remarkable is that again the finite lattice simulations are unable to distinguish

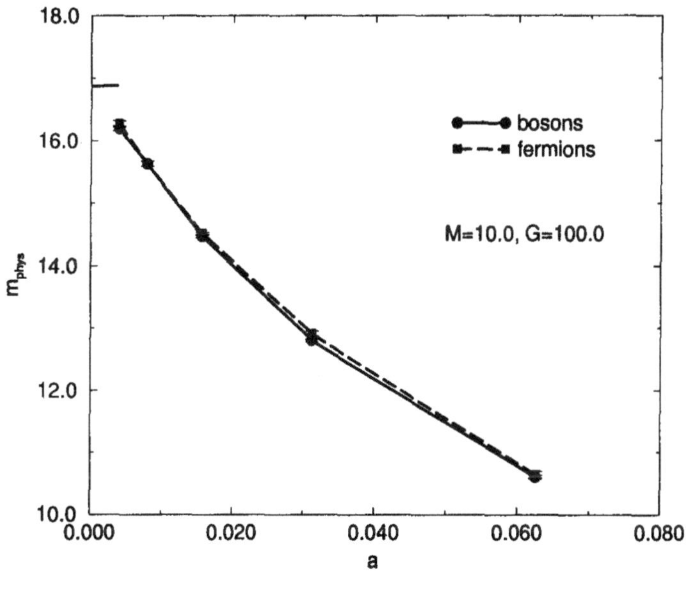

Figure 2. Boson and fermion masses vs. lattice spacing a at $m_{\text{phys}} = 10.0$, $g_{\text{phys}} = 100$, $r = 1$.

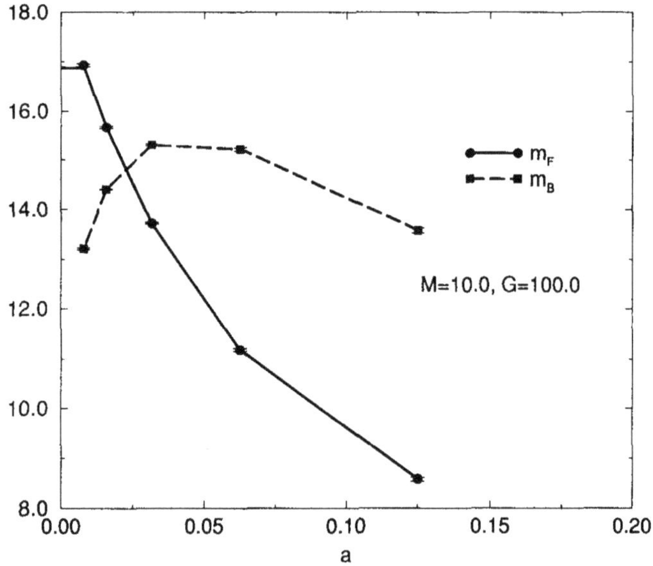

Figure 3. Boson and fermion masses vs. lattice spacing a at $m_{\text{phys}} = 10.0$, $g_{\text{phys}} = 100$, $r = 1$, simulated with "naive" bosonic action.

between the bosonic and fermionic mass gaps (though they are both far from the continuum limit value).

For comparison we include the results of a simulation conducted using the naive bosonic action ($\Box$ rather than D^2) at the same values $m_{\text{phys}} = 10.0$, $g_{\text{phys}} = 100$, $r = 1$. This is shown in Figure 3. It is clear that the finite lattice case does not exhibit supersymmetry, and while the bosonic and fermionic massgaps initially appear to approach each other, as the lattice spacing decreases past 0.05 the bosons and fermions begin to head towards different continuum limits.

5 Conclusions

In the simulations discussed in this report, we have seen empirically that it it possible to construct an interacting lattice action that exhibits supersymmetry in the continuum limit. A more in depth discussion of this work and an analytic discussion can be found in[1]. Further extensions of this numerical study to higher dimensions and related actions also appear promising.

References

1. Simon Catterall and Eric Gregory, hep-lat/0006013, (to appear in *Phys. Lett.* B , (2000)).
2. Hiroto So and Naoya Ukita, *Phys. Lett.* B **457**, 314 (1999)
   Tatsumi Aoyama and Yoshio Kikukawa, *Phys. Rev.* D **59**, (1999).
3. M. Golterman and D. Petcher, *Nucl. Phys.* B **319**, 307 (1989).
4. W. Bietenholz, *Mod. Phys. Lett.* **A14**, 513 (1999).
5. E. Witten, *Nucl. Phys.* B **185**, 513 (1981).
6. S. Duane, A. Kennedy, B. Pendleton, and D. Roweth, *Phys. Lett.* B **195**, 216 (1987).
7. H. Nicolai, *Phys. Lett.* B **89**, 341 (1980)
8. N. Sakai and M. Sakamoto, *Nucl. Phys.* B **229**, 173 (1983)
9. Matteo Beccaria, Giuseppe Curci, Erika D'Ambrosio, *Phys. Rev.* D **58**, 065009 (1998).

ANOMALY, CHARGE QUANTIZATION AND FAMILY

C. Q. GENG
Department of Physics, National Tsing Hua University, Hsinchu, Taiwan

We first review the three known chiral anomalies in four dimensions and then use the anomaly free conditions to study the uniqueness of quark and lepton representations and charge quantizations in the standard model. We also extend our results to theory with an arbitrary number of color. Finally, we discuss the family problem.

Although the standard model [1] of $SU(3)_C \times SU(2)_L \times U(1)_Y$ has been remarkably successful experimentally, there are several puzzles, such as why the electric charges of quarks and leptons are quantized and why there are three fermion families? In this talk I would like to study these two puzzles in the viewpoint of the chiral gauge anomaly cancellations.

It is well-known that the anomaly free conditions arising from the theoretical requirements of renormalizability and self-consistency are the most elegant tool to test the gauge theory. Three anomalies thus far have been identified for chiral gauge theories in four dimensions: (1) The triangular (perturbative) chiral gauge anomaly,[2] which must be canceled to avoid the breakdown of gauge invariance and renormalizability of the theory; we call this the *triangular* anomaly. (2) The global (non-perturbative) $SU(2)$ chiral gauge anomaly,[3] which must be absent in order to define the fermion integral in a gauge invariant way; we call this the *global* anomaly. This anomaly was first pointed out by Witten,[3] and is known as the Witten $SU(2)$ anomaly. He showed in 1982 that any $SU(2)$ gauge theory with an odd number of left-handed fermion (Weyl) doublets is mathematically inconsistent. (3) The mixed (perturbative) chiral gauge-gravitational anomaly,[4,5] which must be canceled in order to ensure general covariance of the theory; we call this the *mixed* anomaly. This anomaly was first discussed by Delburgo and Salam[4] in 1972 and its consequences studied by Alvarez-Gaumé and Witten[5] in 1983, who concluded that a necessary condition for consistency of the theory coupled to grav ity is that the sum of the $U(1)$ charges of the left-handed fermions vanishes, i.e., $TrQ = 0$.

We now review the three chiral anomalies for the simple Lie groups.

1. *The Triangular Anomaly.* It has been shown[6] that the simple Lie groups: $SU(2)$, $SO(2k+1)(k > 2)$, $SO(4k)(k \geq 2)$, $SO(4k+2)(k \geq 2)$, $Sp(2k)$, G_2, F_4, E_6, E_7, and E_8 are safe groups. The only simple groups with possible triangular anomaly are the unitary groups $SU(n)(n \geq 3)$. Therefore, if we start with the groups which do not contain $SU(n)(n \geq 3)$ group, the

theory will be free of triangular anomaly.

2. *The Global Anomaly.* We classify the simple Lie groups G into the following two classes. (I) $Sp(2k)(Sp(2) \simeq SU(2))$. These groups[7] have the property of

$$\mathbf{\Pi}_4(Sp(2k)) = \mathbf{Z}_2 \ , \tag{1}$$

where $\mathbf{\Pi}_4$ is the fourth homotopy group and $\mathbf{Z}_2$ is the two-valued discrete group (like parity). According to Witten,[3] the group $G^{(I)} = Sp(2k)$ has global anomaly if the number of fermion zero modes (for $SU(2)$ group, it is equal to the number of fermion doublets) is odd. (II) $SU(n)(n \geq 3), SO(2k+1)(k > 2), SO(4k)(k \geq 2), SO(4k+2)(k \geq 2), G_2, F_4, E_6, E_7$, and E_8. These groups ($G^{(II)}$) have no global anomaly since their fourth homotopy groups are trivial,[3,7] i.e.,

$$\mathbf{\Pi}_4(G^{(II)}) = 0 \ . \tag{2}$$

However, the interesting question[8] arises as to how one can know at the level of $G^{(II)}$ whether such a theory is global anomaly-free when $G^{(II)}$ breaks down to groups which contain $G^{(I)}$. Recently, we present a sufficient condition[8] that for any simple group G, containing $Sp(2k)$ as a subgroup, and for which $\mathbf{\Pi}_4(G) = 0$, the vanishing of the triangular perturbative anomaly for Weyl representations of G will guarantee the absence of the global non-perturbative $Sp(2k)$ anomaly.

3. *The Mixed Anomaly.* This anomaly is non-trivial only for the theory in which there is $U(1)$ symmetry with non-zero total charges.[5] Obviously, all the simple Lie groups ($G^{(I),(II)}$) are safe groups. Furthermore, when these groups break down to groups which contain $U(1)$, e.g.,

$$G \rightarrow g \times \prod_i U(1)_i \ , \tag{3}$$

unlike the previous case, there is no mixed anomaly since the $U(1)$ operators are the generators of G and must be traceless.

The triangular anomaly-free of the standard model was first noted[9] in 1972 for each quark-lepton family. It was clear that with only the triangular anomaly-free condition[10] one could not explain the empirically determined quark-lepton representations and their quantized hypercharges. We now study[11] the question of the uniqueness of quarks and leptons in the standard model by insisting on all three anomaly-free conditions. With an arbitrary color number N (≥ 3), we begin by allowing an arbitrary number of (left-handed) Weyl representations under the group of $SU(N) \times SU(2) \times U(1)$, i.e.,

$$
\begin{array}{cccl}
SU(N) \times SU(2) \times U(1) & & & \\
N & 2 & Q_i\,, & i = 1, \cdots, j \\
N & 1 & Q'_i\,, & i = 1, \cdots, k \\
\overline{N} & 1 & \overline{Q}_i\,, & i = 1, \cdots, l \\
\overline{N} & 2 & \overline{Q}'_i\,, & i = 1, \cdots, m \\
1 & 2 & q_i\,, & i = 1, \cdots, n \\
1 & 1 & \overline{q}_i\,, & i = 1, \cdots, p
\end{array} \tag{4}
$$

where the integers j, k, l, m, n and p and the $U(1)$ charges are all arbitrary. The triangular anomaly free conditions then lead to the following equations:

$$
\begin{aligned}
[SU(N)]^3 &: \sum_{i=1}^{j} 2 + \sum_{i=1}^{k} 2 - \sum_{i=1}^{l} 1 - \sum_{i=1}^{m} 2 = 0\,, \\
[SU(N)]^2\, U(1) &: 2\sum_{i=1}^{j} Q_i + \sum_{i=1}^{k} Q'_i + \sum_{i=1}^{l} \overline{Q}'_i + 2\sum_{i=1}^{m} \overline{Q}'_i = 0\,, \\
[SU(2)]^2\, U(1) &: N\sum_{i=1}^{j} Q_i + N\sum_{i=1}^{m} \overline{Q}'_i + \sum_{i=1}^{n} q_i = 0\,,
\end{aligned} \tag{5}
$$

$$
[U(1)]^3 : N\sum_{i=1}^{j} Q_i^3 + \frac{N}{2}\sum_{i=1}^{k} Q_i'^3 + \frac{N}{2}\sum_{i=1}^{l} \overline{Q}_i^3 + N\sum_{i=1}^{m} \overline{Q}_i'^3 + \sum_{i=1}^{n} q_i^3 + \frac{1}{2}\sum_{i=1}^{p} \overline{q}_i^3 = 0\,.
$$

The global $SU(2)$ anomaly-free condition is

$$
N\,j + N\,m + n = E\,, \tag{6}
$$

where E is an even integer. Finally the mixed anomaly-free condition is

$$
[U(1)] : N\sum_{i=1}^{j} Q_i + \frac{N}{2}\sum_{i=1}^{k} Q'_i + \frac{N}{2}\sum_{i=1}^{l} \overline{Q}_i + N\sum_{i=1}^{m} \overline{Q}'_i + \sum_{i=1}^{n} q_i + \frac{1}{2}\sum_{i=1}^{p} \overline{q}_i = 0\,. \tag{7}
$$

The requirements of minimality and the three anomaly-free conditions [Eqs. (5)-(7)] lead to the values: (I) if N= even #, $j = 1$, $k = 0$, $l = 2$, $m = n = p = 0$, and

$$
Q_1 = 0\,,\ \overline{Q}_1 = -\overline{Q}_2\,; \tag{8}
$$

and (II) if N= odd #, $j = 1$, $k = 0$, $l = 2$, $m = 0$, $n = 1$, $p = 1$, and to two solutions of $U(1)$ charges

$$Q_1 = \frac{1}{N}\,,\ \overline{Q}_1 = -\frac{N+1}{N}\,,\ \overline{Q}_2 = \frac{N-1}{N}\,,\ \overline{q}_1 = -2q_1 = -2,\,, \tag{9}$$

$$Q_1 = q_1 = \overline{q}_1 = 0\,,\ \overline{Q}_1 = -\overline{Q}_2\,, \tag{10}$$

where we have chosen the normalization $q_1 = -1$ in Eq. (9). For $N = 3$, the solutions in Eqs. (9) and (10) are the "standard model" and the so called "bizarre" ones, respectively. We note that the "inert" state (1,1,0) for the "bizarre" solution is a non-chiral representation and it must be excluded. It is interesting to note that the "bizarre" solution may be viewed as the standard one when $N \to \infty$. Without considering the "bizarre" solution, for the odd number of color, all the $U(1)$ charges are uniquely determined. In this case, the resulting Weyl representations of $SU(N)$ and $SU(2)$ and their $U(1)$ charges are those in the standard model if N=3 (cf. Table 1). The electric charges of quarks and leptons for an arbitrary odd number of color N are given in Table 1 where the electroweak symmetry is spontaneously broken down to $U(1)_{EM}$ by the Higgs mechanism.

Table 1. The quantum numbers of quark and lepton representations under $SU(N)_C \times SU(2)_L \times U(1)_Y$ and $SU(N)_C \times U(1)_{EM}$

Particles $(i = 1,2,3)$	$SU(N)_C \times SU(2)_L \times U(1)_Y$			$\rightarrow$	$SU(N)_C \times U(1)_{EM}$	
$\begin{pmatrix} u \\ d \end{pmatrix}_L^i$	N	2	$\frac{1}{N}$		$\begin{pmatrix} N \\ N \end{pmatrix}$	$\begin{pmatrix} \frac{N+1}{2N} \\ -\frac{N-1}{2N} \end{pmatrix}$
$u_L^{c\ i}$	$\overline{N}$	1	$-\frac{N+1}{N}$		$\overline{N}$	$-\frac{N+1}{2N}$
$d_L^{c\ i}$	$\overline{N}$	1	$\frac{N-1}{N}$		$\overline{N}$	$\frac{N-1}{2N}$
$\begin{pmatrix} \nu \\ e \end{pmatrix}_L^i$	1	2	-1		$\begin{pmatrix} 1 \\ 1 \end{pmatrix}$	$\begin{pmatrix} 0 \\ -1 \end{pmatrix}$
$e_L^{c\ i}$	1	1	2		1	1

For the standard model of $SU(3)_C \times SU(2)_L \times U(1)_Y$, we thus find that the requirements of minimality and freedom from all three chiral gauge anomalies lead to a unique set of Weyl representations (and their $U(1)_Y$ charges) of the standard group that correspond to the observed quarks and leptons of one family. Furthermore, the $U(1)_Y$ charges of these quarks and leptons are quantized and correctly determined by adding the mixed anomaly-free condition and thus a long-standing puzzle of the electric charge quantization of quark and lepton can be solved within the content of the standard model.

In spite of the success of the standard model, it is still a mystery why the three anomaly cancellations, especially the global and the mixed ones, should be satisfied. Naturally one hopes that new physics beyond the standard model can provide us an explanation to this question. ¿From the above studies we see that the three anomaly-free conditions in the standard model may be automatically satisfied if it comes from a large group, especially, a grand unification group. For example, with the E_6 grand unification theory, the triangular, the global, and the mixed anomalies are trivial at the level of E_6 which guarantees their freedom at the standard group level. We thus conclude that the resolution of the question of the uniqueness of the massless fermion representations and $U(1)_Y$ charges for the standard group – when viewed from the standpoint of the perturbative triangular and mixed chiral gauge-gravitational anomalies and the absence of the non-perturbative global $SU(2)$ chiral gauge anomaly in four dimensions – argues strongly for some new physics beyond the standard model.

Finally, we discuss the family issue. It is clear that, as one can see from the above study, the imposition of all three anomaly-free conditions for the standard model does not shed any immediate light on the "generation problem". In fact, the quantum numbers in Table 1 are generation blind. Moreover, if one enlarges the standard group to include an $SU(2)$ or $SU(3)$ group, one can show that the theories are precisely the one family fermion structure of the left-right symmetric model[11] $SU(3)_C \times SU(2)_L \times SU(2)_R \times U(1)$ and the chiral-color model,[12] $SU(3)_{CL} \times SU(3)_{CR} \times SU(2)_L \times U(1)_Y$, respectively, instead of having a family group. Clearly, some new ideas[13] are needed to constrain on the number of families which would be a key to the new physics. We now present a toy model which gives rise to three families of quarks and leptons. In the standard model, in each family there are 15 Weyl spinors. With a right handed neutrino, the number becomes 16. For three families, the total numbers are 48. One may put all these 48 Weyl spinors into a flavor box to form a large global symmetry as $U(48)$.[13] ¿From the study in Eqs. (4)-(10), we can extend the group of $SU(N) \times SU(2) \times U(1)$ with both even and odd numbers of N to a larger group of $SU(N) \times SU(2) \times SU(2)$ in which N has to be an even number as shown in Table 2. For N=4, it is just the Pati-Salam model,[14] which contains a right-handed neutrino. We remark that the representations under $SU(N) \times SU(2) \times SU(2)$ in Table 2 are unique unlike the case with a $U(1)$

symmetry and there is no more "bizarre" solution like the one in Eq. (10).

Table 2. The fermion quantum numbers under $SU(N) \times SU(2) \times SU(2)$

$SU(N)_C$	$\times$ $SU(2)$	$\times$ $SU(2)$
N	2	1
$\overline{N}$	1	2

We now take the global flavor symmetry $U(48)$ and gauge its subgroup $SU(12) \times SU(2) \times SU(2)$ so that the fermions transform according to the representations given in Table 2 with $N = 12$. Thus, the model is a generalized Pati-Salam theory with the color being 12. The symmetry breaking chains by various suitable scalars are given as follows:

$$
\begin{array}{c}
SU(12)_C \times SU(2)_L \times SU(2)_R \\
\begin{array}{ccc} 12 & 2 & 1 \\ \overline{12} & 1 & 2 \end{array} \\
\downarrow \\
SU(12)_C \times SU(2)_L \times SU(2)_R \\
\downarrow \\
SU(8)_C \times SU(4)_{C1} \times SU(2)_L \times SU(2)_R \times U(1) \\
\downarrow \\
SU(4)_{C3} \times SU(4)_{C2} \times SU(4)_{C1} \times SU(2)_L \times SU(2)_R \times U(1) \times U(1) \\
\downarrow \\
\downarrow \\
SU(4)_C \times SU(2)_L \times SU(2)_R \\
\downarrow \\
\downarrow \\
SU(3)_C \times SU(2)_L \times U(1)_Y \\
\underbrace{\textit{three quark and lepton families}}
\end{array}
$$

Therefore, there are three generations of quarks and leptons under the standard group of $SU(3)_C \times SU(2)_L \times U(1)_Y$. However, before taking this model seriously, more works have to be done.

In sum, we have found that the requirements of minimality and freedom from all three chiral gauge anomalies in four dimensions lead to a unique set of Weyl representations of the standard group, corresponding to the observed quarks and leptons of one family. Furthermore, the $U(1)_Y$ charges of these quarks and leptons are quantized and correctly determined by adding the mixed anomaly-free condition and thus a long-standing puzzle of the electric charge quantization of quark and lepton can be solved within the content of the standard model. The determination of the uniqueness of the standard model dup to the anomaly cancellations argues strongly for new physics beyond the standard model, especially some form of the quark-lepton unification. However, there is still no answer to the family problem. Maybe there are possibly some as-yet-unidentified anomalies in four dimensions, or larger symmetries like $SU(12)_C \times SU(2)_L \times SU(2)_R$, or higher dimensions,[13] or presons,[15] or others.

Acknowledgements

This work was supported in part by the National Science Council of Taiwan, China under contract number NSC-89-2112-M-007-013.

1. Glashow, S. L., Nucl. Phys. **22**, 579 (1961); Weinberg, S., Phys. Rev. Lett. **19**, 1264 (1964); Salam, A., " Elementary Particle Theory", ed. Svarthom, N. (Almquist and Wilsell, Stockholm, 1968).
2. Adler, S. Phys. Rev. **117**, 2426 (1969); Bell, J and Jackiw, R., Nuovo Cimento **51A**, 47 (1969); Bardeen, W., Phys. Rev. **184**, 1848 (1969).
3. Witten, E., Phys. Lett. **B117**, 324 (1982).
4. Delbourgo, R. and Salam, A., Phys. Lett. **B40**, 381 (1972); see also Eguchi, T. and Frend, T., Phys. Rev. Lett. **37**, 1251 (1976).
5. Alvarez-Guamé, L. and Witten, E., Nucl. Phys. **B234**, 269 (1983).
6. Georgi, H. and Glashow, S. L. Phys. Rev. **D6**, 1159 (1972); Okubo, S., Phys. Rev. **D16**, 3528 (1977).
7. Atiyah, M. T., Proc. of the *XVIIIth Solvay Conference on Physics* at Univ. of Texas, Austin, Texas, 1982 (Phys. Rep. **104**, 203 (1984)).
8. Geng, C. Q., Marshak, R. E., Zhao, Z. Y., and Okubo, S., Phys. Rev. **D36**, 1953 (1987).
9. Bouchiat, C., Illiopoulos, J. and Meyer, Ph., Phys. Lett. **B73**, 519 (1972); Gross, D. J. and Jackiw, R., Phys. Rev. **D6**, 477 (1972).
10. For reviews, see Glashow, S. L., "Quarks and Leptons", ed. Levy, M., (Cargese, 1979); "Fundamental Interactions", ed. Levy, M., (Cargese, 1981).
11. Geng, C. Q. and Marshak, R. E., Phys. Rev. **D39**, 693 (1989).
12. Geng, C. Q., Phys. Rev. **D39**, 2402 (1989).
13. For a review, see Peccei, R. D., DESY Report No. DESY-88-078.
14. Pati, J. C. and Salam, A, Phys. Rev. **D10**, 275 (1974).
15. Geng, C. Q. and Marshak, R. E., Phys. Rev. **D35**, 2278 (1987).

PARTICIPANTS

Hiroaki Arisue :
Osaka Prefectural College of Technology,
Saiwai-cho, Neyagawa,
Osaka 572, Japan
arisue@las.osaka-pct.ac.jp

Vicente Azcoiti Perez :
Departamento de Física Teórica,
Facultad de Ciencias,
Universidad de Zaragoza,
50009 Zaragoza, Spain
azcoiti@azcoiti.unizar.es

Cheng-Guang Bao :
Department of Physics,
Zhongshan University,
Guangzhou 510275, China

Wolfgang Bietenholz :
NORDITA, Blegdamsvej 17
DK-2100 Copenhagen, Denmark
bietenho@nordita.dk

Rudolph Burkhalter :
Center for Computational Physics and Institute of Physics,
University of Tsukuba,
Tsukuba, Ibaraki 305-8577, Japan
burkhalt@rccp.tsukuba.ac.jp

Hoi Fung Chau :
Department of Physics,
University of Hong Kong,
Pokfulam Road,
Hong Kong
hfchau@hkusua.hku.hk

Ying Chen :
Institute of High Energy Physics,
Chinese Academy of Sciences,
P.O.Box 918-4,
Beijing 100039, China
cheny@hptc5.ihep.ac.cn

Xiao-Ni Cheng :
Department of Physics,
Zhongshan University,
Guangzhou 510275, China

Yi-Zhong Fang :
Department of Physics,
Zhongshan University,
Guangzhou 510275, China

Angelo Galante :
Dipartimento di Fisica
Università di L'Aquila,
67100 L'Aquila, Italy
angelo.galante@lngs.infn.it

Chao-Qiang Geng :
Department of Physics,
National Tsing Hua University,
Hsinchu, Taiwan
geng@phys.nthu.edu.tw

Eric B. Gregory :
Department of Physics,
Zhongshan University,
Guangzhou 510275, China
steric@zsu.edu.cn

Shuo-Hong Guo :
Department of Physics,
Zhongshan University,
Guangzhou 510275, China

Chris J. Hamer :
School of Physics,
University of New South Wales,
Sydney, 2006, Australia
C.Hamer@unsw.edu.au

Chun-Qing Huang :
Department of Physics,
Zhongshan University,
Guangzhou 510275, China

Wolfhard Janke :
Institut für Theoretische Physik,
Universität Leipzig,
04109 Leipzig, Germany
wolfhard.janke@itp.uni-leipzig.de

Yoshio Kikukawa :
Department of Physics,
Nagoya University, 464-8602, Japan
kikukawa@eken.phys.nagoya-u.ac.jp

Dean Lee :
Department of Physics,
University of Massachusetts,
Amherst, MA 01003 USA
dlee@physics.umass.edu

Jia Li :
Department of Physics,
Zhongshan University,
Guangzhou 510275, China

Qi-Ye Li :
Department of Physics,
Zhongshan University,
Guangzhou 510275, China

Zhi-Bin Li :
Department of Physics,
Zhongshan University,
Guangzhou 510275, China

Ruo-Sheng Liao :
Department of Physics,
Zhongshan University,
Guangzhou 510275, China

Qiong-Gui Lin :
Department of Physics,
Zhongshan University,
Guangzhou 510275, China

Chuan Liu :
Department of Physics,
Peking University,
Beijing, 100871, China
liuchuan@mail.phy.pku.edu.cn

Chun Liu :
Institute of Theoretical Physics,
Academia Sinica
P.O. Box 2735,
Beijing, 100080, China
liuc@itp.ac.cn

Jin-Jiang Liu :
Department of Physics,
Zhongshan University,
Guangzhou 510275, China

Liang-Gang Liu :
Department of Physics,
Zhongshan University,
Guangzhou 510275, China

Qin-Yong Liu :
Department of Physics,
Zhongshan University,
Guangzhou 510275, China

Xiao-Meng Liu :
Department of Physics,
Zhongshan University,
Guangzhou 510275, China

Yubin Liu :
Department of Physics,
Hiroshima University,
Higashi–Hiroshima 739–8521, Japan
liu@hirohe.hepl.hiroshima-u.jp

Xiang-Qian Luo :
Department of Physics,
Zhongshan University,
Guangzhou 510275, China
stslxq@zsu.edu.cn

Jian-Ping Ma :
Institute of Theoretical Physics,
Academia Sinica
P.O. Box 2735,
Beijing, 100080, China
majp@itp.ac.cn

Martin Maul :
Department of Theoretical Physics,
Lund University,
Sölvegatan 14A,
S - 223 62 Lund, Sweden
maul@thep.lu.se

Zhong-Hao Mei :
Department of Physics,
Zhongshan University,
Guangzhou 510275, China

Zhi-Gang Pan :
Zhejiang Institute of Modern Physics,
Zhejiang University,
Hangzhou 310027, P.R. China
zgpan@zimp.zju.edu.cn

Wei Pang :
Department of Physics,
Zhongshan University,
Guangzhou 510275, China

Michele Pepe :
Inst. für Theoretische Physik,
ETH Hönggerberg, CH-8093,
Zürich, Switzerland
pepe@itp.phys.ethz.ch

Bengt Petersson :
University of Bielefeld,
Faculty for Physics,
P.O. Box 10 01 31,
D-33501 Bielefeld, Germany
bengt@physik.uni-bielefeld.de

Takuya Saito :
Department of Physics,
Hiroshima University,
Higashi–Hiroshima 739–8521, Japan
tsaito@hirohe.hepl.hiroshima-u.ac.jp

Gerrit Schierholz :
Theory Group,
Deutsches Electronon-Synchrotron DESY,
D-22603 Hamburg, Germany
gsch@x4u2.desy.de

Lothar Schülke :
Fachbereich Physik,
Universität Siegen,
D-57068 Siegen, Germany
schuelke@pollux.physik.uni-siegen.de

Shu-Ping Sito :
Department of Physics,
Zhongshan University,
Guangzhou 510275, China

Tsuneo Suzuki :
Institute for Theoretical Physics,
Kanazawa University,
Kanazawa 920-1192, Japan
suzuki@hep.s.kanazawa-u.ac.jp

Tetsuya Takaishi :
Department of Physics,
Hiroshima University of Economics,
Hiroshima, 731-0192, Japan
tt-taka@hue.ac.jp

Stefan Thurner :
Institut für Kernphysik,
TU Wien, Wiedner Hauptstrasse 8-10,
1040 Vienna, Austria
thurner@univie.ac.at

Matthieu Tissier :
Laboratoire de Physique Théorique et des Hautes Energies,
Université Paris VI-Pierre et Marie Curie - Paris VII-Denis Diderot,
2 Place Jussieu,
F-75252 Paris Cedex 05, France
tissier@lpthe.jussieu.fr

Lei Wang :
Zhejiang Institute of Modern Physics,
Zhejiang University,
Hangzhou 310027, P.R. China
lwang@zimp.zju.edu.cn

Qing Wang :
Fachbereich Physik,
Universität Siegen,
D-57068 Siegen, Germany

Ji-Min Wu :
Institute of High Energy Physics,
Chinese Academy of Sciences,
P.O.Box 918-4,
Beijing 100039, China
wujm@alpha02.ihep.ac.cn

Hao Xu :
Department of Physics,
Zhongshan University
Guangzhou 510275, China

Jia-Rui Xu :
Department of Chemistry,
Zhongshan University
Guangzhou 510275, China

He-Ping Ying :
Zhejiang Institute of Modern Physics,
Zhejiang University,
Hangzhou 310027, China
hpying@zimp.zju.edu.cn

www.ingramcontent.com/pod-product-compliance
Ingram Content Group UK Ltd.
Pitfield, Milton Keynes, MK11 3LW, UK
UKHW020359150926
8624IPUK00004B/240

9 789810 245955